Applied Numerical Methods for Chemical Engineers

Applied Numerical Methods for Chemical Engineers

Navid Mostoufi
School of Chemical Engineering
College of Engineering
University of Tehran

Alkis Constantinides
Professor of Chemical and Biochemical Engineering
Rutgers University
(deceased)

Academic Press is an imprint of Elsevier
125 London Wall, London EC2Y 5AS, United Kingdom
525 B Street, Suite 1650, San Diego, CA 92101, United States
50 Hampshire Street, 5th Floor, Cambridge, MA 02139, United States
The Boulevard, Langford Lane, Kidlington, Oxford OX5 1GB, United Kingdom

Notices
Knowledge and best practice in this field are constantly changing. As new research and experience broaden our understanding, changes in research methods, professional practices, or medical treatment may become necessary.

Practitioners and researchers must always rely on their own experience and knowledge in evaluating and using any information, methods, compounds, or experiments described herein. In using such information or methods they should be mindful of their own safety and the safety of others, including parties for whom they have a professional responsibility.

To the fullest extent of the law, neither the Publisher nor the authors, contributors, or editors, assume any liability for any injury and/or damage to persons or property as a matter of products liability, negligence or otherwise, or from any use or operation of any methods, products, instructions, or ideas contained in the material herein.

ISBN: 978-0-12-822961-3

For Information on all Academic Press publications visit our website at https://www.elsevier.com/books-and-journals

Publisher: Katey Birtcher
Senior Acquisitions Editor: Steve Merken
Editorial Project Manager: Naomi Robertson
Production Project Manager: Kamatchi Madhavan
Cover Designer: Greg Harris

Typeset by MPS Limited, Chennai, India
Printed in India
Last digit is the print number: 9 8 7 6 5 4 3 2 1

Working together to grow libraries in developing countries

www.elsevier.com • www.bookaid.org

Dedicated to my wife Fereshteh and my children Kourosh and Soroush

and

In the memory of Professor Alkis Constantinides
(1941–2018)

Preface

This book emphasizes the derivation of a variety of numerical methods and their application to the solution of engineering problems, with special attention to problems in the chemical engineering field. These algorithms encompass linear and nonlinear algebraic equations, eigenvalue problems, finite difference methods, interpolation, differentiation and integration, ordinary differential equations, boundary value problems, partial differential equations, and linear and nonlinear regression analysis. MATLAB® and Microsoft Excel are adopted as the calculation environment throughout the book because of their ability to perform calculations in matrix form, their large library of built-in functions, their strong structural language, and their rich graphical visualization tools. The reader is expected to have a basic knowledge of using MATLAB and Excel.

Several worked examples are given in each chapter to demonstrate the numerical techniques. Most of these examples require computer programs for their solution. These examples are solved with MATLAB programming or in Excel. This book is accompanied by all the programs and Excel files that appear in the text. In the examples solved by MATLAB it is tried to present a general function that implements the particular method that is given as the worked example. It may be applied to solve problems in the same category of application. The Excel spreadsheets are, however, specific to that example, but the steps for solving a similar problem are shown in detail. The algorithms for many methods are given as flowcharts throughout the text. These flowcharts should be considered as a general guide; there are not flowcharts for all methods. The programs and spreadsheets are given in the source code so that they can be modified easily to suit the needs of the user. In addition, the programs and spreadsheets are described in detail in the text to provide the reader with a thorough background and understanding of how MATLAB is used to implement the numerical methods. For a better understanding of the methods, many examples are solved step by step manually in the text.

It is worth mentioning that the main purpose of this book is to teach the student numerical methods and problem solving rather than be a manual for MATLAB and Excel. Therefore new MATLAB functions and Excel spreadsheets have been developed to demonstrate the numerical methods covered in this text. Admittedly, MATLAB and Excel already have their own built-in functions for some of the methods introduced in this book. The built-in functions are mentioned and discussed, whenever they exist. Also a section at the end of each chapter is dedicated to introducing the functions in MATLAB and Excel that can carry out the calculations presented in the chapter.

Basic and advanced numerical methods are covered in each chapter. Whenever feasible for a specific course, the more advanced techniques are covered in the last few sections of each chapter. A one-semester graduate-level course in applied numerical methods would cover all the material in this book. An undergraduate course (junior or senior level) would cover the more basic methods in each chapter. The professor teaching the course may omit some sections in an undergraduate course. Of course, this choice is left to the discretion of the professor.

I would like to thank Professor Shuguang Deng from Arizona State University for encouraging me to revise my previous book (coauthored with the late Professor Alkis Constantinides) and also supplying several useful after-chapter problems. I deeply appreciate the kind help of Seyed Majid

Amirshahkarami and Ali Sadeghikia for typing the equations and Mr. Sina Kheirabadi for checking the m-files. I am especially grateful to Kamal Nosrati for useful discussions, brilliant suggestions, and for providing many examples. Finally, I would like to extend my gratitude to my colleagues Professors Rahmat Sotudeh-Gharebagh and Reza Zarghami for their support and synergy.

Teaching ancillaries for this book, including solutions manual, downloadable MATLAB code, and image bank, are available online to qualified instructors. Visit https://educate.elsevier.com/book/details/9780128229613 for more information and to register for access.

Navid Mostoufi

Contents

NONLINEAR EQUATIONS

CHAPTER OUTLINE

MOTIVATION

Many problems in engineering and science require the solution of nonlinear equations. Several examples of such problems drawn from the field of chemical engineering are discussed. Many nonlinear equations are in the form of a polynomial. Therefore, the types of roots of polynomial equations and their approximation are discussed first. A number of effective iterative methods for finding roots of a general nonlinear algebraic equation are described next. Some of these methods require only one starting point, while some others need two initial guesses for root finding. Methods of solution for a set of nonlinear algebraic equations are also developed in this chapter.

Applied Numerical Methods for Chemical Engineers. DOI: https://doi.org/10.1016/B978-0-12-822961-3.00001-7

1.1 INTRODUCTION

In thermodynamics, the pressure-volume-temperature relationship of real gases is described by the equation of state. There are several semitheoretical or empirical equations, such as Redlich−Kwong (RK), Soave−Redlich−Kwong (SRK), and the Benedict−Webb−Rubin (BWR) equations, which have been used extensively in chemical engineering. For example, the SRK equation of state has the form:

$$P = \frac{RT}{V-b} - \frac{a\alpha}{V(V+b)} \tag{1.1}$$

where P, V, and T are the pressure, specific volume, and temperature, respectively. R is the gas constant, α is a function of temperature, and a and b are constants, specific for each gas. Eq. (1.1) is a third-degree polynomial in V and can be easily rearranged into the canonical form for a polynomial, which is:

$$Z^3 - Z^2 + (A - B - B^2)Z - AB = 0 \tag{1.2}$$

where $Z = PV/RT$ is the compressibility factor, $A = \alpha aP/R^2T^2$, and $B = bP/RT$. Therefore, the problem of finding the specific volume of a gas at a given temperature and pressure reduces to the problem of finding the appropriate root of a polynomial equation.

In the calculations for multicomponent separations, it is often necessary to estimate the minimum reflux ratio of a multistage distillation column. The shortcut method developed for this purpose requires the solution of the equation [1]:

$$\sum_{j=1}^{n} \frac{\alpha_j z_{jF} F}{\alpha_j - \varphi} - F(1-q) = 0 \tag{1.3}$$

where F is the molar feed flow rate, n is the number of components in the feed, z_{jF} is the mole fraction of each component in the feed, q is the quality of the feed, α_j is the relative volatility of each component at average column conditions, and φ is the root of the equation. The feed flow rate, composition, and quality are usually known, and the average column conditions can be approximated. Therefore, φ is the only unknown in Eq. (1.3). Because this equation is a polynomial in φ of degree n, there are n possible values of φ (roots) that satisfy the equation.

The friction factor f for turbulent flow of an incompressible fluid in a pipe is given by the nonlinear Colebrook equation:

$$\sqrt{\frac{1}{f}} = -2\log\left(\frac{\varepsilon/D}{3.7} + \frac{2.51}{Re\sqrt{f}}\right) \tag{1.4}$$

where ε and D are the roughness and inside diameter of the pipe, respectively, and Re is the Reynolds number. This equation does not readily rearrange itself into a polynomial form; however, it can be arranged so that all the nonzero terms are on the left side of the equation as follows:

$$\sqrt{\frac{1}{f}} + 2\log\left(\frac{\varepsilon/D}{3.7} + \frac{2.51}{Re\sqrt{f}}\right) = 0 \tag{1.5}$$

The method of differential operators is applied in finding analytical solutions of nth-order linear homogeneous differential equations. The general form of an nth-order linear homogeneous differential equation is:

$$a_n \frac{d^n y}{dx^n} + a_{n-1} \frac{d^{n-1} y}{dx^{n-1}} + \ldots + a_1 \frac{dy}{dx} + a_0 y = 0 \qquad (1.6)$$

By defining D as the differentiation operator with respect to x:

$$D = \frac{d}{dx} \qquad (1.7)$$

Eq. (1.6) can be written as:

$$\left[a_n D^n + a_{n-1} D^{n-1} + \ldots + a_1 D + a_0 \right] y = 0 \qquad (1.8)$$

where the bracketed term puts the differential equation in operator form. For Eq. (1.8) to have a nontrivial solution, the differential operator must be equal to zero:

$$a_n D^n + a_{n-1} D^{n-1} + \ldots + a_1 D + a_0 = 0 \qquad (1.9)$$

This, of course, is a polynomial equation in D whose roots must be evaluated in order to construct the complementary solution of the differential equation.

The field of process dynamics and control often requires the location of the roots of transfer functions which usually have the form of polynomial equations. In kinetics and reaction design, the simultaneous solution of rate equations and energy balances results in mathematical models of simultaneous nonlinear and transcendental equations. Methods of solution for these and other such problems are developed in this chapter. It is worth noting that most systems encountered in practice are multivariable in nature, but we will start with single variable problems first to develop the foundation in a simple way.

1.2 TYPES OF ROOTS AND THEIR APPROXIMATION

All the nonlinear equations presented in this chapter can be written in the general form:

$$f(x) = 0 \qquad (1.10)$$

where x is a single variable that can have multiple values (roots) that satisfy this equation. The function $f(x)$ may assume a variety of nonlinear functionalities ranging from that of a polynomial equation whose canonical form is:

$$f(x) = a_n x^n + a_{n-1} x^{n-1} + \ldots + a_1 x + a_0 = 0 \qquad (1.11)$$

to transcendental equations, which involve trigonometric, exponential, or logarithmic terms. The roots of these functions could be

1. Real and distinct
2. Real and repeated
3. Complex conjugates
4. A combination of any or all of the above

The real parts of the roots may be positive, negative, or zero.

Fig. 1.1 graphically demonstrates all the above cases using fourth-degree polynomials. Fig. 1.1a is a plot of the following polynomial equation:

$$x^4 + 6x^3 + 7x^2 - 6x - 8 = 0 \tag{1.12}$$

which has four real and distinct roots at -4, -2, -1, and 1, as indicated by the intersections of the function with the x-axis. Fig. 1.1b is a graph of the following polynomial equation:

$$x^4 + 7x^3 + 12x^2 - 4x - 16 = 0 \tag{1.13}$$

which has two real and distinct roots at -4 and 1, and two real and repeated roots at -2. The point of tangency with the x-axis indicates the presence of the repeated roots. At this point $f(x) = 0$ and $f'(x) = 0$. Fig. 1.1c is a plot of the following polynomial equation:

$$x^4 - 6x^3 + 18x^2 - 30x + 25 = 0 \tag{1.14}$$

which has only complex roots at $1 \pm 2i$ and $2 \pm i$. In this case, no intersection with the x-axis of the Cartesian coordinate system occurs because all of the roots are located in the complex plane.

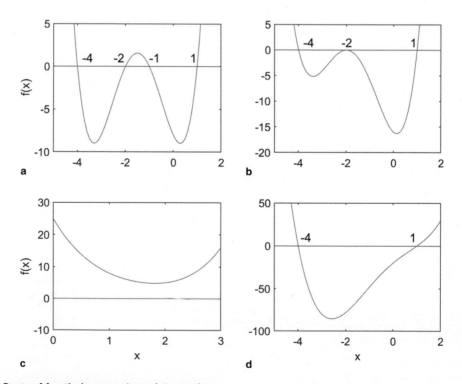

FIG. 1.1 Roots of fourth-degree polynomial equations.

(a) Four real distinct. (b) Two real and two repeated. (c) Four complex. (d) Two real and two complex.

Finally, Fig. 1.1d demonstrates the presence of two real and two complex roots with the following polynomial equation:

$$x^4 + x^3 - 5x^2 + 23x - 20 = 0 \tag{1.15}$$

whose roots are -4, 1, and $1 \pm 2i$. As expected, the function crosses the x-axis only at two points: -4 and 1.

The roots of an nth-degree polynomial such as Eq. (1.11) may be verified using Newton's relations, which are:

Newton's first relation:

$$\sum_{i=1}^{n} x_i = -\frac{a_{n-1}}{a_n} \tag{1.16}$$

where x_i are the roots of the polynomial.

Newton's second relation:

$$\sum_{i,j=1}^{n} x_i x_j = \frac{a_{n-2}}{a_n} \tag{1.17}$$

Newton's third relation:

$$\sum_{i,j,k=1}^{n} x_i x_j x_k = -\frac{a_{n-3}}{a_n} \tag{1.18}$$

Newton's nth relation:

$$x_1 x_2 x_3 \ldots x_n = (-1)^n \frac{a_0}{a_n} \tag{1.19}$$

where $i \neq j \neq k \neq \ldots$ for all the foregoing equations, which contain products of roots.

In certain problems, it may be necessary to locate all the roots of the equation, including the complex roots. This is the case in finding the zeros and poles of transfer functions in process control applications and in formulating the analytical solution of linear nth-order differential equations. On the other hand, some other problems may require the location of only one of the roots. For example, in the solution of the equation of state, the positive real root is the one of interest. In any case, the physical constraints of the problem may dictate the feasible region of search where only a subset of the total number of roots may be indicated. Also, the physical characteristics of the problem may provide an approximate value of the desired root.

The most effective way of finding the roots of nonlinear equations is to devise iterative algorithms that start at an initial estimate of a root and converge to the exact value of the desired root in a finite number of steps. Once a root is located, it may be removed by synthetic division if the equation is of the polynomial form. Otherwise, convergence on the same root may be avoided by initiating the search for subsequent roots in a different region of the feasible space.

For equations of the polynomial form, Descartes' rule of sign may be used to determine the number of positive and negative roots. This rule states: The number of positive roots is equal to the number of sign changes in the coefficients of the equation (or less than that by an even integer); the number of negative roots is equal to the number of sign repetitions in the coefficients (or less than that by an even integer). Zero coefficients are counted as positive [2]. The purpose of the qualifier "less than that by an even integer" is to allow for the existence of conjugate pairs of complex roots. The reader is encouraged to apply Descartes' rule to Eqs. (1.12)–(1.15) to verify the results already shown.

If the problem to be solved is a purely mathematical one, that is, the model whose roots are being sought has no physical origin, then brute-force methods would have to be used to establish approximate starting values of the roots for the iterative technique. Two categories of such methods will be mentioned here. The first one is a truncation method applicable to the equation of the polynomial form. For example, the following polynomial:

$$a_4 x^4 + a_3 x^3 + a_2 x^2 + a_1 x + a_0 = 0 \tag{1.20}$$

may have its lower-powered terms truncated:

$$a_4 x^4 + a_3 x^3 \simeq 0 \tag{1.21}$$

to yield an approximation of one of the roots:

$$x = -\frac{a_3}{a_4} \tag{1.22}$$

Alternatively, if the higher-powered terms are truncated:

$$a_1 x + a_0 \simeq 0 \tag{1.23}$$

the approximate root is:

$$x = -\frac{a_0}{a_1} \tag{1.24}$$

This technique applied to the SRK equation (Eq. 1.2) results in:

$$Z = \frac{PV}{RT} = 1 \tag{1.25}$$

This, of course, is the well-known *ideal gas law*, which is an excellent approximation of the pressure-volume-temperature relationship of real gases at low pressures. On the other end of the polynomial, truncation of the higher-powered terms results in:

$$Z = \frac{AB}{A - B - B^2} \tag{1.26}$$

giving a value of Z very close to zero, which is the case for liquids. In this case, the physical considerations of the problem determine that Eq. (1.25) or Eq. (1.26) should be used for the gas phase or liquid phase respectively to initiate the iterative search technique for the real root starting from the approximate estimate.

Another method of locating initial estimates of the roots is to scan the entire region of search by small increments and to observe the steps in which a change of sign in the function $f(x)$ occurs. This signals that the function $f(x)$ crosses the x-axis within the particular step. This search can be done easily in the MATLAB® environment using *fplot* function. Once the function $f(x)$ is introduced in a MATLAB function *file_name.m*, the statement *fplot('file_name',[a b])* shows the plot of the function from $x = a$ to $x = b$. The values of a and b may be changed until the plot crosses the x-axis.

The scan method may be a rather time-consuming procedure for polynomials whose roots lie in a large region of search. A variation of this search is the method of bisection, which repeatedly divides the interval of search by 2 and always retains that half of the search interval in which the change of sign has occurred. When the range of search has been narrowed down sufficiently, a more accurate search technique would then be applied within that step to refine the value of the root.

More efficient methods based on the rearrangement of the function to $x = g(x)$ (*method of successive substitution* or *fixed-point iteration*), Wegstein, the bisection method (*interval halving*), linear interpolation of the function (*method of false position*), and the tangential descent of the function (*Newton-Raphson method*) will be described in the next four sections of this chapter.

1.3 THE METHOD OF SUCCESSIVE SUBSTITUTION

The simplest single-point iterative root-finding technique can be developed by rearranging the function $f(x)$ so that x is on the left-hand side of the equation:

$$x = g(x) \tag{1.27}$$

The function $g(x)$ is a formula to predict the root. In fact, the root is the intersection of the line $y = x$ with the curve $y = g(x)$. Starting with an initial value of x_1, as shown in Fig. 1.2a, we obtain the value of x_2:

$$x_2 = g(x_1) \tag{1.28}$$

which is closer to the root than x_1 and can be used as an initial value for the next iteration. Therefore, the general iterative formula for this method is:

$$x_{n+1} = g(x_n) \tag{1.29}$$

A sufficient condition for convergence of Eq. (1.29) to the root x^* is that $g'(x) < 1$ for all x in the search interval. Fig. 1.2b shows the case when this condition is not valid and the method

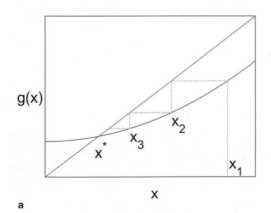
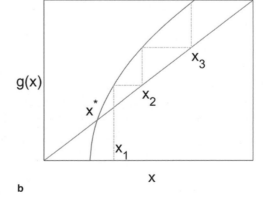

a   b

FIG. 1.2 Use of the method of successive substitution.

When the method (a) converges and (b) diverges.

diverges. This analytical test is often difficult in practice. In a computer program, it is easier to determine whether $x_3 - x_2 < x_2 - x_1$ and therefore, the successive x_n values converge. This method sometimes is referred to as the *successive substitution method*. The advantage of this method is that it can be started with only a single point, without the need for calculating the derivative of the function.

EXAMPLE 1.1 SOLUTION OF THE COLEBROOK EQUATION BY THE SUCCESSIVE SUBSTITUTION METHOD

Calculate the friction factor from the Colebrook equation (Eq. 1.4) for the flow of a fluid in a pipe with $\varepsilon/D = 10^{-4}$ and $Re = 10^5$.

Method of Solution

Eq. (1.29) is used for solving this problem. The iterative procedure stops when the difference between two succeeding approximations of the root is less than the convergence criterion or when the number of iterations reaches a specified value, whichever is satisfied first.

The flowchart of calculations for this example is shown in Fig. E1.1a. First, input data should be determined. These data include values specific to the function (ε/D and Re for the Colebrook equation) and parameters of the solution method (initial guess x_1, convergence value δ, the maximum number of iterations N_{max}). Next, the function for which the root is to be evaluated should be presented in the form of Eq. (1.27). In the case of the Colebrook equation, Eq. (1.4) is rearranged to solve for f:

$$f = \frac{1}{\left[2\log\left(\frac{\varepsilon/D}{3.7} + \frac{2.51}{Re\sqrt{f}} \right) \right]^2} = g(f)$$

The right-hand side of this equation is taken as $g(f)$.

The next section is the main iteration loop, in which the iteration according to Eq. (1.29) takes place and the convergence is checked. Numerical results will be provided when convergence is reached.

Starting with an initial value of $f_1 = 0.01$, the first three iterations are as follows:

$$f_2 = g(f_1) = \frac{1}{\left[2\log\left(\frac{\varepsilon/D}{3.7} + \frac{2.51}{Re\sqrt{f_1}} \right) \right]^2} = \frac{1}{\left[2\log\left(\frac{10^{-4}}{3.7} + \frac{2.51}{10^5\sqrt{0.01}} \right) \right]^2} = 0.0198$$

$$f_3 = g(f_2) = \frac{1}{\left[2\log\left(\frac{\varepsilon/D}{3.7} + \frac{2.51}{Re\sqrt{f_2}} \right) \right]^2} = \frac{1}{\left[2\log\left(\frac{10^{-4}}{3.7} + \frac{2.51}{10^5\sqrt{0.0198}} \right) \right]^2} = 0.0184$$

(Continued)

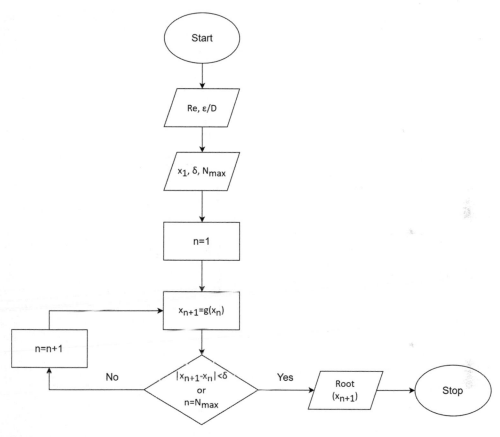

FIG. E1.1a

Flowchart of the method of successive substitution.

EXAMPLE 1.1 (CONTINUED)

$$f_4 = g(f_3) = \cfrac{1}{\left[2\log\left(\cfrac{\varepsilon/D}{3.7} + \cfrac{2.51}{Re\sqrt{f_3}}\right)\right]^2} = \cfrac{1}{\left[2\log\left(\cfrac{10^{-4}}{3.7} + \cfrac{2.51}{10^5\sqrt{0.0184}}\right)\right]^2} = 0.0185$$

It can be seen from these calculations that this method has converged to 0.0185 after three iterations. In fact, the convergence criterion is 10^{-4} in this example.

These calculations can also be performed in Excel. Fig. E1.1b demonstrates the calculation steps in Excel.

(Continued)

EXAMPLE 1.1 (CONTINUED)

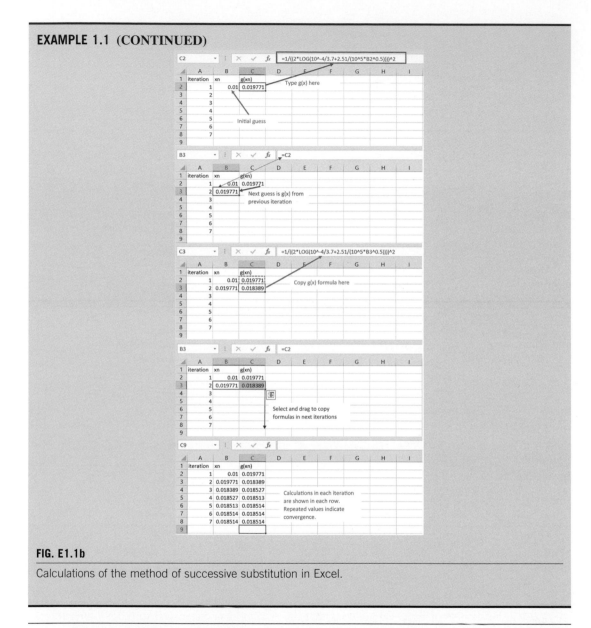

FIG. E1.1b

Calculations of the method of successive substitution in Excel.

1.4 THE WEGSTEIN METHOD

The Wegstein method may also be used for the solution of the equations of the form:

$$x = g(x) \tag{1.27}$$

Starting with an initial value of x_1, we first obtain another estimation of the root from:

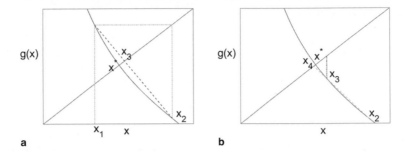

FIG. 1.3

The Wegstein method.

$$x_2 = g(x_1) \tag{1.28}$$

As shown in Fig. 1.3, x_2 does not have to be closer to the root than x_1. At this stage, we estimate the function $g(x)$ with a line passing from the points $[x_1, g(x_1)]$ and $[x_2, g(x_2)]$:

$$\frac{y - g(x_1)}{x - x_1} = \frac{g(x_2) - g(x_1)}{x_2 - x_1} \tag{1.30}$$

and find the next estimation of the root, x_3, from the intersection of the line (1.30) and the line $y = x$:

$$x_3 = \frac{x_1 g(x_2) - x_2 g(x_1)}{x_1 - g(x_1) - x_2 + g(x_2)} \tag{1.31}$$

It can be seen from Fig. 1.3a that x_3 is closer to the root than either x_1 or x_2. In the next iteration, we pass the line from the points $[x_2, g(x_2)]$ and $[x_3, g(x_3)]$ and again evaluate the next estimation of the root from the intersection of this line with $y = x$. Therefore, the general iterative formula for the Wegstein method is:

$$x_{n+1} = \frac{x_{n-1} g(x_n) - x_n g(x_{n-1})}{x_{n-1} - g(x_{n-1}) - x_n + g(x_n)} \qquad n \geq 2 \tag{1.32}$$

The Wegstein method converges, even under conditions in which the method of successive substitution does not. Moreover, it accelerates the convergence when the successive substitution method is stable.

EXAMPLE 1.2 SOLUTION OF THE COLEBROOK EQUATION BY THE WEGSTEIN METHOD

Repeat Example 1.1 and calculate the friction factor from the Colebrook equation (Eq. 1.4) for the flow of a fluid in a pipe with $\varepsilon/D = 10^{-4}$ and $Re = 10^5$.

(Continued)

EXAMPLE 1.2 (CONTINUED)
Method of Solution

Eq. (1.32) is used for solving this problem. Again, the iterative procedure stops when the difference between two succeeding approximations of the root is less than the convergence criterion or when the number of iterations reaches a specified value, whichever is satisfied first.

The flowchart of calculations for this example is shown in Fig. E1.2a. First, input data should be determined. These data include values specific to the function (ε/D and Re for the

(Continued)

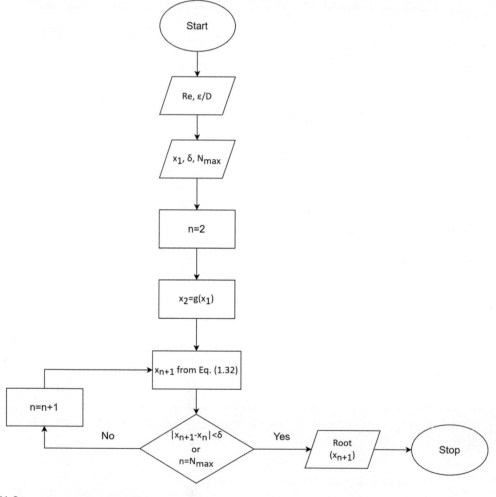

FIG. E1.2a

Flowchart of the Wegstein method.

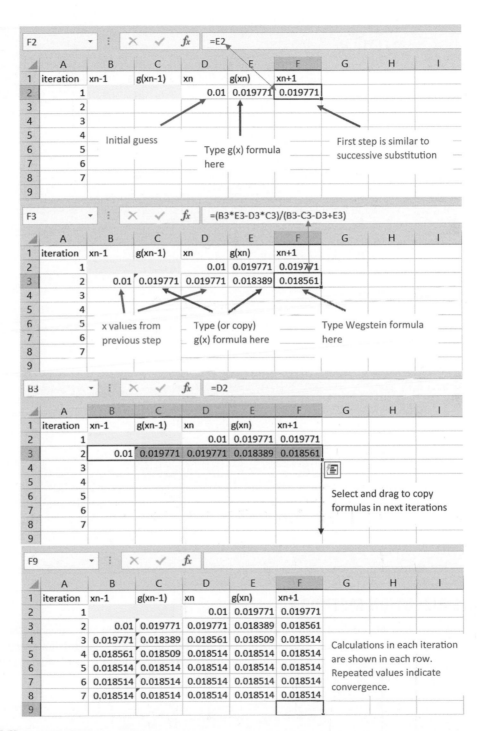

FIG. E1.2b

Calculations of the Wegstein method in Excel.

EXAMPLE 1.2 (CONTINUED)

Colebrook equation) and parameters of the solution method (initial guess x_1, convergence value δ, the maximum number of iterations N_{max}). Next, the function for which the root is to be evaluated should be presented in the form of Eq. (1.27). In this example, the same function as in Example 1.1 can be employed:

$$f = \frac{1}{\left[2\log\left(\frac{\varepsilon/D}{3.7} + \frac{2.51}{Re\sqrt{f}}\right)\right]^2} = g(f)$$

By evaluating this function for the initial guess, the second point in the Wegstein method, x_2, is obtained, based on Eq. (1.28). The next section in the function is the main iteration loop, in which the iteration according to Eq. (1.32) takes place and the convergence is checked. Numerical results will be provided when convergence is reached.

Starting with an initial value of $f_1 = 0.01$, the second point is calculated from:

$$f_2 = \frac{1}{\left[2\log\left(\frac{\varepsilon/D}{3.7} + \frac{2.51}{Re\sqrt{f_1}}\right)\right]^2} = \frac{1}{\left[2\log\left(\frac{10^{-4}}{3.7} + \frac{2.51}{10^5\sqrt{0.01}}\right)\right]^2} = 0.0198$$

The next two iterations are as follows:

$g(f_2) = 0.0184,$

$$f_3 = \frac{f_1 g(f_2) - f_2 g(f_1)}{f_1 - g(f_1) - f_2 + g(f_2)} = \frac{0.01 \times 0.0184 - 0.0198 \times 0.0198}{0.01 - 0.0198 - 0.0198 + 0.0184} = 0.0186$$

$g(f_3) = 0.0188,$

$$f_4 = \frac{f_2 g(f_3) - f_3 g(f_2)}{f_2 - g(f_2) - f_3 + g(f_3)} = \frac{0.0198 \times 0.0185 - 0.0186 \times 0.0184}{0.0198 - 0.0184 - 0.0186 + 0.0185} = 0.0185$$

It can be seen from these calculations that this method has converged to 0.0185 after two iterations, which is faster than in the method of successive substitution (see Example 1.1). The convergence criterion is also 10^{-4} in this example.

These calculations can also be performed in Excel. Fig. E1.2b demonstrates the calculation steps in Excel.

1.5 THE BISECTION METHOD

This method, which is also called interval halving, is very simple and always guarantees convergence. The calculation steps of the bisection method are shown in Fig. 1.4. In this method, we should first scan the domain of interest to find two points with opposite function signs. These two points are shown as x_1 and x_2 in Fig. 1.4. Now, we locate the midpoint:

$$x_3 = \frac{x_1 + x_2}{2} \tag{1.33}$$

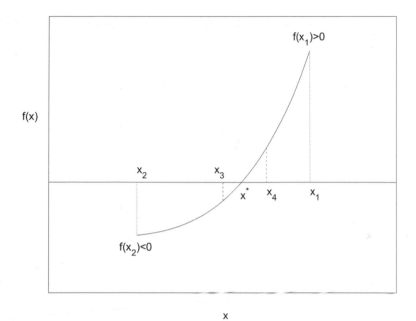

FIG. 1.4

The bisection method.

and evaluate the function value at this point. For the curve shown in Fig. 1.4, the function value at x_3 is negative. Therefore, the root x^* is between x_1 and x_3 and x_3 replaces x_2 in the next iteration. Next, we repeat this operation with x_1 and x_3 to find x_4:

$$x_4 = \frac{x_1 + x_3}{2} \tag{1.34}$$

This procedure continues until the sign changing interval becomes as small as desired.

The general formula for the bisection method is:

$$x_n = \frac{x^+ + x^-}{2} \tag{1.35}$$

in which x^+ is the value at which $f(x^+) > 0$ and x^- is the value at which $f(x^-) < 0$. For the next iteration, x^+ or x^- should be replaced by x_n according to the sign of $f(x_n)$.

Although the bisection method is robust and guarantees convergence, it is relatively slow, especially when near the root. For this reason, it is usually used to obtain a rough approximation of the root and then switch to a faster root finding method.

1.6 THE METHOD OF LINEAR INTERPOLATION

This technique, also known as the method of false position, is based on linear interpolation between two points on the function that have been found by a scan to lie on either side of a root. For example, x_1 and x_2 in Fig. 1.5a are positioned on opposite sides of the root x^* of the nonlinear function $f(x)$. The points $[x, f(x)]$ and $[x_2, f(x_2)]$ are connected by a straight line, which we will call a chord, whose equation is:

$$y(x) = ax + b \tag{1.36}$$

Since this chord passes through the two points $[x_1, f(x_1)]$ and $[x_2, f(x_2)]$, its slope is:

$$a = \frac{f(x_2) - f(x_1)}{x_2 - x_1} \tag{1.37}$$

and its y-intercept is:

$$b = f(x_1) - ax_1 \tag{1.38}$$

Eq. (1.36) then becomes:

$$y(x) = \left[\frac{f(x_2) - f(x_1)}{x_2 - x_1}\right] x + \left\{ f(x_1) - \left[\frac{f(x_2) - f(x_1)}{x_2 - x_1}\right] x_1 \right\} \tag{1.39}$$

Locating x_3 using Eq. (1.39), where $y(x_3) = 0$:

$$x_3 = x_1 - \frac{f(x_1)(x_2 - x_1)}{f(x_2) - f(x_1)} \tag{1.40}$$

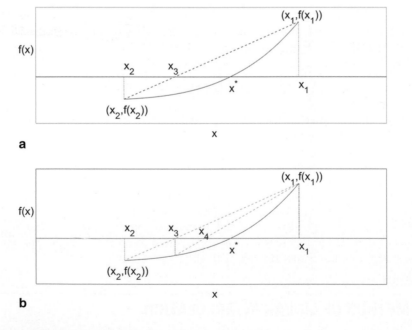

FIG. 1.5

Method of linear interpolation.

Note that for the shape of the curve chosen in Fig. 1.5, x_3 is nearer to the root x^* than either x_1 or x_2. This of course will not always be the case with all functions. A discussion of criteria for convergence will be given in the next section.

According to Fig. 1.5, $f(x_3)$ has the same sign as $f(x_2)$; therefore, x_2 may be replaced by x_3. Now repeating the foregoing operation and connecting the points $[x_1, f(x_1)]$ and $[x_3, f(x_3)]$ with a new chord, as shown in Fig. 1.5b, we obtain the value of x_4:

$$x_4 = x_1 - \frac{f(x_1)(x_3 - x_1)}{f(x_3) - f(x_1)} \tag{1.41}$$

which is nearer to the root than x_3. For the general formulation of this method consider x^+ to be the value at which $f(x^+) > 0$ and x^- to be the value at which $f(x^-) < 0$. The next improved approximation of the root of the function may be calculated by successive application of the general formula:

$$x_n = x^+ - \frac{f(x^+)(x^+ - x^-)}{f(x^+) - f(x^-)} \tag{1.42}$$

For the next iteration, x^+ or x^- should be replaced by x_n according to the sign of $f(x_n)$.

This method is known by several names: method of chords, linear interpolation, false position (*regula falsi*). Its simplicity of calculation (no need for evaluating derivatives of the function) gave it its popularity in the early days of numerical computations. However, its accuracy and speed of convergence are hampered by the choice of x_1, which forms the pivot point for all subsequent iterations.

EXAMPLE 1.3 SOLUTION OF THE COLEBROOK EQUATION BY THE METHOD OF LINEAR INTERPOLATION

Repeat Example 1.1 and calculate the friction factor from the Colebrook equation (Eq. 1.4) for the flow of a fluid in a pipe with $\varepsilon/D = 10^{-4}$ and $Re = 10^5$.

Method of Solution

Eq. (1.42) is used for solving this problem. As before, the iterative procedure stops when the difference between two succeeding approximations of the root is less than the convergence criterion or when the number of iterations reaches a specified value, whichever is satisfied first.

The flowchart of calculations for this example is shown in Fig. E1.3a. First, input data should be determined. These data include values specific to the function (ε/D and Re for the Colebrook equation) and parameters of the solution method (two initial guesses x_1 and x_2, convergence value δ, the maximum number of iterations N_{max}). The function for which the root is to be evaluated is Eq. (1.5):

$$F(f) = \sqrt{\frac{1}{f}} + 2\log\left(\frac{\varepsilon/D}{3.7} + \frac{2.51}{Re\sqrt{f}}\right)$$

(Continued)

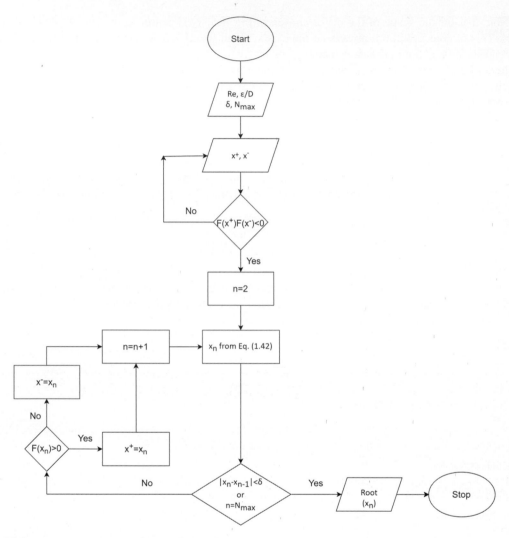

FIG. E1.3a

Flowchart of the method of linear interpolation.

EXAMPLE 1.3 (CONTINUED)

We should find two initial guesses for which the function has opposite signs. As shown in the flowchart, if the function in these values is of the same sign, new guesses should be made until the function values with opposite signs are found.

(Continued)

EXAMPLE 1.3 (CONTINUED)

The next section in the function is the main iteration loop, in which the iteration according to Eq. (1.42) takes place. After the next estimation of the root is calculated, it should be checked whether the function value at this point is positive or negative. This estimation replaces either x^+ or x^- depending on the sign of the function. The convergence is checked in each iteration and numerical results will be provided when convergence is reached.

We start calculations with two initial values of $f_1 = x^+ = 0.01$ and $f_2 = x^- = 0.03$. The function has opposite signs at these values:

$$F(0.01) = \sqrt{\frac{1}{0.01}} + 2\log\left(\frac{10^{-4}}{3.7} + \frac{2.51}{10^5\sqrt{0.01}}\right) = 2.8882 > 0$$

$$F(0.03) = \sqrt{\frac{1}{0.03}} + 2\log\left(\frac{10^{-4}}{3.7} + \frac{2.51}{10^5\sqrt{0.03}}\right) = -1.7557 < 0$$

The iterations are as follows:

$$f_3 = 0.01 - \frac{2.8882(0.01 - 0.03)}{2.8882 - (-1.7557)} = 0.0224$$

The function is negative at this point:

$$F(0.0224) = \sqrt{\frac{1}{0.0224}} + 2\log\left(\frac{10^{-4}}{3.7} + \frac{2.51}{10^5\sqrt{0.0224}}\right) = -0.7460 < 0$$

Therefore, f_3 replaces x^- and Eq. (1.42) is repeated:

$$f_4 = 0.01 - \frac{2.8882(0.01 - 0.0224)}{2.8882 - (-0.7460)} = 0.0199$$

Again, the function is negative at this point and f_4 replaces x^- in the next iteration:

$$F(0.0199) = \sqrt{\frac{1}{0.0199}} + 2\log\left(\frac{10^{-4}}{3.7} + \frac{2.51}{10^5\sqrt{0.0199}}\right) = -0.2850 < 0$$

$$f_5 = 0.01 - \frac{2.8882(0.01 - 0.0199)}{2.8882 - (-0.2850)} = 0.0190$$

Likewise,

$$F(0.0190) = \sqrt{\frac{1}{0.0190}} + 0.86\ln\left(\frac{10^{-4}}{3.7} + \frac{2.51}{10^5\sqrt{0.0190}}\right) = -0.1039 < 0$$

(Continued)

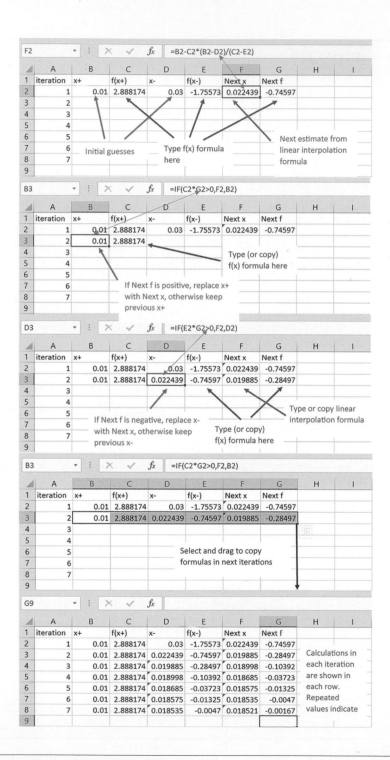

FIG. E1.3b

Calculations of the method of linear interpolation in Excel.

EXAMPLE 1.3 (CONTINUED)

$$f_6 = 0.01 - \frac{2.8882(0.01 - 0.0190)}{2.8882 - (-0.1039)} = 0.0187$$

$$F(0.0187) = \sqrt{\frac{1}{0.0187}} + 0.86\ln\left(\frac{10^{-4}}{3.7} + \frac{2.51}{10^5\sqrt{0.0187}}\right) = -0.0372 < 0$$

$$f_7 = 0.01 - \frac{2.8882(0.01 - 0.0187)}{2.8882 - (-0.0372)} = 0.0186$$

It can be seen from these calculations that this method has converged to 0.0186 after five iterations, with a convergence criterion of 10^{-4}. Although more iterations are needed in the method of linear interpolation, compared to the successive substitution and Wegstein methods (see Examples 1.1 and 1.2), to reach the root of the Colebrook equation, convergence is guaranteed in this method, which is not always the case for the aforementioned other methods.

The calculations of this example can also be performed in Excel. Fig. E1.3b demonstrates the calculation steps in Excel.

1.7 THE NEWTON-RAPHSON METHOD

The best known, and possibly the most widely used, technique for locating the roots of nonlinear equations is the *Newton-Raphson* method. This method is based on a Taylor series expansion of the nonlinear function $f(x)$ around an initial estimate (x_1) of the root:

$$f(x) = f(x_1) + f'(x_1)(x - x_1) + \frac{f''(x_1)}{2!}(x - x_1)^2 + \frac{f'''(x_1)}{3!}(x - x_1)^3 + \ldots \tag{1.43}$$

Because what is being sought is the value of x which forces the function $f(x)$ to assume zero value, the left side of Eq. (1.43) is set to zero and the resulting equation is solved for x. However, the right-hand side is an infinite series. Therefore, a finite number of terms must be retained and the remaining terms must be truncated. Retaining only the first two terms on the right-hand side of the Taylor series is equivalent to linearizing the function $f(x)$. This operation results in:

$$x = x_1 - \frac{f(x_1)}{f'(x_1)} \tag{1.44}$$

that is, the value of x is calculated from x_1 by correcting this initial guess by $f(x_1)/f'(x_1)$. The geometrical significance of this correction is shown in Fig. 1.6a. The value of x is obtained by moving from x_1 to x in the direction of the tangent $f'(x)$ of the function (x).

Because the Taylor series was truncated, retaining only two terms, the new value x will not yet satisfy Eq. (1.10). We will designate this value as x_2 and reapply the Taylor series linearization at

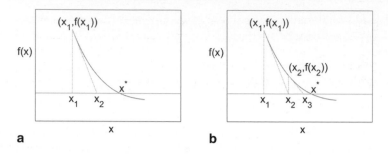

FIG. 1.6

The Newton-Raphson method.

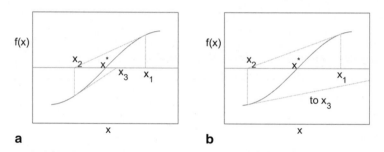

FIG. 1.7 Choice of initial guesses affects the convergence of the Newton-Raphson method.

(a) Convergence. (b) Divergence.

x_2 (shown in Fig. 1.6b) to obtain x_3. Repetitive application of this step converts Eq. (1.44) to an iterative formula:

$$x_{n+1} = x_n - \frac{f(x_n)}{f'(x_n)} \qquad (1.45)$$

In contrast to the method of linear interpolation, the Newton-Raphson method uses the newly found position as the starting point for each subsequent iteration.

In the discussion for both linear interpolation and Newton-Raphson methods, a certain shape of the function was used to demonstrate how these techniques converge toward a root in the space of search. However, the shapes of nonlinear functions may vary drastically and convergence is not always guaranteed. As a matter of fact, divergence is more likely to occur, as shown in Fig. 1.7, unless extreme care is taken in the choice of the initial starting points.

To investigate the convergence behavior of the Newton-Raphson method, one has to examine the term $[f(x_n)/f'(x_n)]$ in Eq. (1.45). This is the error term or correction term applied to the previous estimate of the root at each iteration. A function with a strong vertical trajectory near the root will cause the denominator of the error term to be large; therefore, the convergence will be quite fast. If, however, $f(x)$ is nearly horizontal near the root, the convergence will be slow. If at any point during the search, $f'(x) = 0$, the method will fail due to division by zero. Inflection points on the curve, within the region of search, are also troublesome and may cause the search to diverge.

A sufficient but not necessary condition for convergence of the Newton-Raphson method was stated by Lapidus [3] as follows: "If $f'(x)$ and $f''(x)$ do not change the sign in the interval (x_1, x^*) and if $f(x_1)$ and $f''(x_1)$ have the same sign, the iteration will always converge to x^*." These convergence criteria may be easily programmed as part of the computer program that performs the Newton-Raphson search, and a warning may be issued or other appropriate action may be taken by the computer if the conditions are violated.

A more accurate extension of the Newton-Raphson method is Newton's second-order method, which truncates the right-hand side of the Taylor series (Eq. 1.43) after the third term to yield the equation:

$$\frac{f''(x_1)}{2!}(\Delta x_1)^2 + f'(x_1)\Delta x_1 + f(x_1) = 0 \tag{1.46}$$

where $\Delta x_1 = x - x_1$. This is a quadratic equation in x_1 whose solution is given by:

$$\Delta x_1 = \frac{-f'(x_1) \pm \sqrt{[f'(x_1)]^2 - 2f''(x_1)f(x_1)}}{f''(x_1)} \tag{1.47}$$

The general iterative formula for this method would be:

$$x_{n+1}^+ = x_n - \frac{f'(x_n)}{f''(x_n)} + \frac{\sqrt{[f'(x_n)]^2 - 2f''(x_n)f(x_n)}}{f''(x_n)} \tag{1.48a}$$

or

$$x_{n+1}^- = x_n - \frac{f'(x_n)}{f''(x_n)} - \frac{\sqrt{[f'(x_n)]^2 - 2f''(x_n)f(x_n)}}{f''(x_n)} \tag{1.48b}$$

The choice between Eq. (1.48a) and Eq. (1.48b) will be determined by exploring both values of x^+_{n+1} and x^-_{n+1} and determining which one results in the function $f(x^+_{n+1})$ or $f(x^-_{n+1})$ being closer to zero.

An alternative to the foregoing exploration will be to treat Eq. (1.46) as another nonlinear equation in x and to apply the Newton-Raphson method for its solution:

$$F(\Delta x) = \frac{f''(x_1)}{2!}(\Delta x)^2 + f'(x_1)\Delta x + f(x_1) = 0 \tag{1.49}$$

where

$$\Delta x_{n+1} = \Delta x_n - \frac{F(\Delta x_n)}{F'(\Delta x_n)} \tag{1.50}$$

Two nested Newton-Raphson algorithms would have to be programmed together as follows:

1. Assume a value of x_1.
2. Calculate x_1 from Eq. (1.47).
3. Calculate Δx_2 from Eq. (1.50).
4. Calculate x_2 from $x_2 = x_1 + \Delta x_2$.
5. Repeat steps 2 to 4 until convergence is achieved.

EXAMPLE 1.4 SOLUTION OF THE COLEBROOK EQUATION BY THE NEWTON-RAPHSON METHOD

Repeat Example 1.1 and calculate the friction factor from the Colebrook equation (Eq. 1.4) for the flow of a fluid in a pipe with $\varepsilon/D = 10^{-4}$ and $Re = 10^5$.

Method of Solution

Eq. (1.45) is used for solving this problem. Again, the iterative procedure stops when the difference between two succeeding approximations of the root is less than the convergence criterion or when the number of iterations reaches a specified value, whichever is satisfied first.

The flowchart of calculations for this example is shown in Fig. E1.4a. First, input data should be determined. These data include values specific to the function (ε/D and Re for the Colebrook equation) and parameters of the solution method (initial guess x_1, convergence value δ, the maximum number of iterations N_{max}). The function for which the root is to be evaluated is Eq. (1.5):

$$F(f) = \sqrt{\frac{1}{f}} + 2\log\left(\frac{\varepsilon/D}{3.7} + \frac{2.51}{Re\sqrt{f}}\right)$$

The next section in the function is the main iteration loop, in which the iteration according to Eq. (1.45) takes place. In the Newton-Raphson method, we need to evaluate the derivative of the function in each step. This can be done either analytically or numerically. Because in this example the function is explicitly known, we use the analytical formula of the derivative:

$$F'(f) = -\frac{1}{2f^{3/2}} - \frac{\frac{2.51}{Re}}{\ln 10 f^{3/2}\left(\frac{\varepsilon}{3.7} + \frac{2.51}{Re\sqrt{f}}\right)}$$

The convergence is checked in each iteration and numerical results will be provided when convergence is reached.

Starting with an initial value of $f_1 = 0.01$, the second point is calculated from:

$$F(f_1) = \sqrt{\frac{1}{f_1}} + 2\log\left(\frac{\frac{\varepsilon}{D}}{3.7} + \frac{2.51}{Re\sqrt{f_1}}\right) = \sqrt{\frac{1}{0.01}} + 2\log\left(\frac{10^{-4}}{3.7} + \frac{2.51}{10^5\sqrt{0.01}}\right) = 2.8882$$

$$F'(f_1) = -\frac{1}{2f_1^{3/2}} - \frac{\frac{2.51}{Re}}{\ln 10 f_1^{3/2}\left(\frac{\frac{\varepsilon}{D}}{3.7} + \frac{2.51}{Re\sqrt{f_1}}\right)} = -\frac{1}{2\times 0.01^{3/2}} - \frac{\frac{2.51}{10^5}}{\ln 10 \times 0.01^{3/2}\left(\frac{10^{-4}}{3.7} + \frac{2.51}{10^5\sqrt{0.01}}\right)}$$

$$= -539.2077$$

(*Continued*)

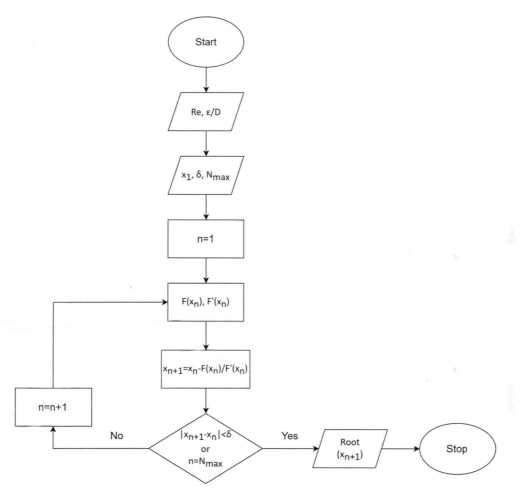

FIG. E1.4a

Flowchart of the Newton-Raphson method.

EXAMPLE 1.4 (CONTINUED)

$$f_2 = f_1 - \frac{F(f_1)}{F'(f_1)} = 0.01 - \frac{2.8882}{-539.2077} = 0.0154$$

The next three iterations are as follows:

$$F(f_2) = \sqrt{\frac{1}{0.0154}} + 2\log\left(\frac{10^{-4}}{3.7} + \frac{2.51}{10^5\sqrt{0.0154}}\right) = 0.7915$$

(Continued)

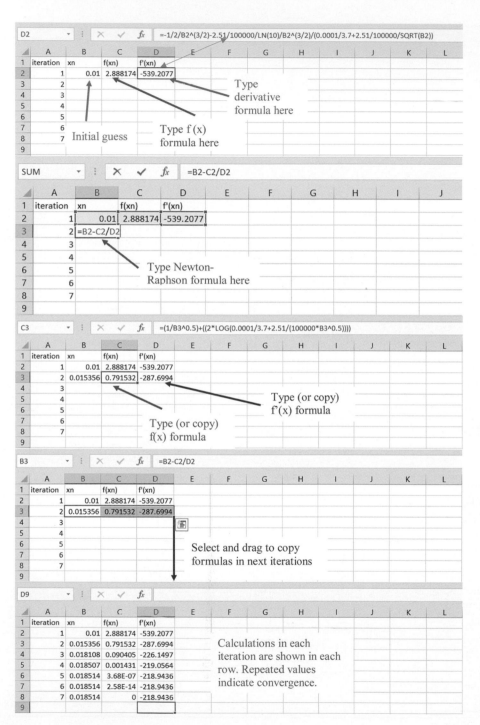

FIG. E1.4b

Calculations of the Newton-Raphson method in Excel.

EXAMPLE 1.4 (CONTINUED)

$$F'(f_2) = -\frac{1}{2 \times 0.0154^{3/2}} - \frac{\frac{2.51}{10^5}}{\ln 10 \times 0.0154^{3/2} \left(\frac{10^{-4}}{3.7} + \frac{2.51}{10^5 \sqrt{0.0154}}\right)} = -287.6994$$

$$f_3 = 0.0155 - \frac{0.7915}{-287.6994} = 0.0181$$

$$F(f_3) = \sqrt{\frac{1}{0.0181}} + 2\log\left(\frac{10^{-4}}{3.7} + \frac{2.51}{10^5 \sqrt{0.0181}}\right) = 0.0904$$

$$F'(f_3) = -\frac{1}{2 \times 0.0181^{3/2}} - \frac{\frac{2.51}{10^5}}{\ln 10 \times 0.0181^{3/2} \left(\frac{10^{-4}}{3.7} + \frac{2.51}{10^5 \sqrt{0.0181}}\right)} = -226.1497$$

$$f_4 = 0.0181 - \frac{0.0904}{-226.1497} = 0.0185$$

$$F(f_4) = \sqrt{\frac{1}{0.0188}} + 2\log\left(\frac{10^{-4}}{3.7} + \frac{2.51}{10^5 \sqrt{0.0188}}\right) = 0.0017$$

It can be seen from these calculations that this method has converged to 0.0185 after three iterations, which is faster than in the method of linear interpolation (see Example 1.3). The convergence criterion is also 10^{-4} in this example.

These calculations can also be performed in Excel. Fig. E1.4b demonstrates the calculation steps in Excel.

EXAMPLE 1.5 FINDING A ROOT OF A NONLINEAR ALGEBRAIC EQUATION BY SUCCESSIVE SUBSTITUTION, WEGSTEIN, LINEAR INTERPOLATION, AND NEWTON-RAPHSON METHODS APPLIED TO THE SRK EQUATION OF STATE

Develop MATLAB functions to solve nonlinear equations by the successive substitution, the Wegstein, the linear interpolation, and the Newton-Raphson root finding techniques. Use these functions to calculate the specific volume of pure gas, at a given temperature and pressure, from the SRK equation of state:

$$P = \frac{RT}{V - b} - \frac{a\alpha}{V(V - b)}$$

The equation constants, a and b, are obtained from:

(Continued)

EXAMPLE 1.5 (CONTINUED)

$$a = \frac{0.4278R^2 T_C^2}{P_C}$$

$$b = \frac{0.0867 R T_C}{P_C}$$

where T_C and P_C are critical temperature and pressure, respectively. The variable α is an empirical function of temperature:

$$\alpha = \left[1 + S \left(1 - \sqrt{\frac{T}{T_C}} \right) \right]^2$$

The value of S is a function of the acentric factor, ω, of the gas:

$$S = 0.48508 + 1.55171\omega - 0.15613\omega^2$$

Calculate the specific volume of n-butane vapor at 500 K and temperatures from 1 to 40 atm. The physical properties of n-butane are:

$$T_C = 425.2 \text{ K}, \ P_C = 3797 \text{ kPa}, \ \omega = 0.1931$$

and the gas constant is:

$$R = 8314 \text{ J/kmol.K}$$

Compare the results graphically with the ones obtained from using the ideal gas law. What conclusion do you draw from this comparison?

Method of Solution

Eqs. (1.29), (1.32), (1.42), and (1.45) are used for the methods of $x = g(x)$, Wegstein, linear interpolation, and Newton-Raphson, respectively. The iterative procedure stops when the difference between two succeeding approximations of the root is less than the convergence criterion (default value is 10^{-6}), or when the number of iterations reaches 100, whichever is satisfied first. The program may show the convergence results numerically, if required, to illustrate how each method arrives at the answer. For finding the gas-specific volume from the SRK equation of state, Eq. (1.2), which is a third-degree polynomial in compressibility factor, is solved. The starting value for the iterative method is $Z = 1$, which is the compressibility factor of the ideal gas.

Program Description

The programs and functions developed in this example can be found in https://www.elsevier.com/books-and-journals/book-companion/9780128229613

Four MATLAB functions called *XGX.m*, *Wegstein.m*, *LI.m*, and *NR.m* are developed to find the root of a general nonlinear equation using successive substitution, Wegstein, linear

(Continued)

EXAMPLE 1.5 (CONTINUED)

interpolation, and Newton-Raphson methods, respectively. The name of the nonlinear function that is subject to root finding is introduced in the input arguments; therefore, these MATLAB functions may be applied to any problem.

Successive substitution method (XGX.m): This function starts with the initialization section in which input arguments are evaluated and initial values for the main iteration are introduced. The first argument is the name of the MATLAB function in which the function $g(x)$ is described. The second argument is a starting value and has to be a scalar. By default, convergence is assumed when two succeeding iterations result in root approximations with less than 10^{-6} indifference. If another value for convergence is desired, it may be introduced to the function by the third argument. A value of 1 as the fourth argument makes the function show the results of each iteration step numerically. The third and fourth arguments are optional. Every additional argument that is introduced after the fourth argument is passed directly to the function $g(x)$. In this case, if it is desired to use the default values for the third and fourth arguments, an empty matrix should be entered in their place. For the solution of the problem, the constants of the SRK are passed to the function by introducing them in the fifth and sixth arguments.

The next section in the function is the main iteration loop, in which the iteration according to Eq. (1.29) takes place and the convergence is checked. In the case of the SRK equation, Eq. (1.2) can be rearranged to solve for Z:

$$Z = \left[Z^2 - \left(A - B - B^3 \right) Z + AB \right]^{1/3}$$

The right-hand side of this equation is taken as $g(Z)$ and is introduced in the MATLAB function *SRKg.m*. Numerical results of the calculations are also shown, if requested, in each iteration of this section.

Wegstein method (Wegstein.m): This function consists of the same parts as the *XGX.m* function. *Wegstein.m* can use the same $g(x)$ function as *XGX.m*. Therefore, the MATLAB function *SRKg.m* is applicable to this problem.

Linear interpolation method (LI.m): This function consists of the same parts as the two previous functions. The number of input arguments is one more than that of *XGX.m* because the linear interpolation method needs two starting points. Special care should be taken to introduce two starting values in which the function has opposite signs. Eq. (1.5) is used without change as the function the root of which is to be located:

$$f(Z) = Z^3 - Z^2 + \left(A - B - B^3 \right) Z - AB$$

This function is contained in a MATLAB function called *SRK.m*.

Newton-Raphson method (NR.m): The structure of this function is the same as that of previous functions. The derivative of the function is taken numerically to reduce the inputs. It is also more applicable to complicated functions. The reader may perform analytical differentiation using the MATLAB Symbolic Toolbox for obtaining a function to evaluate the resulting expressions for the derivative and use it instead of numerical derivation. To solve

(Continued)

EXAMPLE 1.5 (CONTINUED)

this problem, the same MATLAB function *SRK.m*, which represents Eq. (1.5), may be used with this function to calculate the value of the compressibility factor.

MATLAB program *Example1_5.m* solves the SRK equation of state by four different methods described earlier. At the beginning of this program, temperature, pressure range, and the physical properties of *n*-butane are entered. The constants of the SRK equation of state are calculated next. The values of *A* and *B* (used in Eq. 1.2) are also calculated in this section. Next, the user is asked to pick the method of solution. The name of the *m*-file in which the function that is to be used for root finding should be introduced to the program by the user. Evaluation of the root is done in the third part of the program. In this part, the root of the equation, closest to the ideal gas, is determined using the selected MATLAB function based on the choice of the user. The last part of the program *Example1_5.m* plots the results of the calculation both for SRK and ideal gas equations of state. It also shows some of the numerical results.

Results

>> **Example1_5**
Input the vector of pressure range (Pa) = [1:40]*101325
Input temperature (K) = 500
Critical temperature (K) = 425.2
Critical pressure (Pa) = 3797e3
Acentric factor = 0.1931
1) Successive substitution
2) Wegstein
3) Linear Interpolation
4) Newton Raphson
0) Exit
Choose the method of solution: 4
Function containing the Soave-Redlich-Kwong equation: 'SRK'

Results
Pres. = 101325.00 Ideal gas vol. = 41.0264 Real gas vol. = 40.8111
Pres. = 1013250.00 Ideal gas vol. = 4.1026 Real gas vol. = 3.8838
Pres. = 2026500.00 Ideal gas vol. = 2.0513 Real gas vol. = 1.8284
Pres. = 3039750.00 Ideal gas vol. = 1.3675 Real gas vol. = 1.1407
Pres. = 4053000.00 Ideal gas vol. = 1.0257 Real gas vol. = 0.7954

Discussion of Results

In this example, we use the *Example1_5.m* program to calculate the specific volume of a gas using the SRK equation of state. Additional information such as temperature, pressure, and physical properties are entered by the user through the program. Only the run for the

(Continued)

EXAMPLE 1.5 (CONTINUED)

Newton-Raphson method is shown in the foregoing results. However, the reader can try the other three methods, using the functions provided earlier, to solve this example.

Above the critical temperature, the SRK equation of state has only one real root that is of interest, the one located near the value given by the ideal gas law. Therefore, the latter, which corresponds to $Z = 1$, is used as the initial guess of the root.

A direct comparison between the SRK and ideal gas volumes is made in Fig. E1.5. It can be seen from this figure that the ideal gas equation overestimates gas volumes and, as expected from thermodynamic principles, the deviation from ideality increases as the pressure increases.

Let's compare the convergence speed of the root-finding methods considered in this example. For this purpose, we find the root only for $P = 40$ atm. Using the functions developed in this example, we can consider *trace* = 1 and pass it into the root fining function (the fourth input argument in *XGX*, *Wegstein*, and *Newton* and the fifth in *LI*) to show the calculation steps. The runs are not shown here and only the number of iterations for each

(Continued)

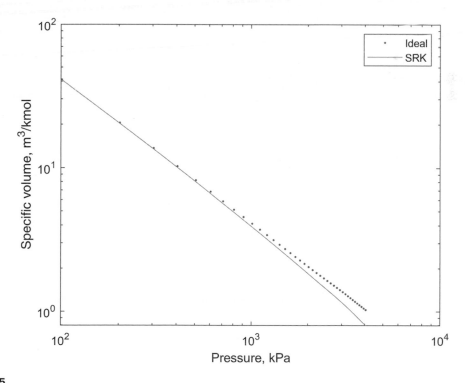

FIG. E1.5

Solution to Example 1.5: Graphical comparison between SRK and ideal gas equations of state.

Table E1.5 Efficiency of convergence of different methods.

Method	No. of iterations
Successive substitution	38
Wegstein	6
Linear interpolation	6
Newton-Raphson	5

EXAMPLE 1.5 (CONTINUED)

method is listed in Table E1.5. It can be seen in this table that successive substitution is the slowest method and Newton-Raphson is the fastest. Note that the default value of tolerance (1×10^{-6}) was used in all cases.

1.8 SYNTHETIC DIVISION ALGORITHM

If the nonlinear equation being solved is of the polynomial form, each real root (located by one of the methods already discussed) can be removed from the polynomial by synthetic division, thus reducing the degree of the polynomial from n to $n - 1$. Each successive application of the synthetic division algorithm will reduce the degree of the polynomial by one until all real roots have been located.

A simple computational algorithm for synthetic division has been given by Lapidus [3]. Consider the fourth-degree polynomial:

$$f(x) = a_4 x^4 + a_3 x^3 + a_2 x^2 + a_1 x + a_0 = 0 \qquad (1.51)$$

whose first real root has been determined to be x^*. This root can be factored out as follows:

$$f(x) = (x - x^*)(b_3 x^3 + b_2 x^2 + b_1 x + b_0) = 0 \qquad (1.52)$$

To determine the coefficients (b_i) of the third-degree polynomial first, multiply out Eq. (1.52) and rearrange in descending power of x:

$$f(x) = b_3 x^4 + (b_2 - b_3 x^*)x^3 + (b_1 - b_2 x^*)x^2 + (b_0 - b_1 x^*)x - b_0 x^* = 0 \qquad (1.53)$$

Equating Eqs. (1.51) and (1.53), the coefficients of like powers of x must be equal to each other; that is:

$$\begin{aligned}
a_4 &= b_3 \\
a_3 &= b_2 - b_3 x^* \\
a_2 &= b_1 - b_2 x^* \\
a_1 &= b_0 - b_1 x^* \\
a_0 &= -b_0 x^*
\end{aligned} \qquad (1.54)$$

Solving Eq. (1.54) for b_0 to b_3 we obtain:

$$b_3 = a_4$$
$$b_2 = a_3 + b_3 x*$$
$$b_1 = a_2 + b_2 x*$$
$$b_0 = a_1 + b_1 x*$$
(1.55)

In general notation, for a polynomial of nth degree, the new coefficients after application of synthetic division are given by:

$$b_{n-1} = a_n$$
$$b_{n-1-r} = a_{n-r} + b_{n-r} x*$$
(1.56)

where $r = 1, 2, (n-1)$. The polynomial is then reduced by one degree:

$$n_{new} = n_{previous} - 1$$
(1.57)

and the newly calculated coefficients are renamed:

$$a_i = b_i \qquad i = 1, 2, \cdots, n-1$$
(1.58)

This procedure is repeated until all real roots are extracted. When this is accomplished, the remainder polynomial will contain the complex roots. The presence of a pair of complex roots will give a quadratic equation that can be easily solved by the quadratic formula. However, two or more pairs of complex roots require the application of more elaborate techniques such as the eigenvalue method, which is developed in the next section.

1.9 THE EIGENVALUE METHOD

The concept of eigenvalues will be discussed in Chapter 2 of this textbook. As a preview of that topic, we will state that a square matrix has a *characteristic polynomial* whose roots are called the *eigenvalues* of the matrix. However, root-finding methods that have been discussed up to here are not efficient techniques for calculating eigenvalues. There are more efficient eigenvalue methods to find the roots of the characteristic polynomial (see Section 2.9).

It can be shown that Eq. (1.11) is the characteristic polynomial of the following $n \times n$ companion matrix A, which contains the coefficients of the original polynomial in the following form:

$$A = \begin{bmatrix} -\dfrac{a_{n-1}}{a_n} & -\dfrac{a_{n-2}}{a_n} & \cdots & -\dfrac{a_2}{a_n} & -\dfrac{a_1}{a_n} & -\dfrac{a_0}{a_n} \\ 1 & 0 & \cdots & 0 & 0 & 0 \\ 0 & 1 & \cdots & 0 & 0 & 0 \\ \vdots & \vdots & \ddots & \vdots & \vdots & \vdots \\ 0 & 0 & \cdots & 1 & 0 & 0 \\ 0 & 0 & \cdots & 0 & 1 & 0 \end{bmatrix}$$
(1.59)

Therefore, finding the eigenvalues of A is equivalent to locating the roots of the polynomial (1.11).

MATLAB has its own function, *roots.m*, for calculating all the roots of a polynomial equation of the form (1.11). This function accomplishes the task of finding the roots of the polynomial Eq. (1.11) by first converting the polynomial to the companion matrix A shown in Eq. (1.59). It then uses the built-in function *eig.m*, which calculates the eigenvalues of a matrix, to evaluate the eigenvalues of the companion matrix that are also the roots of the polynomial (1.11).

EXAMPLE 1.6 SOLUTION OF *N*TH-DEGREE POLYNOMIALS AND TRANSFER FUNCTIONS USING THE NEWTON-RAPHSON METHOD WITH SYNTHETIC DIVISION AND EIGENVALUE METHOD

Consider the isothermal continuous stirred tank reactor (CSTR) shown in Fig. E1.6. Components A and R are fed to the reactor at rates of Q and $(q - Q)$, respectively. The following complex reaction scheme develops in the reactor:

$$A + R \to B$$
$$B + R \to C$$
$$C + R \rightleftarrows D$$
$$D + R \to E$$

This problem was analyzed by Douglas [4] to illustrate the various techniques for designing simple feedback control systems. In his analysis of this system, Douglas made the following assumptions:

1. Component R is present in the reactor in sufficiently large excess so that the reaction rates can be approximated by first-order expressions.
2. The feed compositions of components B, C, D, and E are zero.
3. A particular set of values is chosen for feed concentrations, feed rates, the kinetic rate constant, and reactor volume.
4. Disturbances are due to changes in the composition of component R in the vessel.

The control objective is to maintain the composition of component C in the reactor as close as possible to the steady-state design value, despite the fact that disturbances enter the system. This objective is accomplished by measuring the actual composition of C and using the difference between the desired and measured values to manipulate the inlet flow rate Q of component A.

Douglas developed the following transfer function for the reactor with a proportional control system:

(Continued)

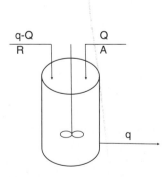

FIG. E1.6

The continuous stirred tank reactor.

EXAMPLE 1.6 (CONTINUED)

$$K_C \frac{2.98(s + 2.25)}{(s + 1.45)(s + 2.85)^2(s + 4.35)} = -1$$

where K_C is the gain of the proportional controller. This control system is stable for values of K_C that yield roots of the transfer function having negative real parts.

Using the Newton-Raphson method with the synthetic division or eigenvalue method, determine the roots of this transfer function for a range of values of the proportional gain K_C and calculate the critical value of K_C above which the system becomes unstable. Write the program so that it can be used to solve nth-degree polynomials or transfer functions of the type shown in the foregoing equation.

Method of Solution

In the Newton-Raphson method with synthetic division, Eq. (1.45) is used for the evaluation of each root. Eqs. (1.56)−(1.58) are then applied to perform synthetic division to extract each root from the polynomial and reduce the latter by one degree. When the nth-degree polynomial has been reduced to a quadratic:

$$a_2 x^2 + a_1 x + a_0 = 0$$

the program uses the quadratic solution formula:

$$x_{1,2} = \frac{-a_1 \pm \sqrt{a_1^2 - 4a_2 a_0}}{2a_2}$$

to check for the existence of a pair of complex roots. In the eigenvalue method, the MATLAB function *roots* may be used directly.

The numerator and the denominator of the transfer function are multiplied out to yield:

$$\frac{(2.98s + 6.705)K_C}{s^4 + 11.50s^3 + 47.49s^2 + 83.0632s + 51.2327} = -1$$

A first-degree polynomial is present in the numerator and a fourth-degree polynomial in the denominator. To convert this to the canonical form of a polynomial, we multiply through by the denominator and rearrange to obtain:

$$s^4 + 11.50s^3 + 47.49s^2 + (83.0632 + 2.98K_C)s + (51.2327 + 6.705K_C) = 0$$

It is obvious that once a value of K_C is chosen, the two bracketed terms of this equation can be added to form a single fourth-degree polynomial whose roots can be evaluated.

When $K_C = 0$, the transfer function has the following four negative real roots that can be found by inspection of the original transfer function:

(Continued)

EXAMPLE 1.6 (CONTINUED)

$$s_1 = -1.45 \quad s_2 = -2.85 \quad s_3 = -2.85 \quad s_4 = -4.35$$

These are called the poles of the open-loop transfer function.

The value that causes one or more of the roots of the transfer function to become positive (or have positive real parts) is called the critical value of the proportional gain, K_C. This critical value is calculated as follows:

1. A range of search for K_C is established.
2. The bisection method is used to search for this range.
3. All the roots of the transfer function are evaluated at each step of the bisection search.
4. The roots are checked for positive real part. The range of K_C, over which the change from negative to positive roots occurs, is retained.
5. Steps 2 to 4 are repeated until successive values of K_C change by less than a convergence criterion, ε.

Program Description

The programs and functions developed in this example can be found in https://www.elsevier.com/books-and-journals/book-companion/9780128229613

The MATLAB function *NRsdivision.m* calculates all roots of a polynomial by the Newton-Raphson method with the synthetic division as described in the Method of Solution. Unlike other functions employing the Newton-Raphson method, this function does not need a starting value as one of the input arguments. Instead, the function generates a starting point at each step according to Eq. (1.22). Only polynomials that have no more than one pair of complex roots can be handled by this function. If the polynomial has more than a pair of complex roots, the function *roots* should be used instead. The function is written in general form and may be used in other programs directly.

The MATLAB program *Example1_6.m* searches for the desired value of K_C by the bisection method. At the beginning of the program, the user is asked to enter the coefficients of the numerator and the denominator of the transfer function (in descending s powers). The numerator and the denominator may be of any degree with the limitation that the numerator cannot have a degree greater than or equal to that of the denominator. The user should also enter the range of search and method of root finding. It is good practice to choose zero for the minimum value of the range; thus poles of the open-loop transfer function are evaluated in the first step of the search. The maximum value must be higher than the critical value; otherwise, the search will not arrive at the critical value.

The stability of the system is examined at the minimum, maximum, and midpoints of the range of search of K_C. That half of the interval in which the change from negative to positive (stable to the unstable system) occurs is retained by the bisection algorithm. This new interval is bisected again and the evaluation of the system stability is repeated until the convergence criterion, which is $\Delta K_C < 0.001$, is met.

To determine whether the system is stable or unstable, the two polynomials are combined, as shown in the Method of Solution, using K_C as the multiplier of the polynomial from the numerator of the transfer function. Function *NRsdivision* (which uses the Newton-Raphson

(*Continued*)

EXAMPLE 1.6 (CONTINUED)

method with synthetic division algorithm) or function *roots* (which uses the eigenvalue algorithm) is called to calculate the roots of the overall polynomial function and the sign of all roots is checked for positive real parts. A flag named stbl indicates that the system is stable (all negative roots; stbl = 1) or unstable (positive root; stbl = 0).

Results

>> Example1_6

Vector of coefficients of the numerator polynomial = [2.98, 6.705]
Vector of coefficients of the denominator polynomial = [1, 11.5, 47.49, 83.0632, 51.2327]
Lower limit of the range of search = 0
Upper limit of the range of search = 100
1) Newton-Raphson with synthetic division
2) Eigenvalue method
Method of root finding = 1
Kc = 0.0000
Roots = − 4.35 − 2.8591 − 2.8409 − 1.45
Kc = 100.0000
Roots = − 9.851 − 2.248 0.2995 + 5.701i 0.2995-5.701i
Kc = 50.0000
Roots = − 8.4949 − 2.2459 − 0.3796 + 4.485i − 0.3796-4.485i
Kc = 75.0000
Roots = − 9.2487 − 2.2473-0.001993 + 5.163i − 0.001993-5.163i
Kc = 87.5000
Roots = − 9.5641 − 2.2477 0.1559 + 5.445i 0.1559-5.445i
Kc = 81.2500
Roots = − 9.4104 − 2.2475 0.07893 + 5.308i 0.07893-5.308i
Kc = 78.1250
Roots = − 9.3306 − 2.2474 0.039 + 5.237i 0.039-5.237i
Kc = 76.5625
Roots = − 9.29 − 2.2473 0.01864 + 5.2i 0.01864- 5.2i
Kc = 75.7813
Roots = − 9.2694 − 2.2473 0.00836 + 5.182i 0.00836-5.182i
Kc = 75.3906
Roots = − 9.2591 − 2.2473 0.003192 + 5.173i 0.003192-5.173i
Kc = 75.1953
Roots = − 9.2539 − 2.2473 0.0006016 + 5.168i 0.0006016-5.168i
Kc = 75.0977
Roots = − 9.2513 − 2.2473 − 0.0006953 + 5.166i − 0.0006953-5.166i
Kc = 75.1465

(Continued)

EXAMPLE 1.6 (CONTINUED)

Roots = − 9.2526 − 2.2473 − 4.667e-05 + 5.167i − 4.667e-05-5.167i
Kc = 75.1709
Roots = − 9.2533 − 2.2473 0.0002775 + 5.167i 0.0002775-5.167i
Kc = 75.1587
Roots = − 9.2529 − 2.2473 0.0001154 + 5.167i 0.0001154-5.167i
Kc = 75.1526
Roots = − 9.2528 − 2.2473 3.438e-05 + 5.167i 3.438e-05-5.167i
Kc = 75.1495
Roots = − 9.2527 − 2.2473 − 6.147e-06 + 5.167i − 6.147e-06-5.167i
Kc = 75.1511
Roots = − 9.2527 − 2.2473 1.412e-05 + 5.167i 1.412e-05-5.167i
Kc = 75.1503
Roots = − 9.2527 − 2.2473 3.985e-06 + 5.167i 3.985e-06-5.167i

Discussion of Results

The range of search for the proportional gain (K_C) is chosen to be between 0 and 100. A convergence criterion of 0.001 is used and may be changed by the user, if necessary. The bisection method evaluates the roots at the low end of the range $(K_C = 0)$ and finds them to have the predicted values of:

$$-4.3500 \quad -2.8591 \quad -28409 \quad \text{and} \quad -1.4500$$

The small difference between the two middle roots and their actual values is due to rounding off the coefficients of the denominator polynomial. This deviation is very small in comparison to the root itself and it can be ignored. At the upper end of the range $(K_C = 100)$ the roots are:

$$-9.8510 \quad -2.2480 \quad \text{and} \quad 0.2995 \pm 5.7011i$$

The system is unstable because of the positive real components of the roots. At the midrange $(K_C = 50)$ the system is still stable because all the real parts of the roots are negative. The bisection method continues its search in the range of 50 to 100. In a total of 19 evaluations, the algorithm arrives at the critical value of K_C in the range:

$$75.1495 < K_C < 75.1503$$

In the event that the critical value of the gain was outside the limits of the original range of search, the program would have detected this early in the search and would have issued a warning and stopped running.

1.10 NEWTON'S METHOD FOR SOLVING SYSTEM OF NONLINEAR EQUATIONS

If the mathematical model involves two (or more) simultaneous nonlinear equations in two (or more) unknowns, the Newton-Raphson method can be extended to solve these equations simultaneously. In what follows we will first develop the Newton-Raphson method for two equations and then expand the algorithm to a system of k equations.

The model for two unknowns will have the general form:

$$f_1(x_1, x_2) = 0$$
$$f_2(x_1, x_2) = 0 \tag{1.60}$$

where f_1 and f_2 are nonlinear functions of variables x_1 and x_2. Both these functions may be expanded in two-dimensional Taylor series around an initial estimate of $x_1^{(1)}$ and $x_2^{(1)}$:

$$f_1(x_1, x_2) = f_1\left(x_1^{(1)}, x_2^{(1)}\right) + \left.\frac{\partial f_1}{\partial x_1}\right|_{x^{(1)}}\left(x_1 - x_1^{(1)}\right) + \left.\frac{\partial f_1}{\partial x_2}\right|_{x^{(1)}}\left(x_2 - x_2^{(1)}\right) + \ldots$$

$$f_2(x_1, x_2) = f_2\left(x_1^{(1)}, x_2^{(1)}\right) + \left.\frac{\partial f_2}{\partial x_1}\right|_{x^{(1)}}\left(x_1 - x_1^{(1)}\right) + \left.\frac{\partial f_2}{\partial x_2}\right|_{x^{(1)}}\left(x_2 - x_2^{(1)}\right) + \ldots \tag{1.61}$$

The superscript (1) will be used to designate the iteration number of the estimate.

Setting the left-hand sides of Eq. (1.61) to zero and truncating the second-order and higher derivatives of the Taylor series, we obtain the following equations:

$$\left.\frac{\partial f_1}{\partial x_1}\right|_{x^{(1)}}\left(x_1 - x_1^{(1)}\right) + \left.\frac{\partial f_1}{\partial x_2}\right|_{x^{(1)}}\left(x_2 - x_2^{(1)}\right) = -f_1\left(x_1^{(1)}, x_2^{(1)}\right)$$

$$\left.\frac{\partial f_2}{\partial x_1}\right|_{x^{(1)}}\left(x_1 - x_1^{(1)}\right) + \left.\frac{\partial f_2}{\partial x_2}\right|_{x^{(1)}}\left(x_2 - x_2^{(1)}\right) = -f_2\left(x_1^{(1)}, x_2^{(1)}\right) \tag{1.62}$$

If we define the correction variable as:

$$\delta_1^{(1)} = x_1 - x_1^{(1)}$$
$$\delta_2^{(1)} = x_2 - x_2^{(1)} \tag{1.63}$$

then Eq. (1.62) simplify to:

$$\left.\frac{\partial f_1}{\partial x_1}\right|_{x^{(1)}}\delta_1^{(1)} + \left.\frac{\partial f_1}{\partial x_2}\right|_{x^{(1)}}\delta_2^{(1)} = -f_1\left(x_1^{(1)}, x_2^{(1)}\right)$$

$$\left.\frac{\partial f_2}{\partial x_1}\right|_{x^{(1)}}\delta_1^{(1)} + \left.\frac{\partial f_2}{\partial x_2}\right|_{x^{(1)}}\delta_2^{(1)} = -f_2\left(x_1^{(1)}, x_2^{(1)}\right) \tag{1.64}$$

Eq. (1.64) are a set of simultaneous linear algebraic equations, where the unknowns are $\delta_1^{(1)}$ and $\delta_2^{(1)}$. These equations can be written in matrix format as follows:

$$\begin{bmatrix} \left.\dfrac{\partial f_1}{\partial x_1}\right|_{x^{(1)}} & \left.\dfrac{\partial f_1}{\partial x_2}\right|_{x^{(1)}} \\ \left.\dfrac{\partial f_2}{\partial x_1}\right|_{x^{(1)}} & \left.\dfrac{\partial f_2}{\partial x_2}\right|_{x^{(1)}} \end{bmatrix} \begin{bmatrix} \delta_1^{(1)} \\ \delta_2^{(1)} \end{bmatrix} = - \begin{bmatrix} f_1^{(1)} \\ f_2^{(1)} \end{bmatrix} \tag{1.65}$$

Because this set contains only two equations in two unknowns, it can be readily solved by the application of Cramer's rule (see Chapter 2) to give the first set of values for the correction vector:

$$
\delta_1^{(1)} = - \frac{\left[f_1 \dfrac{\partial f_2}{\partial x_2} - f_2 \dfrac{\partial f_1}{\partial x_2} \right]}{\left[\dfrac{\partial f_1}{\partial x_1} \dfrac{\partial f_2}{\partial x_2} - \dfrac{\partial f_1}{\partial x_2} \dfrac{\partial f_2}{\partial x_1} \right]}
$$

$$
\delta_2^{(1)} = - \frac{\left[f_2 \dfrac{\partial f_1}{\partial x_1} - f_1 \dfrac{\partial f_2}{\partial x_1} \right]}{\left[\dfrac{\partial f_1}{\partial x_1} \dfrac{\partial f_2}{\partial x_2} - \dfrac{\partial f_1}{\partial x_2} \dfrac{\partial f_2}{\partial x_1} \right]}
$$

(1.66)

The superscripts, indicating the iteration number of the estimate, have been omitted from the right-hand side of Eq. (1.66) to avoid overcrowding.

The new estimate of the solution may now be obtained from the previous estimate by adding to it the correction vector:

$$
x^{(n+1)} = x^{(n)} + \delta^{(n)}
$$

(1.67)

This equation is merely a rearrangement and generalization to the $(n + 1)$st iteration of Eq. (1.63).

The method just described for two nonlinear equations is readily expandable to the case of k simultaneous nonlinear equations in k unknowns:

$$
f_1(x_1, \ldots, x_k) = 0
$$
$$
\vdots
$$
$$
f_k(x_1, \ldots, x_k) = 0
$$

(1.68)

The linearization of this set by the application of the Taylor series expansion produces:

$$
\begin{bmatrix} \left.\dfrac{\partial f_1}{\partial x_1}\right|_{x^{(1)}} & \cdots & \left.\dfrac{\partial f_1}{\partial x_k}\right|_{x^{(1)}} \\ \vdots & \ddots & \vdots \\ \left.\dfrac{\partial f_k}{\partial x_1}\right|_{x^{(1)}} & \cdots & \left.\dfrac{\partial f_k}{\partial x_k}\right|_{x^{(1)}} \end{bmatrix} \begin{bmatrix} \delta_1^{(1)} \\ \vdots \\ \delta_k^{(1)} \end{bmatrix} = - \begin{bmatrix} f_1^{(1)} \\ \vdots \\ f_k^{(1)} \end{bmatrix}
$$

(1.69)

In matrix/vector notation, this equation condenses to:

$$
J\delta = -f
$$

(1.70)

where J is the *Jacobian* matrix containing the partial derivatives, δ is the correction vector, and f is the vector of functions. Eq. (1.70) represents a set of linear algebraic equations whose solution will be discussed in Chapter 2.

Strongly nonlinear equations are likely to diverge rapidly. To prevent this situation, relaxation is used to stabilize the iterative solution process. If δ is the correction vector without relaxation, then relaxed change is $\rho\delta$ where ρ is the relaxation factor:

$$
x^{(n+1)} = x^{(n)} + \rho\,\delta^{(n)}
$$

(1.71)

By applying the relaxation, we do not use the corrections calculated from the previous iteration in full, but by attenuating them with a (fractional) relaxation factor. A typical value for ρ is 0.5.

A value of zero inhibits changes and a value of one is equivalent to no relaxation. Relaxation reduces the correction made to the variable from one iteration to the next and may eliminate the tendency of the solution to diverge.

EXAMPLE 1.7 SOLUTION OF NONLINEAR EQUATIONS IN CHEMICAL EQUILIBRIUM USING NEWTON'S METHOD FOR SIMULTANEOUS NONLINEAR EQUATIONS

Develop a MATLAB function to solve n simultaneous nonlinear equations in n unknowns. Apply this function to find the equilibrium conversion of the following reactions:

$$2A + B \rightleftarrows C \qquad K_1 = \frac{C_C}{C_A^2 C_B} = 5 \times 10^{-4}$$

$$A + D \rightleftarrows C \qquad K_2 = \frac{C_C}{C_A C_D} = 4 \times 10^{-2}$$

Initial concentrations are:

$$C_{A,0} = 40 \quad C_{B,0} = 15 \quad C_{C,0} = 0 \quad C_{D,0} = 10$$

All concentrations are in kmol/m^3.

Method of Solution

Eq. (1.70) is applied to calculate the correction vector in each iteration and Eq. (1.71) is used to estimate the new relaxed variables. The variables of this problem are the conversions, x_1 and x_2, of the reactions in the preceding text. The concentrations of the components can be calculated from these conversions and the initial concentrations:

$$C_A = C_{A,0} - 2x_1 C_{B,0} - x_2 C_{D,0} = 40 - 30x_1 - 10x_2$$
$$C_B = (1 - x_1)C_{B,0} = 15 - 15x_1$$
$$C_C = C_{C,0} + x_1 C_{B,0} + x_2 C_{D,0} = 15x_1 + 10x_2$$
$$C_D = (1 - x_2)C_{D,0} = 10 - 10x_2$$

The set of equations that are functions of x_1 and x_2 are:

$$f_1(x_1, x_2) = \frac{C_C}{C_A^2 C_B} - 5 \times 10^{-4} = 0$$

$$f_2(x_1, x_2) = \frac{C_C}{C_A C_D} - 4 \times 10^{-2} = 0$$

The values of x_1 and x_2 are to be calculated by the program so that $f_1 = f_2 = 0$.

Program Description

The programs and functions developed in this example can be found in https://www.elsevier.com/books-and-journals/book-companion/9780128229613

(Continued)

EXAMPLE 1.7 (**CONTINUED**)

The MATLAB function *Newton.m* solves a set of nonlinear equations by Newton's method. The first part of the program is initialization in which the convergence criterion, the relaxation factor, and other initial parameters needed by the program are set. The main iteration loop comes next. In this part, the components of the Jacobian matrix are calculated by numeric differentiation (Eq. 1.69) and new estimates of the roots are calculated according to Eq. (1.71). This procedure continues until the convergence criterion is met or a maximum iteration limit is reached. By default, the convergence criterion is $\max(|x_i^{(n)} - x_i^{(n+1)}|) < 10^{-6}$. This value may be changed through the fourth input argument of the function.

The MATLAB program *Example1_7.m* is written to solve the particular problem of this example. This program simply takes the required input data from the user and calls *Newton.m* to solve the set of equations. The program allows the user to repeat the calculation and try new initial values and relaxation factor without changing the problem parameters.

The set of equations of this example are introduced in the MATLAB function *Ex1_7_func. m*. It is important to note that the function *Newton* should receive the function values at each point as a column vector. This is considered in the function *Ex1_7_func*.

Results

>> **Example1_7**
Vector of initial concentration of A, B, C, and D = [40, 15, 0, 10]
2 A + B = C K1 = 5E − 4
A + D = C K2 = 4E − 2
Name of the file containing the set of equations = 'Ex1_7_func'
Vector of initial guesses = [0.1, 0.9]
Relaxation factor = 1
Results:
x1 = 0.1203, x2 = 0.4787
Solution reached after 8 iterations.
Repeat the calculations (0 / 1)? 1
Vector of initial guesses = [0.1, 0.1]
Relaxation factor = 1
Warning: Maximum iterations reached.
Results:
x1 = 3212963175567458300000000.0000, x2 = -6973207056420066600000000.0000
Solution reached after 100 iterations.
Repeat the calculations (0 / 1)? 1
Vector of initial guesses = [0.1, 0.1]
Relaxation factor = 0.7

(Continued)

EXAMPLE 1.7 (CONTINUED)

Results:

x1 = 0.1203, x2 = 0.4787
Solution reached after 18 iterations.
Repeat the calculations (0 / 1)? 0

Discussion of Results

Three runs are made to test the sensitivity in the choice of initial guesses and the effectiveness of the relaxation factor. In the first run, initial guesses are set to $x_1^{(1)} = 0.01$ and $x_2^{(1)} = 0.09$. With these guesses, the method converges in 8 iterations.

In the next run, the initial guess for x_2 is changed to 0.1. The maximum number of iterations defined in the function *Newton* is 100, and the method does not converge in this case, even in 100 iterations. This test shows the high sensitivity of Newton's method to the initial guess. Introducing the relaxation factor $\rho = 0.7$ in the next run causes the method to converge in only 18 iterations. In other words, the sensitivity of the method to the starting point is reduced when the relaxation factor is less than 1.

1.11 HOMOTOPY METHOD

Newton and successive substitution methods can be very sensitive to the starting point and diverge easily if a good set of initial guesses is not selected. The homotopy method, however, provides a way to guarantee the convergence; thus the initial guesses can be chosen more loosely.

Let's consider the set of nonlinear algebraic equations shown as:

$$f(x) = 0 \qquad (1.72)$$

Here f represents the set of equations and x is the vector of variables. The homotopy function $H(x, t)$ is defined as:

$$H(x, t) = tf(x) + (1 - t) g(x) \qquad (1.73)$$

in which $g(x)$ is a set of auxiliary functions and t is the convergence control parameter in the interval [0, 1]. In the homotopy method, $g(x)$ is assumed such that its root is known.

Starting from the known solution of $g(x) = 0$ at $t = 0$, we can trace the roots of $H(x, t)$ by gradually increasing t until $t = 1$, where the answer is the desired solution. In other words, the calculations start from $t = 0$ (or a value close to zero) and the roots of Eq. (1.73) are found by a method such as Newton or successive substitution. The parameter t is then increased and roots of Eq. (1.73) are found for this new value of the convergence control parameter. The increase of t is continued until $t = 1$ at which the roots of $f(x) = 0$ are computed. The roots computed in each step are used as the initial guess in the next step. Therefore, by considering small enough increments of t-value, we can be sure that the initial guesses are close enough to the roots and convergence can be guaranteed in this way.

Although any function with known roots can be employed as $g(x)$, the two commonly used homotopy methods are:

$$H(x,t) = tf(x) + (1-t)(x-x_0) \quad \text{Fixed point homotopy} \tag{1.74}$$

and

$$H(x,t) = tf(x) + (1-t)f(x_0) \quad \text{Newton homotopy} \tag{1.75}$$

In these equations, x_0 is a known point.

1.12 USING THE BUILT-IN MATLAB AND EXCEL FUNCTIONS

The function *fzero* is a general function for root finding in MATLAB. The statement *fzero*('*file_name*', x_0) finds the root of the function $f(x)$ introduced in the user-defined MATLAB function *file_name.m*. The second argument x_0 is a starting guess. Starting with this initial value, the function *fzero* searches for change in the sign of the function $f(x)$. The calculation then continues with either bisection or linear interpolation method until the convergence is achieved.

As an example, for finding the friction factor from the Colebrook equation by *fzero*, we should first create the following function containing the Colebrook equation:

```
function y = Colebrook(f)
eD = 1e-4; % Relative roughness
Re = 1e5; % Reynolds number

y = 1/sqrt(f) + 2*log10(eD/3.7 + 2.51/Re/sqrt(f));
```

Then, the friction factor, as a root of the Colebrook equation, can be evaluated in the command window as follows:

```
>> fzero('Colebrook',0.01)
ans =
0.0185
```

Note that the data from Example 1.1 have been used in *Colebrook.m* function.

The roots of a polynomial can be computed by the MATLAB function *roots*. The polynomial is introduced to MATLAB as a vector whose elements are the coefficients of this polynomial in descending order.

$$f(x) = a_k x^k + a_{k-1} x^{k-1} + \ldots + a_1 x + a_0 \tag{1.76}$$

Therefore, the input argument to *roots* is the vector of coefficients of the polynomial for which its roots are to be computed.

Consider Example 1.5 in which the SRK cubic equation of state should be solved for computing the compressibility factor from Eq. (1.2). Using the data provided in this example, the constants A and B are 0.2882 and 0.0787, respectively, at 40 atm. Therefore, the roots of Eq. (1.2) can be calculated in the command window as follows:

```
>> A = 0.2882;
>> B = 0.0787;
```

```
>> c = [1, − 1, (A − B − B^2), − A*B];
>> Z = roots(c)
Z =
0.7756 + 0.0000i
0.1122 + 0.1290i
0.1122 − 0.1290i
```

It can be seen that the function *roots* has returned three values for the roots of this cubic polynomial. In this case, the only real root of the function is acceptable for the compressibility factor. Note that for a superheat gas, a cubic equation of state always has one real and two imaginary roots.

The MATLAB function *solve* provides the symbolic solution of algebraic equations. For the Colebrook equation, this function can be used as follows:

```
>> syms f
>> solve(1/sqrt(f) + 2*log10(1e-4/3.7 + 2.51/1e5/sqrt(f)) == 0)
```

or

```
>> eval(solve(1/sqrt(f) + 2*log10(1e-4/3.7 + 2.51/1e5/sqrt(f)) == 0))
ans =
0.0185
```

For solving systems of nonlinear equations of several variables, the MATLAB function *fsolve* can be employed. For solving the problem posed in Example 1.7 by this function, we first define the set of equations in the following function:

```
function f − equil(x)
c0 = [40, 15, 0, 10]; % Initial concentrations
% Equilibrium constants
K1 = 5e − 4;
K2 = 4e − 2;
ca = c0(1) − 2*x(1)*c0(2) − x(2)*c0(4);
cb = (1 − x(1))*c0(2);
cc = c0(3) + x(1)*c0(2) + x(2)*c0(4);
cd = (1 − x(2))*c0(4);
f(1) = cc / ca^2 / cb − K1;
f(2) = cc / ca / cd − K2;
```

Then this set of equations can be solved by the following command, considering the initial guesses 0.1 and 0.9:

```
>> fsolve('equil',[0.1, 0.9])
ans =
0.1203 0.4787
```

We can use *Goal Seek* command in Excel for calculating the root of an equation. The *Goal Seek* command finds the root by changing the value of a certain cell with the Newton-Raphson

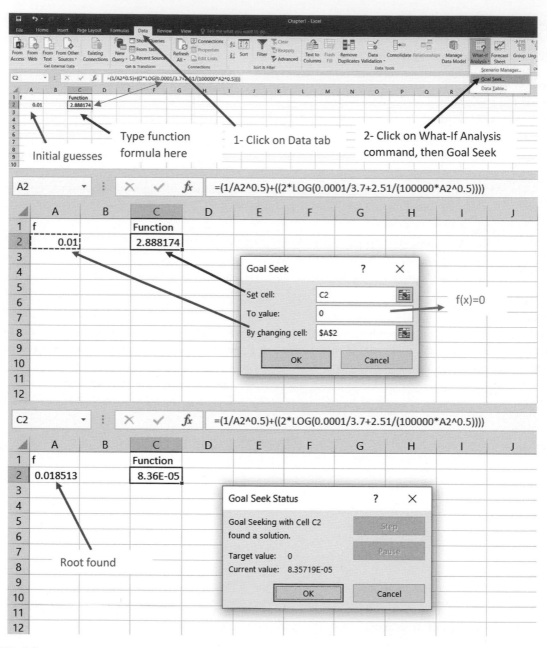

FIG. 1.8

Using Goal Seek command in Excel for calculating friction factor from the Colebrook equation.

method. The reported root depends on the initial guess and we may find different roots, in case they exist, with changing the initial guess. Fig. 1.8 illustrates the required steps for finding the friction factor from the Colebrook equation.

In the end, it should be mentioned that a root fining problem can be converted into a minimization problem. To do so, it is necessary to define the function $\varphi = [f(x)]^2$. Then, the root of the equation $f(x) = 0$ can be computed by determining the value of x that minimizes φ. Such a problem can be solved in MATLAB by *fminbnd* or *fminsearch* and in Excel by *Solver* (which is an add-in program that can be used for *What-If Analysis*).

1.13 SUMMARY

In this chapter, several root-finding methods are developed and specific examples of the solutions are demonstrated using MATLAB and Excel and by hand. Types of roots of polynomial equations and their approximation are discussed first. A general method of finding an initial estimate of roots is scanning the region of search, either graphically or numerically. Then, a more efficient iterative method should be used for obtaining the exact location of the root.

Successive substitution, Wegstein and Newton-Raphson methods need only one starting value while bisection, linear interpolation and secant methods require two initial guesses. Among these methods, the successive substitution is easy to implement, Wegstein is effective, Newton-Raphson is fast, bisection and linear interpolation both guarantee the convergence, while bisection is very slow.

Newton's method is developed for the solution of a set of nonlinear equations. This method is sensitive to the initial guess, especially for highly nonlinear equations. By introducing the homotopy function, the user can use looser initial guesses and still reach convergence.

At the end of the chapter, some built-in MATLAB and Excel functions that can be used for root finding are discussed.

PROBLEMS

1.1 Evaluate all the roots of the following polynomial equations by performing the following steps:
  a. (a)Use Descartes' rule to predict how many positive and how many negative roots each polynomial may have.
  b. Use the Newton-Raphson method with the synthetic division to calculate the numerical values of the roots. To do so, first, apply the MATLAB function *roots.m*, and then the *NRsdivision.m* which was developed in this chapter. Why does the *NRsdivision.m,* program fail to arrive at the answers of some of these polynomials? What is the limitation of this program?
  c. Classify these polynomials according to the four categories described in Section 1.2.
     (i) $x^4 - 16x^3 + 96x^2 - 256x + 256 = 0$
     (ii) $x^4 - 32x^2 + 256 = 0$

$\quad$ **(iii)** $x^4 + 3x^3 + 12x - 16 = 0$

$\quad$ **(iv)** $x^4 + 4x^3 + 18x^2 - 20x + 125 = 0$

$\quad$ **(v)** $x^5 - 8x^4 + 25x^3 - 106x^2 + 170x - 200 = 0$

$\quad$ **(vi)** $x^4 - 10x^3 + 35x^2 - 5x + 24 = 0$

$\quad$ **(vii)** $x^6 - 8x^5 + 11x^4 + 78x^3 - 382x^2 + 800x - 800 = 0$

$\quad$ **(viii)** $x^6 - 8x^5 + 10x^4 + 80x^3 - 371x^2 + 748x - 720 = 0$

1.2 Evaluate the roots of the following transcendental equations.

$\quad$ **(a)** $\sin(x) - 2\exp(-x^2) = 0$

$\quad$ **(b)** $ax - a^x = 0$ for $a = 2, e,$ or 3

$\quad$ **(c)** $\ln(1 + x^2) - \sqrt{|x|} = 0$

$\quad$ **(d)** $e^x/(1 + \cos x) - 1 = 0$

1.3 Repeat Example 1.5 by using the BWR equation of state. BWR equation of state is:

$$P = \frac{RT}{V} + \frac{B_0 RT - A_0 - (C_0/T^2)}{V^2} + \frac{bRT - a}{V^3} + \frac{a\alpha}{V^6} + \frac{c}{V^3 T^2}\left(1 + \frac{\gamma}{V^2}\right)e^{-\gamma/V^2}$$

where A_0, B_0, C_0, a, b, c, α, and γ are constants. When P is in atmosphere, V is in liters per mole, and T is in Kelvin the values of constants for n-butane are:

$$
\begin{array}{lll}
A_0 = 10.0847 & B_0 = 0.124361 & C_0 = 0.992830 \times 10^6 \\
a = 1.88231 & b = 0.0399983 & c = 0.316400 \times 10^6 \\
\alpha = 1.10132 \times 10^{-3} & \gamma = 3.400 \times 10^{-2} & R = 0.08206
\end{array}
$$

1.4 Repeat Example 1.5 by using the Patel-Teja (PT) equations of state. The Patel-Teja equation of state is:

$$P = \frac{RT}{V - b} - \frac{a}{V(V + b) + c(V - b)}$$

where a is a function of temperature, and b and c are constants:

$$
\begin{aligned}
a &= \Omega_a \left(R^2 T_c^2/P_c\right)\left[1 + F\left(1 - \sqrt{T_R}\right)\right]^2 \\
b &= \Omega_b \left(RT_c/P_c\right) \\
c &= \Omega_c \left(RT_c/P_c\right)
\end{aligned}
$$

where

$$\Omega_c = 1 - 3\zeta_c$$

$$\Omega_a = 3\zeta_c^2 + 3\left(1 - 2\zeta_c\right)\Omega_b + \Omega_b^2 + 1 - 3\zeta_c$$

and Ω_b is the smallest positive root of the cubic equation:

$$\Omega_b^3 + \left(2 - 3\zeta_c\right)\Omega_b^2 + 3\zeta_c^2\Omega_b - \zeta_c^3 = 0$$

F and *c* are functions of the acentric factor given by the following quadratic correlations:

$$F = 0.452413 + 1.30982\omega - 0.295937\omega^2$$
$$\zeta_c = 0.329032 - 0.076799\omega + 0.0211947\omega^2$$

Use the data given in Example 1.5 for *n*-butane to calculate the parameters of PT equation.

1.5 The enthalpy of vaporization, ΔH_v, is related to vapor pressure by the thermodynamically exact Clausius-Clapeyron equation:

$$\Delta H_v = -R\Delta Z_v \frac{d\ln P^{sat}}{d(1/T)}$$

where

R = gas constant (=8.314 J/mol K)

$\Delta Z_v = Z_G - Z_L$

Z_G = compressibility factor of saturated vapor

Z_L = compressibility factor of saturated liquid

P^{sat} = vapor pressure

T = absolute temperature

   (a) Employ the SRK equation of state for estimating ΔZ_v for *n*-butane.

   (b) Use the Antoine equation and results of part (a) for evaluating ΔH_v as a function of temperature. The Antoine equation is:

$$\log P^{sat} = A - \frac{B}{T+C}$$

For P^{sat} in bar and T in Kelvin, the constants of this equation for *n*-butane are [5]:

$$A = 4.35576 \quad B = 1175.581 \quad C = -2.071$$

1.6 Riedel equation can be used for evaluating the vapor pressure of pure compounds:

$$\ln P_R^{sat} = A - \frac{B}{T_R} + C\ln T_R + DT_R^6$$

where $T_R = T/T_c$ and $P_R = P/P_c$. Use this equation to calculate the normal boiling point of acetone. Additional data are [6]:

$$A = 10.380 \quad B = 10.677 \quad C = -5.1589 \quad D = 0.29657$$
$$T_c = 508.1 K \quad P_c = 4700 \text{ kPa}$$

1.7 Vapor pressure can be obtained over a wide range of temperature from the Wagner equation:

$$\ln P_R^{sat} = \frac{A\tau + B\tau^{1.5} + C\tau^3 + D\tau^6}{1 \quad \tau}$$

in which $\tau = 1 - T_R$.

Liquid propane is stored in a storage tank for which the design pressure is 10 bar(a). What is the maximum allowable temperature inside the tank?

Additional data are [6]:

$$A = -6.72219 \quad B = 1.33236 \quad C = -2.13868 \quad D = -1.38551$$
$$T_c = 369.8\,\text{K} \quad P_c = 42.72 \text{ bar}$$

1.8 Calculate the terminal velocity of a 0.5 mm iron ball in water from the following force balance equation:

$$(\rho_s - \rho_l)Vg = \frac{1}{2}\rho_l v_T^2 A C_D$$

where ρ_s is the solid density ($7860\,\text{kg/m}^3$), ρ_l is the liquid density ($1000\,\text{kg/m}^3$), V is the volume of the ball, g is the gravitational acceleration ($9.81\,\text{m/s}^2$), v_T is the terminal velocity, A is the cross-sectional area of the ball, and C_D is the drag coefficient.

The drag coefficient for a spherical particle can be obtained from [7]:

$$C_D = \frac{24}{\text{Re}}\left(1 + 0.173\text{Re}^{0.657}\right) + \frac{0.413}{1 + 16300\text{Re}^{-1.09}}$$

in which

$$\text{Re} = \frac{\rho_l v_T d}{\mu}$$

where d is the particle diameter and μ is the viscosity of the liquid ($0.001\,\text{Pa}\cdot\text{s}$).

1.9 F moles per hour of an n-component natural gas stream is introduced as feed to the flash vaporization tank shown in Fig. P1.9. The resulting vapor and liquid streams are withdrawn at the rate of V and L moles per hour, respectively. The mole fractions of the components

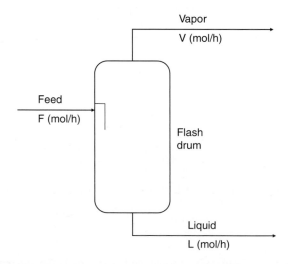

FIG. P1.9

Flash drum.

in the feed, vapor, and liquid are designated by z_i, y_i, and x_i, respectively ($i = 1, 2, \ldots, n$). Assuming vapor-liquid equilibrium and steady state operation, we have:

Overall balance: $F = L + V$

Individual component balances: $z_i F = x_i L + y_i V$ $i = 1, 2, \ldots, n$

Equilibrium relations: $K_i = y_i / x_i$ $i = 1, 2, \ldots, n$

Here, K_i is the equilibrium constant for the ith component at the prevailing temperature and pressure in the tank. From these equations and the fact that:

$$\sum_{i=1}^{n} x_i = \sum_{i=1}^{n} y_i = 1$$

derive the following equation:

$$\sum_{i=1}^{n} \frac{z_i K_i F}{V(K_i - 1) + F} = 1$$

Using the data given in Table P1.9, solve the above equation for V. Also calculate the values of L, the x_i, and the y_i by using the first three equations given above. The test data in Table P1.9 relates to the flashing of a natural gas stream at 11 MPa and 48°C. Assume that $F = 100$ mol/h.

What would be a good value V_0 for starting the iteration? Base this answer on your observations of the data given in Table P1.9.

1.10 The Underwood equation for multicomponent distillation is given as:

$$\left(\sum_{j=1}^{n} \frac{\alpha_j z_{jF} F}{\alpha_j - \varphi} \right) - F(1 - q) = 0$$

Table P1.9 Feed data.

Component	i	z_j	K_j
Methane	1	0.8345	3.09
Carbon dioxide	2	0.0046	1.65
Ethane	3	0.0381	0.72
Propane	4	0.0163	0.39
i-Butane	5	0.0050	0.21
n-Butane	6	0.0074	0.175
Pentanes	7	0.0287	0.093
Hexanes	8	0.0220	0.065
Heptanes +	9	0.0434	0.036
		1.0000	

Table P1.10 Feed data.		
Component in feed	**Mole fraction, z_{jF}**	**Relative volatility, α_j**
C_1	0.05	10
C_2	0.05	5
C_3	0.10	2.05
C_4	0.30	2.0
C_5	0.05	1.5
C_6	0.30	1.0
C_7	0.10	0.9
C_8	0.05	0.1
	1.00	
F = 100 mol/h; q = 1.0 (saturated liquid).		

where

F = molar feed flow rate

n = number of components in the feed

z_{jF} = mole fraction of each component in the feed

q = quality of the feed

α_j = relative volatility of each component at average column conditions

φ = root of the equation

It has been shown by Underwood that $(n - 1)$ of the roots of this equation lie between the values of the relative volatilities as shown below:

$$\alpha_n < \varphi_{n-1} < \alpha_{n-1} < \varphi_{n-2} < \ldots < \alpha_3 < \varphi_2 < \alpha_2 < \varphi_1 < \alpha_1$$

Evaluate the $(n - 1)$ roots of this equation for the case shown in Table P1.10.

1.11 Carbon monoxide from a water gas plant is burned with air in an adiabatic reactor. Both the carbon monoxide and air are being fed to the reactor at 25°C and atmospheric pressure. For the reaction:

$$CO + \frac{1}{2}O_2 \rightleftarrows CO_2$$

the following standard free energy change (at 25°C) has been determined:

$$\Delta G_{T_0}^0 = -257 \text{ kJ/(g mol of CO)}$$

The standard enthalpy change at 25°C has been measured as:

$$\Delta H_{T_0}^0 = -283 \text{ kJ/(g mol of CO)}$$

The standard states for all components are pure gases at 1 atm.

Calculate the adiabatic flame temperature and the conversion of CO for the following two cases:

(a) 0.4 mol of oxygen per mole of CO is provided for the reaction.

(b) 0.8 mol of oxygen per mole of CO is provided for the reaction.

The constant pressure heat capacities for the various constituents in J/(gmol.K) with T in Kelvins are all of the form:

$$C_{Pi} = A_i + B_i T_K + C_i T_K^2$$

For the gases involved here, the constants are as shown in Table P1.11.

Hint: Combine the material balance, enthalpy balance, and equilibrium relationship to form two nonlinear algebraic equations in two unknowns: the temperature and conversion.

1.12 Consider the three-mode feedback control of a stirred-tank heater system (Fig. P1.12). The measured output variable is the feedstream temperature [8]. Using classical methods (i.c., deviation variables, linearization, and Laplace transform) the overall closed-loop transfer function for the control system is given by:

$$\frac{\overline{T}}{\overline{T}_i} = \frac{(\tau_I s)(\tau_v s + 1)(\tau_m s + 1)}{(\tau_I s)(\tau_P s + 1)(\tau_v s + 1)(\tau_m s + 1) + K(\tau_I s + 1 + \tau_D \tau_I s^2)}$$

Table P1.11 Constants of heat capacity equation.

Gas	A	B	C
CO	26.16	8.75×10^{-3}	-1.92×10^{-6}
O_2	25.66	12.5×10^{-3}	-3.37×10^{-6}
CO_2	28.67	35.72×10^{-3}	-10.39×10^{-6}
N_2	26.37	7.61×10^{-3}	-1.44×10^{-6}

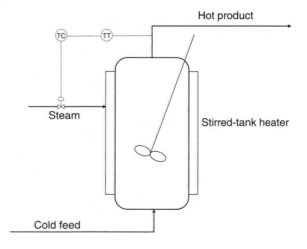

FIG. P1.12

Stirred tank heater.

where

τ_I = reset time constant

τ_D = derivative time constant

$K = K_P K_v K_m K_c$

K_P = first-order process static gain

K_v = first-order valve constant

K_m = first-order measurement constant

K_c = proportional gain for the three-mode controller

$\overline{T}$ = Laplace transform of the output temperature deviation

$\overline{T}_i$ = Laplace transform of the input load temperature deviation

τ_P, τ_m, τ_v = first-order time constants for the process, measurement device, and process valve, respectively.

For a given set of values, the stability of the system can be determined from the roots of the characteristic polynomial (i.e., the polynomial in the denominator of the overall transfer function). Thus,

$$\tau_I \tau_P \tau_m \tau_v s^4 + (\tau_I \tau_P \tau_m + \tau_I \tau_P \tau_v + \tau_I \tau_m \tau_v)s^3 + (K\tau_I \tau_D + \tau_I \tau_P + \tau_I \tau_v + \tau_I \tau_m)s^2 + (\tau_I + K\tau_I)s + K = 0$$

For the following set of parameter values, find the four roots to the characteristic polynomial when K_C is equal to its "critical" value.

$\tau_I = 10$	$\tau_D = 1$	$\tau_P = 10$	$\tau_m = 5$	$\tau_v = 5$
$K_P = 10$	$K_v = 2$	$K_m = 0.09$	$K = 1.8K_c$	

1.13 In the analytical solution of some parabolic partial differential equations in cylindrical coordinates, it is necessary to calculate the roots of the Bessel function first (for example, see Problem 8.15). Find the first N root of the first and the second kind. Use the following approximations for evaluating the initial guesses:

$$J_n(x) \approx \sqrt{\frac{2}{\pi x}} \cos\left(x - \frac{n\pi}{2} - \frac{\pi}{4}\right)$$

$$Y_n(x) \approx \sqrt{\frac{2}{\pi x}} \sin\left(x - \frac{n\pi}{2} - \frac{\pi}{4}\right)$$

Which method of root finding do you recommend?

1.14 A liquid is flowing in a trapezoidal channel at a rate of $Q = 30$ m³/s. The critical depth y for such a channel must satisfy the equation:

$$\frac{Q^2}{gA_c^3}B = 1$$

where g is the gravitational acceleration, A_c is the cross-sectional area of the channel (m²), and B is the width of the channel at the surface (m). For this case, the width and the cross-sectional area can be related to depth y by:

$$B = 3.5 + y$$

and

$$A_c = 3.5y + y^2/2$$

Solve for the critical depth using:

(a) the graphical method

(b) bisection method

(c) false position.

For (b) and (c) use initial guesses of 0.5 and 2.5, and iterate until the approximate error falls below 1% or the number of iterations exceeds 10. Discuss your results.

1.15 The resistivity ρ of doped silicon is given by:

$$\rho = \frac{1}{qn\mu}$$

where q is the electric charge of an electron, n is the electron density, and μ is the electron mobility. The electron density is given in terms of the doping density N and the intrinsic carrier density n_i:

$$n = \frac{1}{2}\left(N + \sqrt{N^2 + 4n_i^2}\right)$$

The electron mobility is a function of temperature, T:

$$\mu = \mu_0 \left(\frac{T}{T_0}\right)^{-2.42}$$

in which T_0 is the reference temperature, and μ_0 is the reference mobility.

Determine N at $T = 900$ K, for which $\rho = 6 \times 10^6$ V $\cdot$ s $\cdot$ cm/C.

Additional data are: $T_0 = 300$ K, $\mu_0 = 1360$ cm^2 (V.s)$^{-1}$, $q = 1.7 \times 10^{-19}$ C, $n_i = 6.2 \times 10^9$ cm^{-3}.

Employ initial guesses of $N = 0$ and 2.5×10^{10}, use

(a) bisection method

(b) false position method.

1.16 The following polynomial relates the specific heat of dry air at low pressure, c_p (kJ/kg K), to temperature (K):

$$c_p = 0.99403 + 1.671 \times 10^{-4}T + 9.7215 \times 10^{-8}T^2 + 9.5838 \times 10^{-11}T^3 + 1.9520 \times 10^{-14}T^4$$

(a) Plot c_p temperature for a range of $T = 0$ to 1000 K.

(b) Determine the temperature at which the specific heat is 1.2 kJ/(kg K).

1.17 The acidity of a saturated solution of magnesium hydroxide in hydrochloric acid is given by:

$$\frac{3.46 \times 10^{-11}}{[H_3O^+]^2} = [H_3O^+] + 3.46 \times 10^{-4}$$

in which $[H_3O^+]$ is the hydronium ion concentration. If we set $x = 10^4[H_3O^+]$, the above equation becomes:

$$x^3 + 3.64x^2 - 36.4 = 0$$

Determine the value of x and $[H_3O^+]$.

1.18 The equilibrium composition (molar fraction) of a mixture of carbon monoxide and oxygen gas at 1 bar is given by the following equation:

$$3.06 = \frac{(1-x)(3+x)^{1/2}}{x(1+x)^{1/2}}$$

Obtain a fixed-point iteration formula for finding the roots of this equation and solve for x. If the iteration does not converge, develop one that does.

1.19 The van der Waals equation of state is:

$$P = \frac{RT}{V-b} - \frac{a}{V^2}$$

where R is the universal gas constant, a and b are positive constants, P is the pressure, T is the absolute temperature, and V is the volume. Show that if one sets:

$$P^* = \frac{27b^2}{a}P, \quad T^* = \frac{27Rb}{8a}T, \quad V^* = \frac{1}{3b}V$$

The equation of state takes the dimensionless form:

$$\left(P^* + \frac{3}{V^{*2}}\right)(3V^* - 1) = 8T^*$$

Find V^* to five decimal places using a numerical method for the following conditions:
  (a) $P^* = 6$, $T^* = 2$,
  (b) $P^* = 4$, $T^* = 1$,
  (c) $P^* = 5$, $T^* = 5$.

1.20 The RK equation of state is given by:

$$P = \frac{RT}{V-b} - \frac{a}{V(V+b)\sqrt{T}}$$

where R is the gas constant [8314 J/(kmol.K)], T is the absolute temperature (K), P is the absolute pressure (Pa) and V is the volume of a kg of gas (m^3/kg). The parameters a and b are calculated by:

$$a = \frac{0.4278R^2T_C^{2.5}}{P_C} \quad b = \frac{0.0867RT_C}{P_C}$$

where $P_C = 49$ bar and $T_C = 305$ K for ethane. What is the amount of ethane that can be held in a 2 m^3 tank at a temperature of 20°C with a pressure of 60 MPa. Use a root-locating method of your choice to calculate V and then determine the mass of methane contained in the tank.

1.21 If heated to sufficiently high temperatures, water vapor dissociates to form oxygen and hydrogen:

$$H_2O \rightleftarrows H_2 + 0.5O_2$$

If it is assumed that this is the only reaction involved, the mole fraction x of H_2O that dissociates can be represented by:

$$K = \frac{x}{1-x}\sqrt{\frac{2P}{2+x}}$$

where K is the equilibrium constant of the reaction and P is the total pressure of the mixture. If $P = 4$ atm and $K = 0.05$, determine the value of x that satisfies the above equation.

1.22 The volume of liquid V in a hollow horizontal cylinder of radius r and length L is related to the depth of the liquid h by:

$$V = \left[r^2\cos^{-1}\left(\frac{r-h}{r}\right) - (r-h)\sqrt{2rh - h^2} \right]L$$

Determine h given $r = 2$ m, $L = 4$ m, and $V = 6$ m^3.

1.23 A direct-fired tubular reactor is used in the thermal cracking of light hydrocarbons or naphthas for the production of olefins, such as ethylene (see Fig. P1.23). The reactants are preheated in the convection section of the furnace, mixed with steam, and then subjected to high temperatures in the radiant section of the furnace. Heat transfer in the radiant section of the furnace takes place through three mechanisms: radiation, conduction, and convection. Heat is transferred by radiation from the walls of the furnace to the surface of the tubes which carry the reactants and it is transferred through the walls of the tubes by conduction and finally to the fluid inside the tubes by convection.

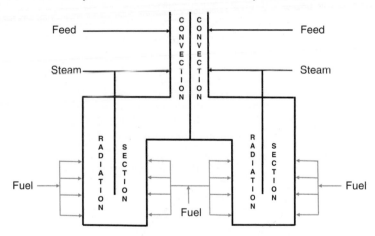

FIG. P1.23

Pyrolysis reactor.

The three heat-transfer mechanisms are quantified as follows:

a. *Radiation*: The Stefan-Boltzmann law of radiation may be written as:

$$\frac{dQ}{dA_o} = \sigma\phi\left(T_R^4 - T_o^4\right)$$

where dQ/dA_o is the rate for heat transfer per unit outside surface area of the tubes, T_R is the "effective" furnace radiation temperature, and T_o is the temperature on the outside surface of the tube. In furnaces with tube banks irradiated from both sides, a reasonable approximation is:

$$T_R = T_G$$

where T_G is the temperature of the flue gas in the reactor. Therefore, the Stefan-Boltzmann equation is revised to:

$$\frac{dQ}{dA_o} = \sigma\phi\left(T_G^4 - T_o^4\right)$$

σ is the Stefan-Boltzmann constant and ϕ is the tube geometry emissivity factor, which depends on the tube arrangement and tube surface emissivity. For single rows of tubes irradiated from both sides:

$$\frac{1}{\phi} = \frac{1}{\varepsilon} - 1 + \frac{\pi}{2\Omega}$$

$$\Omega = \frac{S}{D_o} + \tan^{-1}\sqrt{\left(\frac{S}{D_o}\right)^2 - 1} - \sqrt{\left(\frac{S}{D_o}\right)^2 - 1}$$

where ε is the emissivity of the outside surface of the tube and S is the spacing (pitch) of the tubes (center-to-center) and D_o is the outside diameter of the tubes.

b. *Conduction*: Conduction through the tube wall is given by Fourier's equation:

$$\frac{dQ}{dA_o} = \frac{k_t}{t_t}(T_o - T_i)$$

where T_i is the temperature on the inside surface of the tube, k_t is the thermal conductivity of the tube material, and t_t is the thickness of the tube wall.

c. *Convection*: Convection through the fluid film inside the tube is expressed by:

$$\frac{dQ}{dA_o} = h_i\left(\frac{D_i}{D_o}\right)(T_i - T_f)$$

where D_i is the inside diameter of the tube, T_f is the temperature of the fluid in the tube, and h_i is the heat-transfer film coefficient on the inside of the tube. The film coefficient may be approximated from the Dittus-Boelter equation [9]:

$$h_i = 0.023\left(\frac{k_f}{D_i}\right)Re_f^{0.8}Pr_f^{0.4}$$

where Re_f is the Reynolds number, Pr_f is the Prandtl number, and k_f is the thermal conductivity of the fluid.

Conditions vary drastically along the length of the tube, as the temperature of the fluid inside the tube rises rapidly. The rate of heat transfer is the highest at the entrance conditions and lowest at the exit conditions of the fluid.

Calculate the rate of heat transfer (dQ/dA_o), the temperature on the outside surface of the tube (T_o), and the temperature on the inside surface of the tube (T_i) at a point along the length of the tube where the following conditions exist:

$$T_G = 1200°C \qquad T_f = 800°C$$
$$\varepsilon = 0.9 \qquad \sigma = 5.7 \times 10^{-8} \, W/m^2.K^4$$
$$S = 0.20\,m \qquad D_i = 0.10\,m \qquad\qquad D_o = 0.11\,m$$
$$t_t = 0.006\,m \qquad k_f = 0.175\,W/m.K$$
$$Re_f = 388,000 \qquad Pr_f = 0.660$$

1.24 The elementary reaction A $\rightarrow$ B $+$ C is carried out in a continuous stirred tank reactor (CSTR). Pure A enters the reactor at a flow rate of 12 mol/s and a temperature of 25°C. The reaction is exothermic and cooling water at 50°C is used to absorb the heat generated. The energy balance for this system, assuming constant heat capacity and equal heat capacity of both sides of the reaction, can be written as:

$$-F_{A_0} X \Delta H_R = F_{A_0} C_{P_A}(T - T_0) + UA(T - T_a)$$

where
$F_{A0} =$ molar flow rate, mol/s
$X =$ conversion
$\Delta H_R =$ heat of reaction, J/mol A
$C_{PA} =$ heat capacity of A, J/mol.K
$T =$ reactor temperature, °C
$T_0 =$ reference temperature, 25°C
$T_a =$ cooling water temperature, 20°C
$U =$ overall heat transfer coefficient, W/m².K
$A =$ heat transfer area, m²

For a first-order reaction, the conversion can be calculated from:

$$X = \frac{\tau k}{1 + \tau k}$$

where τ is the residence time of the reactor in seconds and k is the specific reaction rate in s^{-1} defined by the Arrhenius formula:

$$k = 650 \exp\left[-3800/(T + 273)\right]$$

Solve the energy balance equation for temperature and find the steady state operating temperatures of the reactor and the conversions corresponding to these temperatures. Additional data are:

$$\Delta H_R = 1500\,kJ/mol \quad \tau = 10\,s \quad C_{PA} = 4500 \text{ J/mol.K} \quad UA/F_{A0} = 700 \text{ W.s/mol.K}$$

1.25 A chemical reaction A$\rightarrow$B takes place in a series of three continuous stirred tank reactors arranged as shown in Fig. P1.25. The rate of consumption of component A in each reactor is given by:

$$r_A = \frac{0.2C_A^{1.5}}{1 + 0.05C_A} \text{ mol}/L.min.$$

Compute the concentration of A in these reactors at the steady state operation of the reactors. All reactors are of an equal volume of 1000 L. The concentration of the feed is $C_{Ain} = 1.5$ mol/L.

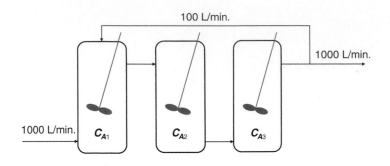

FIG. P1.25

Series of three continuous stirred tank reactors.

1.26 The following equations can be shown to relate the temperatures and pressures on either side of a detonation wave that is moving into a zone of unburned gas [10]:

$$\frac{\gamma_2 m_2 T_1}{m_1 T_2}\left(\frac{P_2}{P_1}\right)^2 - (\gamma_2 + 1)\left(\frac{P_2}{P_1}\right) + 1 = 0$$

$$\frac{\Delta H_{R1}}{c_{p2} T_1} + \frac{T_2}{T_1} - 1 = \frac{(\gamma_2 - 1)m_2}{2\gamma_2 m_1}\left(\frac{P_2}{P_1} - 1\right)\left(1 + \frac{m_1 T_2 P_1}{m_2 T_1 P_2}\right)$$

Here T = absolute temperature, P = absolute pressure, γ_2 = ratio of specific heat at constant pressure to that at constant volume, m = mean molecular weight, ΔH_{R1} = heat of reaction, c_{p2} = specific heat, and the subscripts 1 and 2 refer to the unburned and burned gas, respectively.

Write a program that accepts values for m_1, m_2, γ_2, ΔH_{R1}, c_{p2}, T_1, and P_1 as data, and that will proceed to compute and print values for T_2 and P_2. Run the program with the following data, which apply to the detonation of a mixture of hydrogen and oxygen:

$m_1 = 12$ g/g mol $\qquad\qquad$ $m_2 = 18$ g/g mol $\qquad\qquad$ $T_1 = 300$ K $\qquad$ $\gamma_2 = 1.31$

$\Delta H_{R1} = -58,300$ cal/g mol $\qquad$ $c_{p2} = 9.86$ cal/(g mol $\cdot$ K) $\qquad$ $P_1 = 1$ atm

1.27 Aniline is being removed from water by solvent extraction using toluene [8]. The unit is a 10-stage countercurrent tower, shown in Fig. P1.27. The equilibrium relationship valid at each stage is:

$$m = \frac{Y_i}{X_i} = 9 + 20X_i$$

where

Y_i = (kg of aniline in the toluene phase) / (kg of toluene in the toluene phase)

X_i = (kg of aniline in the water phase) / (kg of water in the water phase)

    (a) The solution to this problem is a set of 10 simultaneous nonlinear equations. Derive these equations from material balances around each stage.

    (b) Solve the above set of equations to find the concentration in both the aqueous and organic phases leaving each stage of the system (X_i and Y_i).

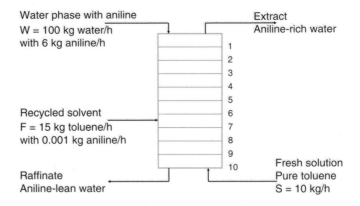

FIG. P1.27

Extraction column

REFERENCES

1. Wankat, P. C. *Separation Process Engineering*, 4th ed.; Pearson: Upper Saddle River, NJ, 2016.
2. Grabiner, D. J. Descartes' Rule of Signs: Another Construction. *Amer. Math. Monthly* **1999,** *106*, 854–855.
3. Lapidus, I. *Digital Computation for Chemical Engineering;* McGraw-Hill: New York, 1962.
4. Douglas, J. M. *Process Dynamics and Control*, vol. 2. Prentice Hall: Englewood Cliffs, NJ, 1972.
5. Das, T. R., Reed, C. O., Eubank, P. T. PVT Surface and Thermodynamic Properties of *n*-Butane. *J. Chem. Eng. Data* **1973,** *18* (3), 244–253.
6. Poling, B., Prausnitz, J. M., O'Connell, J. P. *The Properties of Gases and Liquids*, 5th ed.; McGraw Hill: New York, 2000.
7. Turton, R., Levenspiel, O. A Short Note on the Drag Correlation for Spheres. *Powder Technol.* **1986,** *47*, 83–86.
8. Constantinides, A., Mostoufi, N. *Numerical Methods for Chemical Engineers with MATLAB Applications;* Prentice Hall: Upper Saddle River, NJ, 1999.
9. Incropera, F. P., DeWitt, D. P. *Fundamentals of Heat and Mass Transfer*, 6th ed.; John Wiley & Sons: New York, 2007.
10. Carnahan, B., Luther, H. A., Wilkes, J. O. *Applied Numerical Methods;* John Wiley & Sons: New York, 1969.

SIMULTANEOUS LINEAR ALGEBRAIC EQUATIONS

CHAPTER OUTLINE

MOTIVATION

The mathematical analysis of linear physicochemical systems often results in models consisting of sets of linear algebraic equations. In addition, methods of solution of nonlinear systems and

Applied Numerical Methods for Chemical Engineers. DOI: https://doi.org/10.1016/B978-0-12-822961-3.00002-9

differential equations use the technique of linearization of the models, thus requiring the repetitive solution of sets of linear algebraic equations. These problems may range in complexity from a set of two simultaneous linear algebraic equations to a set involving 1000 or even 10,000 equations. The solution of a set of two to three linear algebraic equations can be obtained easily by the algebraic elimination of variables or by the application of Cramer's rule. However, for systems involving five or more equations, the algebraic elimination method becomes too complex, and Cramer's rule requires a rapidly escalating number of arithmetic operations, too large even for today's high-speed digital computers. Various methods of solution of a set of linear algebraic equations are presented in this chapter. In developing systematic methods for the solution of linear algebraic equations and the evaluation of eigenvalues and eigenvectors of linear systems we will make extensive use of matrix-vector notation. For this reason, and for the benefit of the reader, a review of the selected matrix and vector operations is also given.

2.1 INTRODUCTION

Material and energy balances are the primary tools of chemical engineers. Such balances applied to multistage or multicomponent processes result in sets of equations that can be either differential or algebraic. Often the systems under analysis are nonlinear, thus resulting in sets of nonlinear equations. However, as mentioned earlier, many procedures have been developed which linearize the equations and apply iterative convergence techniques to arrive at the solution of the nonlinear systems.

A classic example of the use of these techniques is in the analysis of distillation columns, such as the one shown in Fig. 2.1. Steady-state material balances applied to the rectifying section of the column yield the following equations:

$$\text{Balance around condenser: } V_1 y_{1i} = L_0 x_{0i} + D x_{Di} \tag{2.1}$$

$$\text{Balance above the } j\text{th stage: } V_j y_{ji} = L_{j-1} x_{j-1,i} + D x_{Di} \tag{2.2}$$

Assuming that the stages are equilibrium stages and that the column uses a total condenser, the following equilibrium relations apply:

$$y_{ji} = K_{ji} x_{ji} \tag{2.3}$$

Substituting Eq. (2.3) in (2.1) and (2.2) and dividing through by Dx_{Di} we get:

$$\frac{V_1 y_{1i}}{Dx_{Di}} = \frac{L_0}{Dx_{Di}} x_{0i} + 1 \tag{2.4}$$

$$\frac{V_j y_{1i}}{Dx_{Di}} = \left(\frac{L_{j-1}}{K_{j-1,i} V_{j-1}}\right) \left(\frac{V_{j-1} y_{j-1,i}}{Dx_{Di}}\right) + 1 \tag{2.5}$$

The molar flow rates of the individual components are defined as:

$$v_{ji} = V_j y_{ji} \tag{2.6}$$

$$d_i = Dx_{Di} \tag{2.7}$$

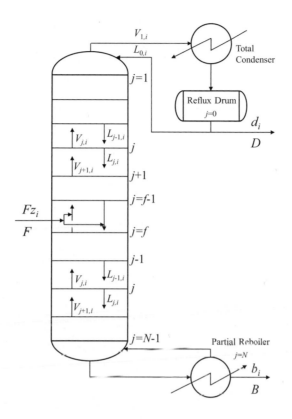

FIG. 2.1

Conventional distillation column.

For any stage j, the absorption ratio is defined as:

$$A_{ji} = \frac{L_j}{K_{ji}V_j} \tag{2.8a}$$

and for the total condenser as:

$$A_{0i} = \frac{L_0}{D} \tag{2.8b}$$

Substitution of Eqs. (2.6)–(2.8) in (2.4) and (2.5) yields:

$$\left(\frac{v_{1i}}{d_i}\right) = A_{0i} + 1 \tag{2.9}$$

$$\left(\frac{v_{ji}}{d_i}\right) = A_{j-1,i}\left(\frac{v_{j-1,i}}{d_i}\right) + 1 \quad 2 \le j \le f - 1 \tag{2.10}$$

For any given trial calculation, the A's are regarded as constants. The unknowns in the above equations are the groups of terms v_{ji}/d_i. If these are replaced by x_{ji} and the subscript i, designating component i, is dropped, the following set of equations can be written for a column containing five equilibrium stages above the feed stage:

$$\begin{aligned}
x_1 &= A_0 + 1 \\
-A_1 x_1 + x_2 &= 1 \\
-A_2 x_2 + x_3 &= 1 \\
-A_3 x_3 + x_4 &= 1 \\
-A_4 x_4 + x_5 &= 1
\end{aligned} \tag{2.11}$$

This is a set of simultaneous linear algebraic equations. It is actually a special set that has nonzero terms only on the diagonal and one adjacent element. It is a bidiagonal set.

The general formulation of a set of simultaneous linear algebraic equations is:

$$\begin{aligned}
a_{11}x_1 + a_{12}x_2 + \ldots + a_{1n}x_n &= c_1 \\
a_{21}x_1 + a_{22}x_2 + \ldots + a_{2n}x_n &= c_2 \\
&\ldots \\
a_{n1}x_1 + a_{n2}x_2 + \ldots + a_{nn}x_n &= c_n
\end{aligned} \tag{2.12}$$

where all of the coefficients a_{ij} could be nonzero. This set is usually condensed in vector-matrix notation as:

$$Ax = c \tag{2.13}$$

where A is the coefficient matrix,

$$A = \begin{bmatrix}
a_{11} & a_{12} & \cdots & a_{1n} \\
a_{21} & a_{22} & \cdots & a_{2n} \\
\vdots & \vdots & \ddots & \vdots \\
a_{n1} & a_{n2} & \cdots & a_{nn}
\end{bmatrix} \tag{2.14}$$

x is the vector of unknown variables,

$$x = \begin{bmatrix} x_1 \\ x_2 \\ \vdots \\ x_n \end{bmatrix} \tag{2.15}$$

and c is the vector of constants:

$$c = \begin{bmatrix} c_1 \\ c_2 \\ \vdots \\ c_n \end{bmatrix} \tag{2.16}$$

When the vector c is the zero vector, the set of equations is called *homogeneous*.

Another example requiring the solution of linear algebraic equations comes from the analysis of complex reaction systems which have monomolecular kinetics. Fig. 2.2 considers a chemical

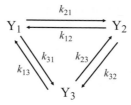

FIG. 2.2

System of chemical reactions.

reaction between the three species, whose concentrations are designated by Y_1, Y_2, Y_3, taking place in a batch reactor.

The equations describing the dynamics of this chemical reaction scheme are:

$$\frac{dY_1}{dt} = -(k_{21} + k_{31})Y_1 + k_{12}Y_2 + k_{13}Y_3$$
$$\frac{dY_2}{dt} = k_{21}Y_1 - (k_{12} + k_{32})Y_2 + k_{23}Y_3 \qquad (2.17)$$
$$\frac{dY_3}{dt} = k_{31}Y_1 + k_{32}Y_2 - (k_{13} + k_{23})Y_3$$

The above set of linear ordinary differential equations may be condensed into matrix notation:

$$\dot{y} = Ky \qquad (2.18)$$

where $\dot{y}$ is the vector of derivatives,

$$\dot{y} = \begin{bmatrix} \dfrac{dY_1}{dt} \\ \dfrac{dY_2}{dt} \\ \dfrac{dY_3}{dt} \end{bmatrix} \qquad (2.19)$$

y is the vector of concentrations of the components,

$$y = \begin{bmatrix} Y_1 \\ Y_2 \\ Y_3 \end{bmatrix} \qquad (2.20)$$

and K is the matrix of kinetic rate constants:

$$K = \begin{bmatrix} -(k_{21} + k_{31}) & k_{12} & k_{13} \\ k_{21} & -(k_{12} + k_{32}) & k_{23} \\ k_{31} & k_{32} & -(k_{13} + k_{23}) \end{bmatrix} \qquad (2.21)$$

The solution of the dynamic problem, which is modeled by Eq. (2.18), would require the evaluation of the characteristic values (eigenvalues) λ_i and characteristic vectors (eigenvectors) x_i of the matrix K. It is shown in Chapter 5 that the solution of a set of linear ordinary differential equations can be obtained from:

$$y = \left[Xe^{\Lambda t}X^{-1}\right]y_0 \tag{5.53}$$

where X is a matrix whose columns are the eigenvectors x_i of K, and $e^{\Lambda t}$ is a matrix with $e^{\lambda_i t}$ on the diagonal and zero elsewhere; X^{-1} is the inverse of X, and y_0 is the vector of initial values of the variables Y_i. Methods for calculating eigenvalues and eigenvectors of matrices are developed later in this chapter.

When a chemical reaction reaches the steady state, the vector of derivatives in Eq. (2.18) becomes zero and Eq. (2.18) simplifies to:

$$Ky = 0 \tag{2.22}$$

This is a set of homogeneous linear algebraic equations whose solution describes the steady-state situation of the aforementioned chemical reaction problem.

Comparison of Eqs. (2.13) and (2.22) reveals that the difference between nonhomogeneous and homogeneous sets of equations is that in the latter the vector of constants c is the zero vector. The steady-state solution of the chemical reaction problem requires finding a unique solution to the set of homogeneous algebraic equations represented by Eq. (2.22).

The solution of some partial differential equations is sometimes carried out by solving sets of simultaneous finite difference equations (see Chapter 7). These equations are often linear in nature and can be solved by the methods which will be discussed in this chapter.

Optimization of a complex assembly of unit operations, such as a chemical plant, or of a cluster of interrelated assemblies, such as a group of refineries, can be accomplished by techniques of linear programming which handle large sets of simultaneous linear equations.

The application of linear and nonlinear regression analysis to fit mathematical models to experimental data and to evaluate the unknown parameters of these models (see Chapter 8) requires the repetitive solution of sets of linear algebraic equations. In addition, the ellipse formed by the correlation coefficient matrix in the parameter hyperspace of these systems must be searched in the direction of the major and minor axes. The directions of these axes are defined by the eigenvectors of the correlation coefficient matrix, and the relative lengths of the axes are measured by the eigenvalues of the correlation coefficient matrix.

In developing systematic methods for the solution of linear algebraic equations and the evaluation of eigenvalues and eigenvectors of linear systems we will make extensive use of matrix-vector notation. For this reason, and for the benefit of the reader, a review of the selected matrix and vector operations is given in the next section. While there are numerous references and textbooks on the basics of matrix operations, the description given below highlights features specific to MATLAB®.

2.2 REVIEW OF SELECTED MATRIX AND VECTOR OPERATIONS

2.2.1 MATRICES AND DETERMINANTS

A matrix is an array of elements arranged in rows and columns as:

$$A = \begin{bmatrix} a_{11} & a_{12} & \cdots & a_{1n} \\ a_{21} & a_{22} & \cdots & a_{2n} \\ \vdots & \vdots & \ddots & \vdots \\ a_{m1} & a_{m2} & \cdots & a_{mn} \end{bmatrix} \tag{2.23}$$

The elements a_{ij} of the matrix may be real numbers, complex numbers, or functions of other variables. Matrix A has m rows and n columns and is said to be of order ($m \times n$). If the number of rows of a matrix is equal to the number of columns, that is, if $m = n$, then the matrix is a square matrix of nth order. A special matrix containing only a single column is called a vector:

$$x = \begin{bmatrix} x_1 \\ x_2 \\ \vdots \\ x_n \end{bmatrix} \tag{2.15}$$

Define another matrix B with k rows and l columns:

$$B = \begin{bmatrix} b_{11} & b_{12} & \cdots & b_{1l} \\ b_{21} & b_{22} & \cdots & b_{2l} \\ \vdots & \vdots & \ddots & \vdots \\ b_{k1} & b_{k2} & \cdots & b_{kl} \end{bmatrix} \tag{2.24}$$

The two matrices A and B can be added to (or subtracted from) each other if they have the same number of rows ($m = k$) and the same number of columns ($n = l$). For example, if both A and B are (3×2) matrices, their sum (or difference) can be written as:

$$A \pm B = \begin{bmatrix} a_{11} \pm b_{11} & a_{12} \pm b_{12} \\ a_{21} \pm b_{21} & a_{22} \pm b_{22} \\ a_{31} \pm b_{31} & a_{32} \pm b_{32} \end{bmatrix} = C \tag{2.25}$$

Matrix C is also a (3×2) matrix. The commutative and associative laws for addition and subtraction apply.

Two matrices can multiply each other if they are conformable. Matrices A and B would be conformable in the order AB if A had the same number of columns as B has rows ($n = k$). If A is of order (4×2) and B is of order (2×3), then the product AB is:

$$AB = \begin{bmatrix} a_{11} & a_{12} \\ a_{21} & a_{22} \\ a_{31} & a_{32} \\ a_{41} & a_{42} \end{bmatrix} \begin{bmatrix} b_{11} & b_{12} & b_{13} \\ b_{21} & b_{22} & b_{23} \end{bmatrix}$$

$$= \begin{bmatrix} a_{11}b_{11} + a_{12}b_{21} & a_{11}b_{12} + a_{12}b_{22} & a_{11}b_{13} + a_{12}b_{23} \\ a_{21}b_{11} + a_{22}b_{21} & a_{21}b_{12} + a_{22}b_{22} & a_{21}b_{13} + a_{22}b_{23} \\ a_{31}b_{11} + a_{32}b_{21} & a_{31}b_{12} + a_{32}b_{22} & a_{31}b_{13} + a_{32}b_{23} \\ a_{41}b_{11} + a_{42}b_{21} & a_{41}b_{12} + a_{42}b_{22} & a_{41}b_{13} + a_{42}b_{23} \end{bmatrix} = E \tag{2.26}$$

The resulting matrix E is of order (4×3). The general equation for performing matrix multiplication is:

$$e_{ij} = \sum_{p=1}^{n} a_{ip}b_{pj} \qquad \begin{cases} i = 1, 2, \ldots, m \\ j = 1, 2, \ldots, l \end{cases} \tag{2.27}$$

The resulting matrix would be of order ($m \times l$).

The commutative law is not usually valid for matrix multiplication, that is,

$$AB \neq BA \tag{2.28}$$

even if the matrices are conformable. The distributive law for multiplication applies to matrices, provided that conformability exists:

$$A(B + C) = AB + AC \tag{2.29}$$

The associative law of multiplication is also valid for matrices:

$$A(BC) = (AB)C \tag{2.30}$$

When working with MATLAB, it should be noted that there is also an element-by-element multiplication for matrices which is completely different from the ordinary multiplication of matrices described earlier. The element-by-element multiplication, whose operator is ".*" (a dot before the ordinary multiplication operator), may be applied only to matrices of the same order and it simply multiplies corresponding elements of the two matrices. For example, if A and B are of order (3×2), then the element-by-element product $A.*B$ is:

$$A.*B = \begin{bmatrix} a_{11} & a_{12} \\ a_{21} & a_{22} \\ a_{31} & a_{32} \end{bmatrix} .* \begin{bmatrix} b_{11} & b_{12} \\ b_{21} & b_{22} \\ b_{31} & b_{32} \end{bmatrix} = \begin{bmatrix} a_{11}b_{11} & a_{12}b_{12} \\ a_{21}b_{21} & a_{22}b_{22} \\ a_{31}b_{31} & a_{32}b_{32} \end{bmatrix} \tag{2.31}$$

The user should pay special attention to the difference between the element-by-element and regular multiplication of matrices since these two operations result in different results. Especially when both matrices are square, the command executes in MATLAB without any error message.

If the rows of an $(m \times n)$ matrix are written as columns, a new matrix of order $(n \times m)$ is formed. This new matrix is called the *transpose* of the original matrix. For example, if matrix A is:

$$A = \begin{bmatrix} 1 & 2 \\ 3 & 4 \\ 5 & 6 \end{bmatrix} \tag{2.32}$$

then the *transpose* A' is:

$$A' = \begin{bmatrix} 1 & 3 & 5 \\ 2 & 4 & 6 \end{bmatrix} \tag{2.33}$$

The transpose of the matrix A is sometimes shown as A^T. The transpose of the sum of two matrices is given by:

$$(A + B)' = A' + B' \tag{2.34}$$

The transpose of the product of two matrices is given by:

$$(AB)' = B'A' \tag{2.35}$$

In MATLAB the transpose of a matrix is simply obtained by adding a prime sign (') after the matrix.

The following definitions apply to square matrices only: A *symmetric* matrix is one that obeys the equation:

$$A = A' \tag{2.36}$$

If the symmetrically situated elements of a matrix are complex conjugates of each other, the matrix is called *Hermitian*. A *diagonal* matrix is one with nonzero elements on the principal diagonal and zero elements everywhere else:

$$\boldsymbol{D} = \begin{bmatrix} d_{11} & 0 & \cdots & 0 \\ 0 & d_{22} & \cdots & 0 \\ \vdots & \vdots & \ddots & \vdots \\ 0 & 0 & \cdots & d_{nn} \end{bmatrix} \tag{2.37}$$

The built-in MATLAB function *diag(x)* creates a diagonal matrix whose main diagonal elements are the components of the vector $\boldsymbol{x}$. If $\boldsymbol{x}$ is a matrix, *diag(x)* is a column vector formed from the elements of the diagonal of $\boldsymbol{x}$.

A *unit* matrix (or *identity* matrix) is a diagonal matrix whose nonzero elements are unity:

$$\boldsymbol{I} = \begin{bmatrix} 1 & 0 & \cdots & 0 \\ 0 & 1 & \cdots & 0 \\ \vdots & \vdots & \ddots & \vdots \\ 0 & 0 & \cdots & 1 \end{bmatrix} \tag{2.38}$$

Multiplication of a matrix (or a vector) by the identity matrix does not alter the matrix (or vector):

$$\boldsymbol{IA} = \boldsymbol{A} \quad \boldsymbol{Ix} = \boldsymbol{x} \tag{2.39}$$

In MATLAB the function *eye(n)* returns an $(n \times n)$ unit matrix.

A *tridiagonal* matrix is one that has nonzero elements on the principal diagonal and its two adjacent diagonals, which we will refer to as the subdiagonal (below) and superdiagonal (above), and zero elements everywhere else:

$$\boldsymbol{T} = \begin{bmatrix} t_{11} & t_{12} & 0 & 0 & & & & 0 & 0 \\ t_{21} & t_{22} & t_{23} & 0 & \cdots & \cdots & \cdots & 0 & 0 \\ 0 & t_{32} & t_{33} & t_{34} & & & & 0 & 0 \\ \vdots & \vdots & \vdots & \vdots & \ddots & & & \vdots & \vdots \\ 0 & 0 & & & & t_{n-2n-3} & t_{n-2n-2} & t_{n-2n-1} & 0 \\ 0 & 0 & \cdots & \cdots & \cdots & 0 & t_{n-1n-2} & t_{n-1n-1} & t_{n-1n} \\ 0 & 0 & & & & 0 & 0 & t_{nn-1} & t_{nn} \end{bmatrix} \tag{2.40}$$

An *upper triangular* matrix is one that has all zero elements below the principal diagonal:

$$\boldsymbol{U} = \begin{bmatrix} u_{11} & u_{12} & u_{13} & \cdots & u_{1n-1} & u_{1n} \\ 0 & u_{22} & u_{23} & \cdots & u_{2n-1} & u_{2n} \\ \vdots & \vdots & \vdots & \ddots & \vdots & \vdots \\ 0 & 0 & 0 & \cdots & u_{n-1n-1} & u_{n-1n} \\ 0 & 0 & 0 & \cdots & 0 & u_{nn} \end{bmatrix} \tag{2.41}$$

In MATLAB the function *triu(A)* constructs an upper triangular matrix out of matrix $\boldsymbol{A}$; that is, it keeps the elements on the main diagonal and above that unchanged and replaces the elements located under the main diagonal with zeros.

A *lower triangular* matrix is one that has all zero elements above the principal diagonal:

$$
L = \begin{bmatrix}
l_{11} & 0 & 0 & \cdots & 0 & 0 \\
l_{21} & l_{22} & 0 & \cdots & 0 & 0 \\
\vdots & \vdots & \vdots & \ddots & \vdots & \vdots \\
l_{n-11} & l_{n-12} & l_{n-13} & \cdots & l_{n-1\,n-1} & 0 \\
l_{n1} & l_{n1} & l_{n3} & \cdots & l_{nn} & l_{nn}
\end{bmatrix}
\tag{2.42}
$$

In MATLAB the function *tril(A)* constructs a lower triangular matrix out of matrix A; that is, it keeps the elements on the main diagonal and below that unchanged and replaces the elements located above the main diagonal with zeros.

Outputs of the MATLAB function *lu(A)* are an upper triangular matrix U and a "psychologically lower triangular matrix," that is, a product of lower triangular and permutation matrices, in L so that $LU = A$.

A *supertriangular* matrix, also called a *Hessenberg* matrix, is one that has all zero elements below the subdiagonal, such as the upper Hessenberg matrix:

$$
H_U = \begin{bmatrix}
h_{11} & h_{12} & h_{13} & \cdots & h_{1n-2} & h_{1n-1} & h_{1n} \\
h_{21} & h_{22} & h_{23} & \cdots & h_{2n-2} & h_{2n-1} & h_{2n} \\
0 & h_{32} & h_{33} & \cdots & h_{3n-2} & h_{3n-1} & h_{3n} \\
\vdots & \vdots & \vdots & \ddots & \vdots & \vdots & \vdots \\
0 & 0 & 0 & \cdots & h_{n-1n-2} & h_{n-1n-1} & h_{n-1n} \\
0 & 0 & 0 & \cdots & 0 & h_{nn-1} & h_{nn}
\end{bmatrix}
\tag{2.43}
$$

or above the superdiagonal, such as the lower Hessenberg matrix:

$$
H_L = \begin{bmatrix}
h_{11} & h_{12} & 0 & \cdots & 0 & 0 \\
h_{21} & h_{22} & h_{23} & \cdots & 0 & 0 \\
\vdots & \vdots & \vdots & \ddots & \vdots & \vdots \\
h_{n-21} & h_{n-22} & h_{n-23} & & h_{n-2\,n-1} & 0 \\
h_{n-11} & h_{n-12} & h_{n-13} & \cdots & h_{n-1\,n-1} & h_{n-1n} \\
h_{n1} & h_{n2} & h_{n3} & & h_{n\,n-1} & h_{nn}
\end{bmatrix}
\tag{2.44}
$$

Tridiagonal, triangular, and Hessenberg matrices are called *banded* matrices.

Matrices can be divided into two general categories: *dense* and *sparse*. The dense matrices are usually of low order and may have only a few zero elements. The sparse matrices may be of high order with many zero elements. A special subcategory of sparse matrices is the group of banded matrices described earlier.

The sum of the elements on the main diagonal of a square matrix is called the *trace*:

$$
tr\, A = \sum_{i=1}^{n} a_{ii}
\tag{2.45}
$$

The sum of the *eigenvalues* of a square matrix is equal to the trace of that matrix:

$$
\sum_{i=1}^{n} \lambda_i = tr\, A
\tag{2.46}
$$

The MATLAB function *trace(A)* calculates the trace of matrix A.

Matrix division is not defined in the normal algebraic sense. Instead, an inverse operation is defined, which uses multiplication to achieve the same results. If a square matrix A and another square matrix B, of the same order as A, lead to the identity matrix I when multiplied together:

$$AB = I \tag{2.47}$$

then B is called the *inverse* of A and is written as A^{-1}. It follows then that:

$$AA^{-1} = A^{-1}A = I \tag{2.48}$$

There are several different ways in MATLAB to calculate the inverse of a square matrix. The function $inv(A)$ gives the inverse of A. Also, the inverse of the matrix may be obtained by $A^\wedge(-1)$. The expression A/B in MATLAB is equivalent to AB^{-1} and the expression $A\backslash B$ is equivalent to $A^{-1}B$. Note that the expression $A./B$ (putting "." before division operator) is an element-by-element division of the elements of the two matrices and the expression $A.^\wedge(-1)$ results in a matrix whose elements are the reciprocals of the elements of the original matrix.

The inverse of the product of two matrices is the product of the inverses of these matrices multiplied in reverse order:

$$(AB)^{-1} = B^{-1}A^{-1} \tag{2.49}$$

This can be generalized to products of more than two matrices:

$$(ABC\ldots KLM)^{-1} = M^{-1}L^{-1}K^{-1}\ldots C^{-1}B^{-1}A^{-1} \tag{2.50}$$

A matrix is singular if the determinant of the matrix is zero. Only *nonsingular* matrices have an inverse.

The value of the determinant, which exists for square matrices only, can be calculated from Laplace's expansion theorem, which involves *minors* and *cofactors* of square matrices. If the row and column containing an element a_{ij} in a square matrix A are deleted, the determinant of the remaining square array is called the *minor* of a_{ij} and is denoted by M_{ij}. The *cofactor* of a_{ij}, denoted by A_{ij}, is given by:

$$A_{ij} = (-1)^{i+j}M_{ij} \tag{2.51}$$

Laplace's expansion theorem states that the determinant of a square matrix A, shown as $|A|$, is equal to the sum of products of the elements of any row (or column) and their respective cofactors:

$$|A| = \sum_{k=1}^{n} a_{ik}A_{ik} \tag{2.52}$$

for any row i, or:

$$|A| = \sum_{k=1}^{n} a_{kj}A_{kj} \tag{2.53}$$

for any column j.

Determinants have the following properties:

Property 1. If all the elements of any row or column of a matrix are zero, its determinant is equal to zero.

Property 2. If the corresponding rows and columns of a matrix are interchanged, its determinant is unchanged.

Property 3. If two rows or two columns of a matrix are interchanged, the sign of the determinant changes.

Property 4. If the elements of two rows or two columns of a matrix are equal, the determinant of the matrix is zero.

Property 5. If the elements of any row or column of a matrix are multiplied by a scalar, this is equivalent to multiplying the determinant by the scalar.

Property 6. Adding the product of a scalar and any row (or column) to any other row (or column) of a matrix leaves the determinant unchanged.

Property 7. The determinant of a triangular matrix is equal to the product of its diagonal elements:

$$|U| = \prod_{i=1}^{n} a_{ii} \tag{2.54}$$

Calculating determinants by the expansion of cofactors is a very algebra-intensive task. Each determinant has $n!$ groups of terms and each group is the product of n elements; thus the total number of multiplications is $(n-1)(n!)$. Evaluating the determinant of a matrix of order (10×10) would require 32,659,200 multiplications. More efficient methods have been developed for evaluating determinants. It will be shown in Section 2.5 that the Gauss elimination method can be used to calculate the determinant of a matrix in addition to finding the solution of simultaneous linear algebraic equations. To evaluate the determinant of a square matrix A in MATLAB, the built-in function $det(A)$ may be used.

The inverse of a matrix cannot always be determined accurately. There are many ill-conditioned matrices. An ill-conditioned matrix can be identified using the following criteria:

a. When the determination of a matrix is very small, the matrix is ill conditioned.

b. When the ratio of the absolute values of the largest and smallest eigenvalues of the matrix is very large, the matrix is ill conditioned.

The *rank r* of matrix A is defined as the order of the largest nonsingular square matrix within A. Consider the $(m \times n)$ matrix:

$$A = \begin{bmatrix} a_{11} & a_{12} & \cdots & a_{1n} \\ a_{21} & a_{22} & \cdots & a_{2n} \\ \vdots & \vdots & \ddots & \vdots \\ a_{m1} & a_{m2} & \cdots & a_{mn} \end{bmatrix} \tag{2.23}$$

where $n \geq m$. The largest square submatrix within A is of order $(m \times m)$. If the determinant of this $(m \times m)$ submatrix is nonzero, then the rank of A is m $(r = m)$. However, if the determinant of the $(m \times m)$ submatrix is equal to zero, then the rank of A is less than m $(r < m)$. The order of the next largest nonsingular submatrix that can be located within A would determine the value of the rank. In MATLAB the function $rank(A)$ gives the value of r, the rank of matrix A.

As an example, let us look at the following (3 × 4) matrix:

$$A = \begin{bmatrix} 3 & 1 & 2 & -4 \\ 5 & 2 & 1 & 3 \\ 6 & 2 & 4 & -8 \end{bmatrix} \tag{2.55}$$

There are four submatrices of order (3 × 3), whose determinants are evaluated below using Laplace's expansion theorem:

$$\begin{vmatrix} 3 & 1 & 2 \\ 5 & 2 & 1 \\ 6 & 2 & 4 \end{vmatrix} = (3)(-1)^2 \begin{vmatrix} 2 & 1 \\ 2 & 4 \end{vmatrix} + (1)(-1)^3 \begin{vmatrix} 5 & 1 \\ 6 & 4 \end{vmatrix} + (2)(-1)^4 \begin{vmatrix} 5 & 2 \\ 6 & 2 \end{vmatrix}$$
$$= (3)(8 - 2) - (1)(20 - 6) + (2)(10 - 12) = 0 \tag{2.56}$$

Similarly,

$$\begin{vmatrix} 3 & 1 & -4 \\ 5 & 2 & 3 \\ 6 & 2 & -8 \end{vmatrix} = 0 \begin{vmatrix} 3 & 2 & -4 \\ 5 & 1 & 3 \\ 6 & 4 & -8 \end{vmatrix} = 0 \begin{vmatrix} 1 & 2 & -4 \\ 2 & 1 & 3 \\ 2 & 4 & -8 \end{vmatrix} = 0 \tag{2.57}$$

Because all the above (3 × 3) submatrices are singular, the rank of A is less than 3. It is easy to find several (2 × 2) submatrices that are nonsingular; therefore $r = 2$.

The same conclusion regarding the singularity of the (3 × 3) submatrices could have been reached via the application of Properties 4 and 5, which were mentioned earlier in this section. Property 4 states, "If the elements of two rows or two columns of a matrix are equal, the determinant of the matrix is zero." Property 5 states, "If the elements of any row or column of a matrix are multiplied by a scalar, this is equivalent to multiplying the determinant by a scalar." Careful inspection of the four (3 × 3) submatrices shows that the first and third rows are multiples of each other. In accordance with Properties 4 and 5, the determinants are zero.

2.2.2 MATRIX TRANSFORMATIONS

It is often desirable to transform a matrix into a different form that is more amenable to a solution. Several such transformations convert matrices without significantly changing their properties. We will divide these transformations into two categories: *elementary* transformations and *similarity* transformations.

Elementary transformations usually change the shape of the matrix but preserve the value of its determinant. In addition, if the matrix represents a set of linear algebraic equations, the solution of the set is not affected by the elementary transformation. The following series of matrix multiplications:

$$L_{n-1}L_{n-2}\ldots L_2L_1A = U \tag{2.58}$$

represents an elementary transformation of matrix A to an upper triangular matrix U. This operation can be shown in condensed form as:

$$LA = U \tag{2.59}$$

where the transformation matrix L is the product of the lower triangular matrices L_i. The form of L_i matrices will be defined in Section 2.5, in conjunction with the development of the Gauss elementary transformation procedure.

Similarity transformations are of the form:

$$Q^{-1}AQ = B \tag{2.60}$$

where Q is a nonsingular square matrix. In this operation matrix A is transformed to matrix B, which is said to be *similar* to A. Similarity, in this case, implies that

1. The determinant of A and B are equal:

$$|A| = |B| \tag{2.61}$$

2. The trace of A and B are the same:

$$trA = trB \tag{2.62}$$

3. The eigenvalues of A and B are identical:

$$|A - \lambda I| = |B - \lambda I| \tag{2.63}$$

If columns of matrix Q are real mutually orthogonal unit vectors, then Q is an orthogonal matrix, and the following relations are true:

$$Q'Q = I \tag{2.64}$$

and

$$Q' = Q^{-1} \tag{2.65}$$

In this case the similarity transformation, represented by Eq. (2.60), can be written as:

$$Q'AQ = B \tag{2.66}$$

and is called an *orthogonal* transformation. Since an orthogonal transformation is a similarity transformation, the three identities (Eqs. 2.61−2.63) pertaining to determinants, traces, and eigenvalues of A and B are equally valid. In MATLAB the function *orth*(A) gives the matrix Q described earlier.

2.2.3 MATRIX POLYNOMIALS AND POWER SERIES

The definition of a scalar polynomial is given as:

$$f(x) = a_n x^n + a_{n-1} x^{n-1} + \ldots + a_1 x + a_0 \tag{2.67}$$

Similarly, a matrix polynomial can be defined as:

$$P(A) = \alpha_n A^n + \alpha_{n-1} A^{n-1} + \ldots + \alpha_1 A + \alpha_0 I \tag{2.68}$$

where A is a square matrix, A^n is the product of A by itself n times, and $A^0 = I$.

Matrices can be used in infinite series, such as the exponential, trigonometric, and logarithmic series. For example, the matrix exponential function is defined as:

$$e^A = I + A + \frac{A^2}{2!} + \frac{A^3}{3!} + \ldots \tag{2.69}$$

and the matrix trigonometric functions as:

$$\sin(A) = A - \frac{A^3}{3!} + \frac{A^5}{5!} - \ldots \tag{2.70}$$

$$\cos(A) = I - \frac{A^2}{2!} + \frac{A^4}{4!} - \ldots \tag{2.71}$$

Note that in MATLAB these functions, for example, $exp(A)$, $cos(A)$, and $sin(A)$, are element-by-element functions and do not obey the above definitions. The MATLAB functions $expm(A)$, $expm1(A)$, $expm2(A)$, and $expm3(A)$ calculate the exponential of the matrix A by different algorithms. The function $expm2(A)$ calculates the exponential of the matrix A as in Eq. (2.69).

2.2.4 VECTOR OPERATIONS

Consider two vectors x and y:

$$x = \begin{bmatrix} x_1 \\ x_2 \\ \vdots \\ x_n \end{bmatrix} \qquad y = \begin{bmatrix} y_1 \\ y_2 \\ \vdots \\ y_n \end{bmatrix} \tag{2.72}$$

and their transposes:

$$x' = [x_1 \; x_2 \; \ldots \; x_n] \qquad y' = [y_1 \; y_2 \; \ldots \; y_n] \tag{2.73}$$

The *scalar* product (or *inner* product) of these two vectors is defined as:

$$x'y = [x_1 \; x_2 \; \ldots \; x_n] \begin{bmatrix} y_1 \\ y_2 \\ \vdots \\ y_n \end{bmatrix} = x_1 y_1 + x_2 y_2 + \ldots + x_n y_n \tag{2.74}$$

As the name implies, this is a scalar quantity. The scalar product is sometimes called the *dot* product since it is also shown by $x.y$.

The *dyadic* product of these two vectors is defined as:

$$xy' = \begin{bmatrix} x_1 \\ x_2 \\ \vdots \\ x_n \end{bmatrix} [y_1 \; y_2 \; \ldots \; y_n] = \begin{bmatrix} x_1 y_1 & x_1 y_2 & \ldots & x_1 y_n \\ x_2 y_1 & x_2 y_2 & \ldots & x_2 y_n \\ \vdots & \vdots & \ddots & \vdots \\ x_n y_1 & x_n y_2 & \ldots & x_n y_n \end{bmatrix} \tag{2.75}$$

This is a matrix of order ($n \times n$). the dyadic product is sometimes shown with no multiplication sign between the two vectors, xy.

The *cross* product of two vectors is a vector:

$$\boldsymbol{xy} = \begin{bmatrix} x_1 \\ x_2 \\ \vdots \\ x_n \end{bmatrix} \begin{bmatrix} y_1 \\ y_2 \\ \vdots \\ y_n \end{bmatrix} = \begin{bmatrix} x_1 y_1 \\ x_2 y_2 \\ \vdots \\ x_n y_n \end{bmatrix} \tag{2.76}$$

The cross product is also called *vector* product and usually, the cross sign is used for showing this product: $\boldsymbol{x} \times \boldsymbol{y}$.

Two nonzero vectors are *orthogonal* if their scalar product is zero:

$$\boldsymbol{x'y} = 0 \tag{2.77}$$

The length of a vector can be calculated from:

$$|\boldsymbol{x}| = \sqrt{\boldsymbol{x'x}} \tag{2.78}$$

A *unit vector* is a vector whose length is unity.

A set of vectors $\boldsymbol{x}, \boldsymbol{y}, \boldsymbol{z},...$ is linearly dependent if there exists a set of scalars $c_1, c_2, c_3,...$ so that:

$$c_1 \boldsymbol{x} + c_2 \boldsymbol{y} + c_3 \boldsymbol{z} + ... = 0 \tag{2.79}$$

Otherwise, the vectors are *linearly independent*.

2.3 CONSISTENCY OF EQUATIONS AND EXISTENCE OF SOLUTIONS

Consider the set of simultaneous linear algebraic equations represented by:

$$\begin{aligned} a_{11}x_1 + a_{12}x_2 + ... + a_{1n}x_n &= c_1 \\ a_{21}x_1 + a_{22}x_2 + ... + a_{2n}x_n &= c_2 \\ &\vdots \\ a_{n1}x_1 + a_{n2}x_2 + ... + a_{nn}x_n &= c_n \end{aligned} \tag{2.80}$$

The coefficient matrix is $\boldsymbol{A}$, the vector of unknowns is $\boldsymbol{x}$, and the vector of constants is $\boldsymbol{c}$.

The *augmented* matrix $\boldsymbol{A_a}$ is defined as the matrix resulting from joining the vector $\boldsymbol{c}$ to the columns of matrix $\boldsymbol{A}$ as follows:

$$\boldsymbol{A_a} = \begin{bmatrix} a_{11} & a_{12} & ... & a_{1n} & c_1 \\ a_{21} & a_{22} & ... & a_{2n} & c_2 \\ \vdots & \vdots & \ddots & \vdots & \vdots \\ a_{n1} & a_{n2} & ... & a_{nn} & c_n \end{bmatrix} \tag{2.81}$$

The set of equations has a solution if, and only if, the rank of the augmented matrix is equal to the rank of the coefficient matrix. If, in addition, the rank is equal to n $(r = n)$, the solution is unique. If the rank is less than n $(r < n)$, there are more unknowns in the set than there are independent equations. In that case the set of equations can be reduced to r independent equations. The remaining $(n - r)$ unknowns must be assigned arbitrary values. This implies that the system of n equations has an infinite number of possible solutions, since the values of $(n - r)$ unknowns are given arbitrary values, and the rest of unknowns depend on these $(n - r)$ values.

A special subcategory of linear algebraic equations is the set whose vector of constants c is the zero vector:

$$Ax = 0 \qquad (2.82)$$

This is called the *homogeneous* set of linear algebraic equations. This set always has the solution:

$$x_1 = x_2 = \ldots = x_n = 0 \qquad (2.83)$$

It is called the trivial solution because it is not of any particular interest. The coefficient matrix and the augmented matrix of a homogeneous set always have the same rank, since the vector c is the zero vector. As stated earlier, if the rank of A is equal to n $(r = n)$, then the set of equations has a unique solution. However, in the case of the homogeneous equations this unique solution is none other than the trivial one. For a homogeneous set to have nontrivial solutions, the determinant of A must be zero; that is, A must be singular.

In summary, the nonhomogeneous set has a *unique nontrivial* solution if the matrix of coefficients A is nonsingular. It has an infinite number of solutions if the matrix A is singular and the ranks of A and A_a are equal to each other. It has no solution at all if the rank of A is lower than the rank of A_a.

The homogeneous set has a *unique*, but trivial, solution if the matrix of coefficients A is nonsingular. It has an infinite number of solutions if the matrix A is singular. The rank of A_a is always equal to the rank of A for a homogeneous system since the vector of constants is the zero vector (Table 2.1.)

2.4 CRAMER'S RULE

Cramer's rule calculates the solution of nonhomogeneous linear algebraic equations of the form:

$$Ax = c \qquad (2.13)$$

Table 2.1 Existence of solutions.

Condition	Nonhomogeneous set $Ax = c$	Homogeneous set $Ax = 0$
rank $A = n$	Unique solution	Unique, but trivial, solution
rank $A < n$		Infinite number of solutions
rank $A < n$ *and* *rank A = rank A_a*	Infinite number of solutions	
rank $A < n$ *and* *rank A < rank A_a*	No solution	

using the determinants of the coefficient matrix A and the substituted matrix A_j as follows:

$$x_j = \frac{|A_j|}{|A|} \quad j = 1, 2, \ldots, n \tag{2.84}$$

The substituted matrix A_j is obtained by replacing column j of matrix A with the vector c:

$$A_j = \begin{bmatrix} a_{11} & \cdots & a_{1j-1} & c_1 & a_{1j+1} & \cdots & a_{1n} \\ a_{21} & \cdots & a_{2j-1} & c_2 & a_{2j+1} & \cdots & a_{2n} \\ & & \vdots & \vdots & \vdots & & \vdots \\ a_{n1} & \cdots & a_{nj-1} & c_n & a_{nj+1} & \cdots & a_{nn} \end{bmatrix} \tag{2.85}$$

The set of equations must be nonhomogeneous because the determinant of A appears in the denominator of Eq. (2.84); the determinant cannot be zero; that is, matrix A must be nonsingular.

For a system of n equations, Cramer's rule evaluates $(n + 1)$ determinants and performs n divisions. Since the calculation of each determinant requires $(n - 1)(n!)$ multiplications, the total number of multiplications and divisions is:

$$(n + 1)(n - 1)(n!) + n \tag{2.86}$$

Table 2.2 illustrates how the number of operations required by Cramer's rule increases as the value of n increases. For $n = 3$, a total of 51 multiplications and divisions are needed. However, where $n = 10$, this number climbs to 359,251,210. For this reason, Cramer's rule is rarely used for systems with $n > 3$. The Gauss elimination, Gauss-Jordan reduction, and Gauss-Seidel methods, to be described in the next three sections of this chapter, are much more efficient methods of solution of linear equations than Cramer's rule.

2.5 GAUSS ELIMINATION METHOD

The most widely used method for the solution of simultaneous linear algebraic equations is the Gauss elimination method. This is based on the principle of converting the set of n equations in n unknowns:

$$\begin{aligned} a_{11}x_1 + a_{12}x_2 + \ldots + a_{1n}x_n &= c_1 \\ a_{21}x_1 + a_{22}x_2 + \ldots + a_{2n}x_n &= c_2 \\ &\vdots \\ a_{n1}x_1 + a_{n2}x_2 + \ldots + a_{nn}x_n &= c_n \end{aligned} \tag{2.12}$$

Table 2.2 Number of operations needed by Cramer's rule.

n	$(n + 1)(n - 1)(n!) + n$	n	$(n + 1)(n - 1)(n!) + n$
3	51	7	241,927
4	364	8	2,540,168
5	2885	9	29,030,409
6	25,206	10	359,251,210

to a triangular set of the form:

$$a_{11}x_1 + a_{12}x_2 + a_{13}x_3 + a_{14}x_4 + \ldots + a_{1n}x_n = c_1$$
$$a'_{22}x_2 + a'_{23}x_3 + a'_{24}x_4 + \ldots + a'_{2n}x_n = c'_2$$
$$a'_{33}x_3 + a'_{34}x_4 + \ldots + a'_{3n}x_n = c'_3$$
$$\vdots$$
$$a'_{n-1n-1}x_{n-1} + a'_{n-1n}x_n = c'_{n-1}$$
$$a'_{nn}x_n = c'_n$$

$$(2.87)$$

whose solution is the same as that of the original set of equations.

The process is essentially that of converting the set:

$$\mathbf{Ax} = \mathbf{c} \tag{2.13}$$

to the equivalent triangular set:

$$\mathbf{Ux} = \mathbf{c}' \tag{2.88}$$

where $\mathbf{U}$ is an upper triangular matrix and $\mathbf{c}'$ is the modified vector of constants. Once triangularization is achieved, the solution of the set can be obtained easily by back substitution starting with variable n and working backward to variable 1.

2.5.1 GAUSS ELIMINATION IN FORMULA FORM

Gauss elimination is accomplished by a series of elementary operations which do not alter the solution of the equation. These operations are based on the following facts:

1. Any equation in the set can be multiplied (or divided) by a nonzero scalar, without affecting the solution.
2. Any equation in the set can be added to (or subtracted from) another equation, without affecting the solution.
3. Any two equations can interchange positions within the set, without affecting the solution.

Two matrices that can be obtained from each other by successive application of the above elementary operations are said to be *equivalent* matrices. The rank and determinant of these matrices are unaltered by the application of elementary operations.

The overall Gauss elimination procedure applied on the $(n) \times (n + 1)$ augmented matrix is condensed into a three-part mathematical formula for initialization, elimination, and back substitution as follows:

Initialization formula:

$$\left.\begin{array}{ll} a_{ij}^{(0)} = a_{ij} & j = 1, 2, \ldots, n \\ a_{ij}^{(0)} = c_{ij} & j = n + 1 \end{array}\right\} i = 1, 2, \ldots, n \tag{2.89}$$

Elimination formula:

$$a_{ij}^{(k)} = a_{ij}^{(k-1)} - \frac{a_{ik}^{(k-1)}}{a_{kk}^{(k-1)}} a_{kj}^{(k-1)} \quad \left\{\begin{array}{l} j = n+1, n, \ldots, k \\ i = k+1, k+2, \ldots, n \end{array}\right. \quad \left\{\begin{array}{l} k = 1, 2, \ldots, n-1 \\ a_{kk}^{(k-1)} \neq 0 \end{array}\right. \tag{2.90}$$

where the initialization step places the elements of the coefficient matrix and the vector of constants into the augmented matrix, and the elimination formula reduces to zero the elements below the diagonal. The counter k is the iteration counter of the outside loop in a set of nested loops that perform the elimination.

It should be noted that the element a_{kk} in the denominator of Eq. (2.90) is always the diagonal element. It is called the *pivot* element. This pivot element must not be zero; otherwise, the computer program will result in an overflow. The computer program can be written so that it rearranges the equations at each step to attain *diagonal dominance* in the coefficient matrix; that is, the row with the largest absolute value pivot element is chosen. This strategy is called *partial pivoting*, and it serves two purposes in the Gaussian elimination procedure: it reduces the possibility of division by zero and increases the accuracy of the Gauss elimination method by using the largest pivot element. If, in addition to rows, the columns are also searched for maximum available pivot element, then the strategy is called *complete pivoting*. If pivoting cannot locate a nonzero element to place on the diagonal, the matrix must be singular. When two columns are interchanged, the corresponding variables must also be interchanged. A program that performs complete pivoting must keep track of the column interchanges to interchange the corresponding variables.

When triangularization of the coefficient matrix has been completed, the algorithm transfers the calculation to the back-substitution formula:

$$
\begin{aligned}
x_n &= \frac{a_{n,n+1}}{a_{nn}} \\
x_i &= \frac{a_{i,n+1} - \sum_{j=i+1}^{n} a_{ij}x_j}{a_{ii}}, \qquad i = n-1, n-2, \ldots, 1
\end{aligned}
\tag{2.91}
$$

These formulas complete the solution of the equations by the Gauss elimination method by calculating all the unknowns from x_n to x_1. The Gauss elimination algorithm requires $n^3/3$ multiplications to evaluate the vector $\boldsymbol{x}$. The flowchart for this method is illustrated in Fig. 2.3.

EXAMPLE 2.1 STEADY STATE TWO-DIMENSIONAL HEAT TRANSFER IN A FLAT PLATE

The steady-state heat conduction equation in the square flat plate shown in Fig. E2.1 is

$$
\frac{\partial^2 T}{\partial x^2} + \frac{\partial^2 T}{\partial y^2} = 0
$$

The temperature on the grid points ($\Delta x = \Delta y$) shown in Fig. E2.1 can be calculated by solving the following set of equations obtained from discretization of the heat conduction equation by the finite difference method (see Chapter 7 for details):

(Continued)

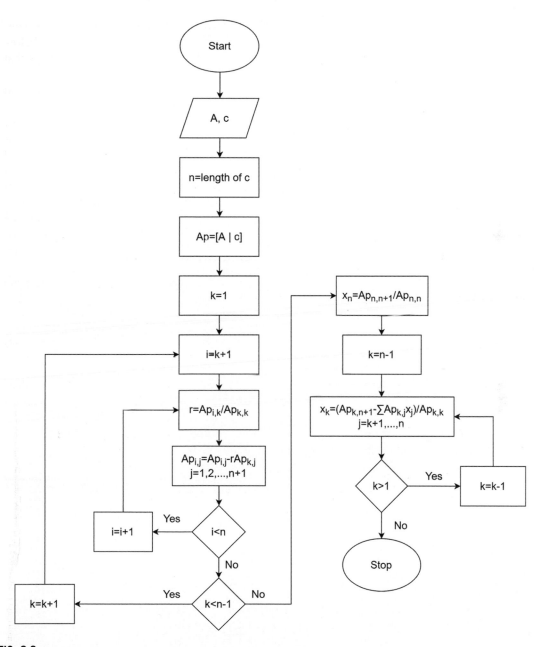

FIG. 2.3

Flowchart of the Gauss elimination method in formula form.

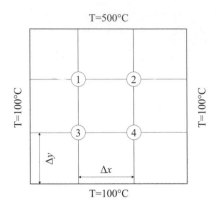

FIG. E2.1

Conduction heat transfer in a plate.

EXAMPLE 2.1 (CONTINUED)

$$T_1 = \frac{T_2 + T_3 + 100 + 500}{4}$$

$$T_2 = \frac{T_1 + T_4 + 100 + 500}{4}$$

$$T_3 = \frac{T_1 + T_4 + 100 + 100}{4}$$

$$T_4 = \frac{T_2 + T_3 + 100 + 100}{4}$$

Use the Gauss elimination method for the solution of this set of linear algebraic equations and determine the steady state temperature profile in this plate.

Method of solution

The set of equations can be rearranged as the following:

$$4T_1 - T_2 - T_3 = 600$$
$$-T_1 + 4T_2 - T_4 = 600$$
$$-T_1 + 4T_3 - T_4 = 200$$
$$-T_2 - T_3 + 4T_4 = 200$$

or

$$\begin{bmatrix} 4 & -1 & -1 & 0 \\ -1 & 4 & 0 & -1 \\ -1 & 0 & 4 & -1 \\ 0 & -1 & -1 & 4 \end{bmatrix} \begin{bmatrix} T_1 \\ T_2 \\ T_3 \\ T_4 \end{bmatrix} = \begin{bmatrix} 600 \\ 600 \\ 200 \\ 200 \end{bmatrix}$$

We will apply the triangularization procedure to obtain the solution of this set of equations. First, form the (4 × 5) augmented matrix of coefficients and constants:

(Continued)

EXAMPLE 2.1 (CONTINUED)

$$\begin{bmatrix} 4 & -1 & -1 & 0 & | & 600 \\ -1 & 4 & 0 & -1 & | & 600 \\ -1 & 0 & 4 & -1 & | & 200 \\ 0 & -1 & -1 & 4 & | & 200 \end{bmatrix}$$

Each complete row of the augmented matrix represents one of the equations of the linear set. Therefore any operations performed on a row of the augmented matrix are automatically performed on the corresponding equation.

To obtain the solution, divide the first row by 4, multiply it by -1, and subtract it from the second row to obtain:

$$\begin{bmatrix} 4 & -1 & -1 & 0 & | & 600 \\ 0 & \dfrac{15}{4} & -\dfrac{1}{4} & -1 & | & 750 \\ -1 & 0 & 4 & -1 & | & 200 \\ 0 & -1 & -1 & 4 & | & 200 \end{bmatrix}$$

Divide the first row by 4, multiply it by -1, and subtract it from the third row to obtain:

$$\begin{bmatrix} 4 & -1 & -1 & 0 & | & 600 \\ 0 & \dfrac{15}{4} & -\dfrac{1}{4} & -1 & | & 750 \\ 0 & -\dfrac{1}{4} & \dfrac{15}{4} & -1 & | & 350 \\ 0 & -1 & -1 & 4 & | & 200 \end{bmatrix}$$

Note that the coefficients in the first column below the diagonal have become zero. Continue the elimination by dividing the second row by 15/4, multiply it by $-1/4$, and subtract it from the third row to obtain:

$$\begin{bmatrix} 4 & -1 & -1 & 0 & | & 600 \\ 0 & \dfrac{15}{4} & -\dfrac{1}{4} & -1 & | & 750 \\ 0 & 0 & \dfrac{56}{15} & -\dfrac{16}{15} & | & 400 \\ 0 & -1 & -1 & 4 & | & 200 \end{bmatrix}$$

Divide the second row by 15/4, multiply it by -1, and subtract it from the fourth row to obtain:

(Continued)

EXAMPLE 2.1 (CONTINUED)

$$\begin{bmatrix} 4 & -1 & -1 & 0 & | & 600 \\ 0 & \dfrac{15}{4} & -\dfrac{1}{4} & -1 & | & 750 \\ 0 & 0 & \dfrac{56}{15} & -\dfrac{16}{15} & | & 400 \\ 0 & 0 & -\dfrac{16}{15} & \dfrac{56}{15} & | & 400 \end{bmatrix}$$

In this stage coefficients in the second column below the diagonal have become zero. Continue the elimination by dividing the third row by 56/15, multiply it by $-16/15$, and subtract it from the fourth row to obtain:

$$\begin{bmatrix} 4 & -1 & -1 & 0 & | & 600 \\ 0 & \dfrac{15}{4} & -\dfrac{1}{4} & -1 & | & 750 \\ 0 & 0 & \dfrac{56}{15} & -\dfrac{16}{15} & | & 400 \\ 0 & 0 & 0 & \dfrac{24}{7} & | & \dfrac{3600}{7} \end{bmatrix}$$

The triangularization of the coefficient part of the augmented matrix is complete and this matrix represents the triangular set of equations:

$$4T_1 - T_2 - T_3 = 600$$
$$\frac{15}{4}T_2 - \frac{1}{4}T_3 - T_4 = 750$$
$$\frac{56}{15}T_3 - \frac{16}{15}T_4 = 400$$
$$\frac{24}{7}T_4 = \frac{3600}{7}$$

whose solution is identical to that of the original set. The solution is obtained by back substitution. From the fourth equation we can compute T_4:

$$T_4 = 150$$

Substitution of the value of T_4 in the third equation and rearrangement gives:

$$T_3 = 150$$

Substitution of the values of T_4 and T_3 in the second equation and rearrangement gives:

$$T_2 = 250$$

(*Continued*)

EXAMPLE 2.1 (CONTINUED)

Finally, the substitution of the values of T_4, T_3, and T_2 in the first equation and rearrangement yields:

$$T_1 = 250$$

2.5.2 GAUSS ELIMINATION IN MATRIX FORM

The Gauss elimination procedure, which was described earlier in formula form, can also be accomplished by a series of matrix multiplications. Two types of special matrices are involved in this operation. Both these matrices are modifications of the identity matrix. The first type, which we designate as P_{ij}, is the identity matrix with the following changes: the unity at position ii switches places with the zero at position ij, and the unity at position jj switches places with the zero at position ji. For example, P_{23} for a fifth-order system is:

$$P_{23} = \begin{bmatrix} 1 & 0 & 0 & 0 & 0 \\ 0 & 0 & 1 & 0 & 0 \\ 0 & 1 & 0 & 0 & 0 \\ 0 & 0 & 0 & 1 & 0 \\ 0 & 0 & 0 & 0 & 1 \end{bmatrix} \tag{2.92}$$

Premultiplication of matrix A by P_{ij} has the effect of interchanging rows i and j. Postmultiplication causes interchange of columns i and j. By definition, $P_{ii} = I$, and multiplication of A by P_{ii} causes no interchanges. The inverse of P_{ij} is identical to P_{ij}.

The second type of matrices used by the Gauss elimination method is unit lower triangular matrices of the form:

$$L_1 = \begin{bmatrix} 1 & 0 & 0 & 0 & 0 \\ -\dfrac{a_{21}^{(0)}}{a_{11}^{(0)}} & 1 & 0 & 0 & 0 \\ -\dfrac{a_{31}^{(0)}}{a_{11}^{(0)}} & 0 & 1 & 0 & 0 \\ -\dfrac{a_{41}^{(0)}}{a_{11}^{(0)}} & 0 & 0 & 1 & 0 \\ -\dfrac{a_{51}^{(0)}}{a_{11}^{(0)}} & 0 & 0 & 0 & 1 \end{bmatrix} \tag{2.93}$$

where the superscript (0) indicates that each L_k matrix uses the elements $a_{ik}^{(k-1)}$ of the previous transformation step. Premultiplication of matrix A by L_k has the effect of reducing to zero the elements below the diagonal in column k. The inverse of L_k has the same form as L_k but with the signs of the off-diagonal elements reversed.

The entire Gauss elimination method, which reduces a nonsingular matrix A to an upper triangular matrix U, can be represented by the following series of matrix multiplications:

$$L_{n-1}L_{n-2}\ldots P_{ij}\ldots L_2L_1P_{ij}A = U \tag{2.94}$$

where the multiplications by P_{ij} cause pivoting, if and when needed, and the multiplications by L_k cause elimination. If pivoting is not performed, Eq. (2.94) simplifies to:

$$L_{n-1}L_{n-2}\ldots L_2L_1A = U \tag{2.58}$$

The matrices L_k are unit lower triangular and their product, defined by matrix L, is also unit lower triangular. With this definition of L, Eq. (2.58) condenses to:

$$LA = U \tag{2.59}$$

Since matrix L is unit lower triangular, it is nonsingular. Its inverse exists and is also a unit lower triangular matrix. If we premultiply both sides of Eq. (2.59) by L^{-1}, we obtain:

$$A = L^{-1}U \tag{2.95}$$

This equation represents the *decomposition* of a nonsingular matrix A into a unit lower triangular matrix and an upper triangular matrix. Furthermore, this decomposition is unique [1]. Therefore the matrix operation of Eq. (2.59) when applied to the augmented matrix $[A \mid c]$ yields the unique solution:

$$L[A|c] \Rightarrow [U|c'] \tag{2.96}$$

of the system of linear algebraic equations:

$$Ax = c \tag{2.13}$$

CALCULATION OF DETERMINANTS BY THE GAUSS METHOD

The Gauss elimination method is also very useful in the calculation of determinants of matrices. The elementary operations used in the Gauss method are consistent with the properties of determinants listed in Section 2.2. Therefore the reduction of a matrix to the equivalent triangular matrix by the Gauss elimination procedure would not alter the value of the determinant of the matrix. The determinant of a triangular matrix is equal to the product of its diagonal elements:

$$|U| = \prod_{i=1}^{n} a_{ii} \tag{2.54}$$

Therefore a matrix whose determinant is to be evaluated should first be converted to the triangular form using the Gauss method and then its determinant should be calculated from the product of the diagonal elements of the triangular matrix.

Example 2.2 demonstrates the Gauss elimination method with the complete pivoting strategy in solving a set of simultaneous linear algebraic equations and in calculating the determinant of the matrix of coefficients.

EXAMPLE 2.2 HEAT TRANSFER IN A PIPE USING THE GAUSS ELIMINATION METHOD FOR SIMULTANEOUS LINEAR ALGEBRAIC EQUATIONS

Write a general MATLAB function that implements the Gauss elimination method with complete pivoting for the solution of nonhomogeneous linear algebraic equations. The function should identify singular matrices and give their rank. Use this function to calculate the interface temperatures in the following problem:

Saturated steam at 130°C is flowing inside a steel pipe having an ID of 20 mm (D_1) and an OD of 25 mm (D_2). The pipe is insulated with 40 mm [($D_3 - D_2$)/2] of insulation on the outside. The convective heat transfer coefficients for the inside steam and outside of the lagging are estimated as $h_i = 1700$ W/m^2.K and $h_o = 3$ W/m^2.K, respectively. The mean thermal conductivity of the metal is $k_s = 45$ W/m.K, and that of the insulation is $k_i = 0.064$ W/m.K. The ambient air temperature is 25°C (Fig. E2.2).

There are three interfaces in this problem and by writing the energy balance at each interface, there will be three linear equations and three unknown temperatures.

Heat transfer from steam to pipe:

$$h_i \pi D_1 (T_s - T_1) = \frac{T_1 - T_2}{\ln(D_2/D_1)/(2\pi k_s)}$$

Heat transfer from pipe to insulation:

$$\frac{T_1 - T_2}{\ln(D_2/D_1)/(2\pi k_s)} = \frac{T_2 - T_3}{\ln(D_3/D_2)/(2\pi k_i)}$$

Heat transfer from insulation to air:

$$\frac{T_2 - T_3}{\ln(D_3/D_2)/(2\pi k_i)} = h_o \pi D_3 (T_3 - T_a)$$

(Continued)

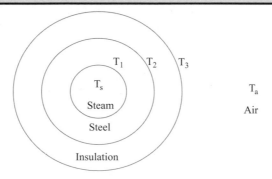

FIG. E2.2

Insulated pipe.

EXAMPLE 2.2 (CONTINUED)

where

T_s = temperature of steam = 130°C

T_1 = temperature of inside wall of pipe (unknown)

T_2 = temperature of outside wall of pipe (unknown)

T_3 = temperature of outside of insulation (unknown)

T_a = ambient temperature = 25°C

Rearranging the above three energy balance equations yields the set of linear algebraic equations, shown below, which can be solved to find the three unknowns T_1, T_2, and T_3.

$$\left[\frac{2k_s}{\ln(D_2/D_1)} + h_iD_1\right]T_1 - \left[\frac{2k_s}{\ln(D_2/D_1)}\right]T_2 = h_iD_1T_s$$

$$\left[\frac{k_s}{\ln(D_2/D_1)}\right]T_1 - \left[\frac{k_s}{\ln(D_2/D_1)} + \frac{k_i}{\ln(D_3/D_2)}\right]T_2 + \left[\frac{k_i}{\ln(D_3/D_2)}\right]T_3 = 0$$

$$\left[\frac{2k_i}{\ln(D_3/D_2)}\right]T_2 - \left[\frac{2k_i}{\ln(D_3/D_2)} + h_oD_3\right]T_3 = -h_oD_3T_a$$

Method of solution

The function is written based on Gauss elimination in matrix form. It applies a complete pivoting strategy by searching rows and columns for the maximum pivot element. It keeps track of column interchanges, which affect the positions of the unknown variables. The function applies the back-substitution formula (Eq. 2.91) to calculate the unknown variables and interchanges their order to correct for column pivoting.

In the beginning, the program checks the determinant of the matrix of coefficients to see if the matrix is singular. If it is singular, the program gives the rank of the matrix and terminates calculations.

Program description

The programs and functions developed in this example can be found in: https://www.elsevier.com/books-and-journals/book-companion/9780128229613

The MATLAB function *Gauss.m* consists of three main sections. In the beginning, it checks the sizes of input arguments (coefficient matrix and vector of constants) to see if they are consistent. It also checks the coefficient matrix for singularity.

The second part of the function is Gauss elimination. In each iteration the program finds the location of the maximum pivot element. It interchanges the row and the column of this pivot element to bring it to diagonal and meanwhile keeps track of the interchanged column. At the end of the loop, the program reduces the elements below the diagonal at which the pivot element is placed.

(Continued)

EXAMPLE 2.2 (CONTINUED)

Finally, the function *Gauss.m* applies the back-substitution formula to evaluate all the unknowns. It also interchanges the order of the unknowns at the same time to correct for any column interchanges that took place during complete pivoting.

The program *Example2_2.m* is written to solve the problem of Example 2.2. It mainly acts as an input file which then builds the coefficient matrix and the vector of constants and finally calls the function *Gauss.m* to solve the set of equations for the unknown temperatures.

Input and results

$>>$ **Example2_2**
Temperature of steam (deg C) = 130
Temperature of air (deg C) = 25
Pipe ID (mm) = 20
Pipe OD (mm) = 25
Insulation thickness (mm) = 40
Inside heat transfer coefficient (W/m^2.K) = 1700
Outside heat transfer coefficient (W/m^2.K) = 3
Heat conductivity of steel (W/m.K) = 45
Heat conductivity of insulation (W/m.K) = 0.064

Results
T1 = 129.79
T2 = 129.77
T3 = 48.12

Discussion of results

The Gauss elimination method finds the interface temperatures as $T_1 = 129.79°C$, $T_2 = 129.77°C$, and $T_3 = 48.12°C$. These values are quite predictable because the heat transfer coefficient of steam and the heat conductivity of steel is very high. Therefore the temperatures at the steam-pipe interface and pipe-insulation interface are very close to the steam temperature. The main resistance to heat transfer is due to insulation.

The values obtained from the function *Gauss.m* may be verified easily in MATLAB by using the original method of solution of the set of linear equations in matrix form, that is, $T = A\backslash c$.

2.6 GAUSS-JORDAN REDUCTION METHOD

The Gauss-Jordan reduction method is an extension of the Gauss elimination method. It reduces a set of n equations from its canonical form of:

$$Ax = c \tag{2.13}$$

to the diagonal set of the form:

$$Ix = c' \qquad (2.97)$$

where I is the unit matrix. Eq. (2.97) is identical to:

$$x = c' \qquad (2.98)$$

In other words, the solution vector is given by the c' vector.

The Gauss-Jordan reduction method applies the same series of elementary operations that are used by the Gauss elimination method. It applies these operations both below and above the diagonal to reduce all the off-diagonal elements of the matrix to zero. Besides, it converts the elements on the diagonal to unity.

2.6.1 GAUSS-JORDAN REDUCTION IN FORMULA FORM

The Gauss-Jordan reduction procedure applied to the $(n) \times (n + 1)$ augmented matrix can be given in a three-part mathematical formula for the initialization, normalization, and reduction steps as follows:

Initialization formula:

$$\left.\begin{array}{ll} a_{ij}^{(0)} = a_{ij}, & j = 1, 2, \ldots, n \\ a_{ij}^{(0)} = c_i, & j = n + 1 \end{array}\right\} \quad i = 1, 2, \ldots, n \qquad (2.99)$$

Normalization formula:

$$a_{kj}^{(k)} = \frac{a_{kj}^{(k-1)}}{a_{kk}^{(k-1)}}, \qquad j = n + 1, n, \ldots, k \qquad \begin{cases} k = 1, 2, \ldots, n \\ a_{kk}^{(k-1)} \neq 0 \end{cases} \qquad (2.100)$$

Reduction formula:

$$a_{ij}^{(k)} = a_{ij}^{(k-1)} - a_{ik}^{(k-1)} a_{kj}^{(k)}, \qquad j = n + 1, n, \ldots, k \qquad \begin{cases} i = 1, 2, \ldots, n \\ i \neq k \end{cases} \qquad \begin{cases} k = 1, 2, \ldots, n \\ a_{kk}^{(k-1)} \neq 0 \end{cases} \qquad (2.101)$$

The initialization formula places the elements of the coefficient matrix in columns 1 to n and the vector of constants in column $(n + 1)$ of the augmented matrix. The normalization formula divides each row of the augmented matrix by its pivot element and makes this change permanent, thus causing the diagonal elements of the coefficient segment of the augmented matrix to become unity. Finally, the reduction formula reduces to zero the off-diagonal elements in each row and column in the coefficient segment of the augmented matrix and converts column $(n + 1)$ to the solution vector. The flowchart for this method is illustrated in Fig. 2.4.

EXAMPLE 2.3 STEADY-STATE TWO-DIMENSIONAL HEAT TRANSFER IN A FLAT PLATE

Repeat Example 2.1, but this time use the Gauss-Jordan reduction method for the solution of this set of linear algebraic equations and determine the steady state temperature profile in the plate.

(Continued)

EXAMPLE 2.3 (CONTINUED)

Method of solution

We will apply the Gauss-Jordan procedure, without pivoting, to the set of linear equations in Example (2.1) to observe the difference between the Gauss-Jordan and the Gauss method. Starting with the augmented matrix,

$$
\begin{bmatrix}
4 & -1 & -1 & 0 & | & 600 \\
-1 & 4 & 0 & -1 & | & 600 \\
-1 & 0 & 4 & -1 & | & 200 \\
0 & -1 & -1 & 4 & | & 200
\end{bmatrix}
$$

Normalize the first row by dividing it by 4:

$$
\begin{bmatrix}
1 & -\dfrac{1}{4} & -\dfrac{1}{4} & 0 & | & 150 \\
-1 & 4 & 0 & -1 & | & 600 \\
-1 & 0 & 4 & -1 & | & 200 \\
0 & -1 & -1 & 4 & | & 200
\end{bmatrix}
$$

Multiply the normalized first row by -1 and subtract it from the second row:

$$
\begin{bmatrix}
1 & -\dfrac{1}{4} & -\dfrac{1}{4} & 0 & | & 150 \\
0 & \dfrac{15}{4} & -\dfrac{1}{4} & -1 & | & 700 \\
-1 & 0 & 4 & -1 & | & 200 \\
0 & -1 & 1 & 4 & | & 200
\end{bmatrix}
$$

Multiply the normalized first row by -1 and subtract it from the third row:

$$
\begin{bmatrix}
1 & -\dfrac{1}{4} & -\dfrac{1}{4} & 0 & | & 150 \\
0 & \dfrac{15}{4} & -\dfrac{1}{4} & -1 & | & 750 \\
0 & -\dfrac{1}{4} & \dfrac{15}{4} & -1 & | & 350 \\
0 & -1 & -1 & 4 & | & 200
\end{bmatrix}
$$

Normalize the second row by dividing it by 15/4:

(Continued)

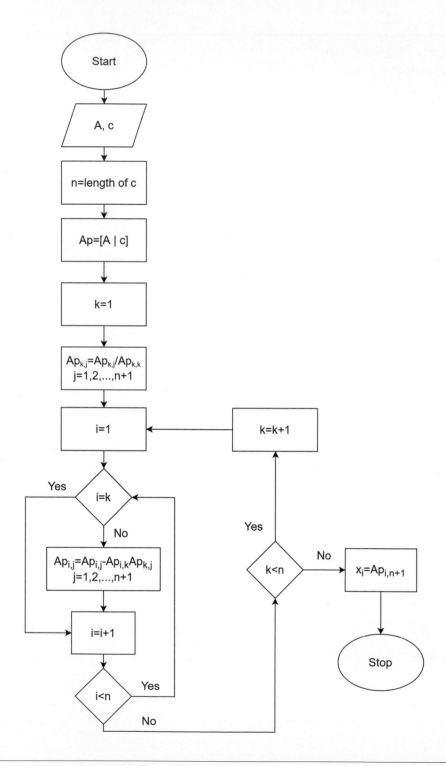

FIG. 2.4

Flowchart of the Gauss-Jordan reduction method in formula form.

EXAMPLE 2.3 (CONTINUED)

$$
\begin{bmatrix}
1 & -\dfrac{1}{4} & -\dfrac{1}{4} & 0 & | & 150 \\[2ex]
0 & 1 & -\dfrac{1}{15} & -\dfrac{4}{15} & | & 200 \\[2ex]
0 & -\dfrac{1}{4} & \dfrac{15}{4} & -1 & | & 350 \\[2ex]
0 & -1 & -1 & 4 & | & 200
\end{bmatrix}
$$

Multiply the normalized second row by $-1/4$ and subtract it from the first row:

$$
\begin{bmatrix}
1 & 0 & -\dfrac{4}{15} & -\dfrac{1}{15} & | & 200 \\[2ex]
0 & 1 & -\dfrac{1}{15} & -\dfrac{4}{15} & | & 200 \\[2ex]
0 & -\dfrac{1}{4} & \dfrac{15}{4} & -1 & | & 350 \\[2ex]
0 & -1 & -1 & 4 & | & 200
\end{bmatrix}
$$

Multiply the normalized second row by $-1/4$ and subtract it from the third row:

$$
\begin{bmatrix}
1 & 0 & -\dfrac{4}{15} & -\dfrac{1}{15} & | & 200 \\[2ex]
0 & 1 & -\dfrac{1}{15} & -\dfrac{4}{15} & | & 200 \\[2ex]
0 & 0 & \dfrac{56}{15} & -\dfrac{4}{15} & | & 400 \\[2ex]
0 & -1 & -1 & 4 & | & 200
\end{bmatrix}
$$

Multiply the normalized second row by -1 and subtract it from the fourth row:

$$
\begin{bmatrix}
1 & 0 & -\dfrac{4}{15} & -\dfrac{1}{15} & | & 200 \\[2ex]
0 & 1 & -\dfrac{1}{15} & -\dfrac{4}{15} & | & 200 \\[2ex]
0 & 0 & \dfrac{56}{15} & -\dfrac{16}{15} & | & 400 \\[2ex]
0 & 0 & -\dfrac{16}{15} & \dfrac{56}{15} & | & 400
\end{bmatrix}
$$

Normalize the third row by dividing it by 56/15:

(Continued)

EXAMPLE 2.3 (CONTINUED)

$$
\begin{bmatrix}
1 & 0 & -\dfrac{4}{15} & -\dfrac{1}{15} & \Big|200 \\[2ex]
0 & 1 & -\dfrac{1}{15} & -\dfrac{4}{15} & \Big|200 \\[2ex]
0 & 0 & 1 & -\dfrac{2}{7} & \Big|\dfrac{750}{7} \\[2ex]
0 & 0 & -\dfrac{16}{15} & \dfrac{56}{15} & \Big|400
\end{bmatrix}
$$

Multiply the normalized third row by $-4/15$ and subtract it from the first row:

$$
\begin{bmatrix}
1 & 0 & 0 & -\dfrac{1}{7} & \Big|\dfrac{1600}{7} \\[2ex]
0 & 1 & -\dfrac{1}{15} & -\dfrac{4}{15} & \Big|200 \\[2ex]
0 & 0 & 1 & -\dfrac{2}{7} & \Big|\dfrac{750}{7} \\[2ex]
0 & 0 & -\dfrac{16}{15} & \dfrac{56}{15} & \Big|400
\end{bmatrix}
$$

Multiply the normalized third row by $-1/15$ and subtract it from the second row:

$$
\begin{bmatrix}
1 & 0 & 0 & -\dfrac{1}{7} & \Big|\dfrac{1600}{7} \\[2ex]
0 & 1 & 0 & -\dfrac{2}{7} & \Big|\dfrac{1450}{7} \\[2ex]
0 & 0 & 1 & -\dfrac{2}{7} & \Big|\dfrac{750}{7} \\[2ex]
0 & 0 & -\dfrac{16}{15} & \dfrac{56}{15} & \Big|400
\end{bmatrix}
$$

Multiply the normalized third row by $-16/15$ and subtract it from the fourth row:

$$
\begin{bmatrix}
1 & 0 & 0 & -\dfrac{1}{7} & \Big|\dfrac{1600}{7} \\[2ex]
0 & 1 & 0 & -\dfrac{2}{7} & \Big|\dfrac{1450}{7} \\[2ex]
0 & 0 & 1 & -\dfrac{2}{7} & \Big|\dfrac{750}{7} \\[2ex]
0 & 0 & 0 & \dfrac{24}{7} & \Big|\dfrac{3600}{7}
\end{bmatrix}
$$

Normalize the fourth row by dividing it by 24/7:

(Continued)

EXAMPLE 2.3 (CONTINUED)

$$\begin{bmatrix} 1 & 0 & 0 & -\dfrac{1}{7} & \left| \dfrac{1600}{7} \right. \\ 0 & 1 & 0 & -\dfrac{2}{7} & \left| \dfrac{1450}{7} \right. \\ 0 & 0 & 1 & -\dfrac{2}{7} & \left| \dfrac{750}{7} \right. \\ 0 & 0 & 0 & 1 & | 150 \end{bmatrix}$$

Multiply the normalized fourth row by $-1/7$ and subtract it from the first row:

$$\begin{bmatrix} 1 & 0 & 0 & 0 & | 250 \\ 0 & 1 & 0 & -\dfrac{2}{7} & \left| \dfrac{1450}{7} \right. \\ 0 & 0 & 1 & -\dfrac{2}{7} & \left| \dfrac{750}{7} \right. \\ 0 & 0 & 0 & 1 & | 150 \end{bmatrix}$$

Multiply the normalized fourth row by $-2/7$ and subtract it from the second row:

$$\begin{bmatrix} 1 & 0 & 0 & 0 & | 250 \\ 0 & 1 & 0 & 0 & | 250 \\ 0 & 0 & 1 & -\dfrac{2}{7} & \left| \dfrac{750}{7} \right. \\ 0 & 0 & 0 & 1 & | 150 \end{bmatrix}$$

Multiply the normalized fourth row by $-2/7$ and subtract it from the third row:

$$\begin{bmatrix} 1 & 0 & 0 & 0 & | 250 \\ 0 & 1 & 0 & 0 & | 250 \\ 0 & 0 & 1 & 0 & | 150 \\ 0 & 0 & 0 & 1 & | 150 \end{bmatrix}$$

This reduced matrix is equivalent to the set of equations $Ix = c'$ and the vector c', which is the last column of the reduced matrix, is the solution of the original set of equations. There is no need for back substitution since the solution is obtained in its final form in vector c'.

2.6.2 GAUSS-JORDAN REDUCTION IN MATRIX FORM

The Gauss-Jordan reduction procedure can also be accomplished by a series of matrix multiplications, similar to those performed in the Gauss elimination method (Section 2.5.2). The matrix P_{ij}, which causes pivoting, is identical to that defined by Eq. (2.93). The matrix L_k must have additional terms above the diagonal to cause the reduction to zero of elements above, as well as below, the diagonal, and a term on the diagonal in order to normalize the element on the diagonal of the original matrix. We will designate this matrix $\overline{L}_k$ and give an example for a fourth-order system with $k = 2$, where the superscript (1) indicates that each $\overline{L}_k$ matrix uses the elements $a_{ik}^{(k-1)}$ of the previous transformation step:

$$\overline{L}_2 = \begin{bmatrix} 1 & -\dfrac{a_{12}^{(1)}}{a_{22}^{(1)}} & 0 & 0 \\ 0 & \dfrac{1}{a_{22}^{(1)}} & 0 & 0 \\ 0 & -\dfrac{a_{32}^{(1)}}{a_{22}^{(1)}} & 1 & 0 \\ 0 & -\dfrac{a_{42}^{(1)}}{a_{22}^{(1)}} & 0 & 1 \end{bmatrix} \tag{2.102}$$

The Gauss-Jordan algorithm reduces a nonsingular matrix A to the identity matrix I by the following series of matrix multiplications:

$$\overline{L}_n \overline{L}_{n-1} \ldots P_{ij} \ldots \overline{L}_2 \overline{L}_1 P_{ij} A = I \tag{2.103}$$

where the multiplications by P_{ij} cause pivoting, if and when needed, and the multiplications by $\overline{L}_k$ cause normalization and reduction. If pivoting is not performed, Eq. (2.103) simplifies to:

$$\overline{L}_n \overline{L}_{n-1} \ldots \overline{L}_2 \overline{L}_1 A = I \tag{2.104}$$

By defining the product of all the $\overline{L}_k$ matrices as $\overline{L}$, we can condense Eq. (2.104) to:

$$\overline{L} A = I \tag{2.105}$$

The matrix operation of Eq. (2.105), when applied to the augmented matrix $[A \mid c]$, yields the unique solution:

$$\overline{L}[A \mid c] \Rightarrow [I \mid c'] \tag{2.106}$$

of the system of linear algebraic equations:

$$A x = c \tag{2.13}$$

whose matrix of coefficients A is nonsingular.

2.6.3 GAUSS-JORDAN REDUCTION WITH MATRIX INVERSION

Matrix $\overline{L}$, in Eq. (2.105), is a nonsingular matrix; therefore its inverse exists. Premultiplying both sides of Eq. (2.105) by $\overline{L}^{-1}$, we obtain:

$$A = \overline{L}^{-1} I \tag{2.107}$$

Taking the inverse of both sides of Eq. (2.107) results in:

$$A^{-1} = \overline{L} I \tag{2.108}$$

This simply states that the inverse of A is equal to $\overline{L}$. This has very important implications in numerical methods because it shows that the Gauss-Jordan reduction method is essentially a matrix inversion algorithm. Eq. (2.108), when rearranged, clearly shows that the application of the reduction operation $\overline{L}$ on the identity matrix yields the inverse of A:

$$\overline{L}I = A^{-1} \tag{2.108}$$

This observation can be used to extend the formula form of the Gauss-Jordan algorithm to give the inverse of matrix A every time it calculates the solution to the set of equations:

$$Ax = c \tag{2.13}$$

This is done by forming the augmented matrix of order $(n) \times (2n + 1)$:

$$[A|c|I] \tag{2.109}$$

and applying the Gauss-Jordan reduction to the augmented matrix. In this case the three-part mathematical formula for the initialization, normalization, and reduction steps is the following:

Initialization formula:

$$\left. \begin{array}{ll} a_{ij}^{(0)} = a_{ij}, & i = 1, 2, \ldots, n \\ a_{ij}^{(0)} = c_i, & i = 1, 2, \ldots, n \\ a_{ij}^{(0)} = 0, & i \neq j \\ a_{ij}^{(0)} = 1, & i = j \end{array} \right\} \quad j = 1, 2, \ldots, n \tag{2.110}$$

Normalization formula:

$$a_{kj}^{(k)} = \frac{a_{kj}^{(k-1)}}{a_{kk}^{(k-1)}}, \qquad j = 2n + 1, 2n, \ldots, k \qquad \begin{cases} k = 1, 2, \ldots, n \\ a_{kk}^{(k-1)} \neq 0 \end{cases} \tag{2.111}$$

Reduction formula:

$$a_{ij}^{(k)} = a_{ij}^{(k-1)} - a_{ik}^{(k-1)} a_{kj}^{(k)}, \qquad j = 2n + 1, 2n, \ldots, k \qquad \begin{cases} i = 1, 2, \ldots, n \\ i \neq k \end{cases} \quad \begin{cases} k = 1, 2, \ldots, n \\ a_{kk}^{(k-1)} \neq 0 \end{cases} \tag{2.112}$$

The first two parts of the initialization formula place the elements of the coefficient matrix in columns 1 to n and the vector of constants in column $(n + 1)$ of the augmented matrix. The last two parts of the initialization step expand the augmented matrix to include the identity matrix in columns $(n + 2)$ to $(2n + 1)$. The normalization formula divides each row of the entire matrix by its pivot element, thus causing the diagonal elements of the coefficient segment of the augmented matrix to become unity. Finally, the reduction formula reduces to zero the off-diagonal elements in each row and column in the coefficient segment of the augmented matrix, converts column $(n + 1)$ to the solution vector, and converts the identity matrix in columns $(n + 2)$ to $(2n + 1)$ to the inverse of A.

Example 2.4 demonstrates the use of the Gauss-Jordan reduction method for the solution of simultaneous linear algebraic equations.

2.7 GAUSS-SEIDEL SUBSTITUTION METHOD

Certain engineering problems yield sets of simultaneous linear algebraic equations, which are *predominantly diagonal* systems. A *predominantly diagonal* system of linear equations has coefficients on the diagonal which are larger in absolute value than the sum of the absolute values of the other coefficients.

EXAMPLE 2.4 SOLUTION OF A STEAM DISTRIBUTION SYSTEM USING THE GAUSS-JORDAN REDUCTION METHOD FOR SIMULTANEOUS LINEAR ALGEBRAIC EQUATIONS

Aniline is being removed from water by solvent extraction using toluene [2]. The unit is a 10-stage countercurrent tower, shown in Fig. E2.4a. The equilibrium relationship valid at each stage is, to a first approximation,

$$m = \frac{Y_i}{X_i} = 9$$

where

Y_i = (kg of aniline in the toluene phase) / (kg of toluene in the toluene phase)
X_i = (kg of aniline in the water phase) / (kg of water in the water phase)

1. The solution to this problem is a set of 10 simultaneous equations. Derive these equations from material balances around each stage.
2. Solve the foregoing set of equations to find the concentration in both the aqueous and organic phases leaving each stage of the system (X_i and Y_i).

Method of solution

We make the following assumptions:

a. The solubility of toluene in water is small and can be neglected.
b. The solubility of water in toluene is small and can be neglected.

(Continued)

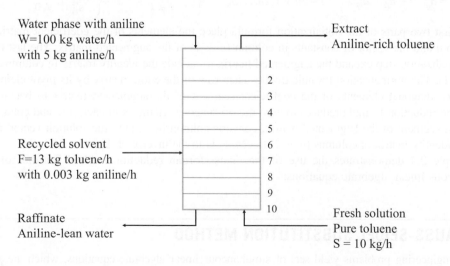

FIG. E2.4a

Extraction column.

EXAMPLE 2.4 (CONTINUED)

With these assumptions, the flow rate of the water phase throughout the tower is $W = 100$ kg/h. The flow rate of the toluene below stage 6 is $S = 10$ kg/h and above stage 6 is $\overline{S} = 23$ kg/h (obtained from a solvent balance around stage 6).

In the following derivation we denote the number of stages with N and the location of the feed tray with f ($N = 10$ and $f = 6$ in this example).

Part (a): Material balance around stage i above the feed tray:

$$WX_{i-1} + \overline{S}Y_{i+1} = WX_i + \overline{S}Y_i \quad i = 1, 2, \cdots, \quad f - 1$$

Material balance around the feed tray (stage f):

$$WX_{f-1} + SY_{f+1} + FY_F = WX_f + \overline{S}Y_f$$

Material balance around stage i below the feed tray:

$$WX_{i-1} + SY_{i+1} = WX_i + SY_i \qquad i = f + 1, f + 2, \cdots, N$$

Here, Y_F is kg aniline/kg toluene in the feed.

Part (b): The equilibrium relationship:

$$Y_i = mX_i$$

can be used in the foregoing set of equations to eliminate the variables Y_i. With this substitution, the system of equations becomes:

$$(W + S)X_1 - mSX_2 = WX_0$$
$$- WX_{i-1} + (W + mS)X_i - mSX_{i+1} = 0 \qquad i = 2, 3, \cdots, f - 1$$
$$- WX_{f-1} + (W + mS)X_f - mSX_{f+1} = FY_F$$
$$- WX_{i-1} + (W + mS)X_i - mSX_{i+1} = 0 \qquad i = f + 1, f + 2, \cdots, N - 1$$
$$- WX_{N-1} + (W + mS)X_N = 0$$

This is a tridiagonal set of simultaneous linear algebraic equations that can be represented by:

$$Ax = c$$

The Gauss or Gauss-Jordan methods may be used for the solution of this problem. The computer program, which is described in the next section, implements the Gauss-Jordan algorithm in matrix form. The program uses a complete pivoting strategy.

Program description

The programs and functions developed in this example can be found in: https://www.elsevier.com/books-and-journals/book-companion/9780128229613

($Continued$)

EXAMPLE 2.4 (CONTINUED)

The MATLAB function *Jordan.m* consists of three main sections. In the beginning it checks the sizes of input arguments (coefficient matrix and vector of constants) to see if they are consistent. It also checks the coefficient matrix for singularity.

The second part of the function is the Gauss-Jordan algorithm with the application of a complete pivoting strategy. In each iteration the program finds the location of the largest pivot element. It interchanges the row and the column of this pivot element to bring it to the diagonal position and keeps track of the interchanged columns. At the end of the loop, the program reduces the elements below and above the diagonal position at which the pivot element is placed.

Finally, the function *Jordan.m* sets the unknowns equal to the elements of the last column of the modified augmented matrix. It also interchanges the order of the unknowns at the same time to correct for any column interchanges that took place during complete pivoting.

The program *Example2_4.m* is written to solve the problem of Example 2.4. It builds the coefficient matrix and the vector of constants and then calls the function *Jordan.m* to solve the set of equations for the unknown flow rates. In the end the program displays the results (X and Y in each stage) both numerically and graphically.

Input and Results

>> **Example2_4**
Number of stages = 10
Equilibrium constant (m) = 9
Water with aniline from top:
W (kg water/hr) = 100
kg aniline/kg water = 0.05
Fresh solvent from bottom:
S (kg toluene/hr) = 10
Recycled solvent:
F (kg toluene/hr) = 13
kg aniline/kg toluene = 0.003
Feed stage no. = 6

***** **Results** *****

Stage	X	Y
1	0.02425	0.21822
2	0.01181	0.10626
3	0.00580	0.05216
4	0.00289	0.02603
5	0.00149	0.01341
6	0.00081	0.00731

(Continued)

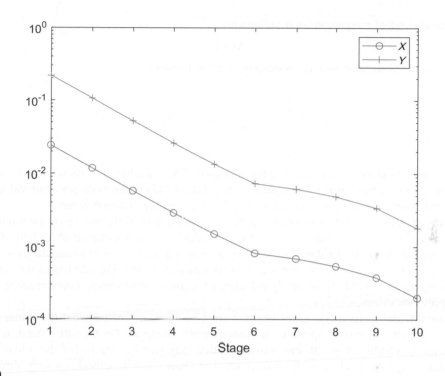

FIG. E2.4b

Profiles of aniline concentration in toluene and water in the extraction column.

EXAMPLE 2.4 (CONTINUED)

7	0.00068	0.00614
8	0.00054	0.00484
9	0.00038	0.00339
10	0.00020	0.00179

Discussion of results

The results are listed numerically and also shown graphically in Fig. E2.4b. It can be seen that aniline concentration in water decreases from the top (stage 1) to the bottom (stage 10) and its concertation in toluene increases from the bottom (stage 10) to the top (stage 1) of the column. Also, note that the slope of the concentration curves changes in the feed stage due to the addition of solvent in this stage.

For a general set of n equations in n unknowns:

$$Ax = c \qquad (2.13)$$

the Gauss-Seidel substitution method corresponds to the formula:

$$x_i = \frac{1}{a_{ii}} \left[c_i - \sum_{\substack{j=1 \\ j \neq i}}^{n} a_{ij} x_j \right], \quad i = 1, 2, \ldots, n \qquad (2.113)$$

Eq. (2.113) is the Gauss-Seidel method in formula form. The calculation starts with an initial guess of the values x_2 to x_n. Each newly calculated x_i from Eq. (2.113) replaces its previous value in subsequent calculations. Substitution continues until the convergence criterion is met.

The Gauss-Seidel substitution method requires an initial guess of the values of the unknowns x_2 to x_n. The initial guesses are used in Eq. (2.113) to calculate a new estimate of x_1 first. This estimate of x_1 and other guessed values are then replaced in Eq. (2.113) to evaluate the new estimate of x_2. The new estimate of x_3 to x_n are successively calculated from Eq. (2.113) in the same way. The iteration continues until all the newly calculated x's converge to within a convergence criterion ε of their previous values.

The Gauss-Seidel method converges to the correct solution, no matter what the initial estimate is, provided that the system of equations is predominantly diagonal. On the other hand, if the system is not predominantly diagonal, the correct solution may still be obtained if the initial estimate of the values of x_2 to x_n is close to the correct set. The Gauss-Seidel method is a very simple algorithm to program and it is computationally very efficient, in comparison with the other methods described in this chapter, provided that the system is predominantly diagonal. These advantages account for this method's wide use in the solution of engineering problems.

EXAMPLE 2.5 STEADY-STATE TWO-DIMENSIONAL HEAT TRANSFER IN A FLAT PLATE

Repeat Example 2.1, but this time use the Gauss-Seidel substitution method for the solution of this set of linear algebraic equations and determine the steady state temperature profile in the plate.

Method of solution

We provided the set of equations obtained in Example 2.1 as follows:

$$4T_1 - T_2 - T_3 = 600$$
$$-T_1 + 4T_2 - T_4 = 600$$
$$-T_1 + 4T_3 - T_4 = 200$$
$$-T_2 - T_3 + 4T_4 = 200$$

This set of equations is a predominantly diagonal set because in all four equations:

(Continued)

EXAMPLE 2.5 (CONTINUED)

$$|4| > |-1| + |-1| + |0|$$

Each equation in the set can be solved for the unknown on its diagonal:

$$T_1 = \frac{T_2 + T_3 + 600}{4}$$

$$T_2 = \frac{T_1 + T_4 + 600}{4}$$

$$T_3 = \frac{T_1 + T_4 + 200}{4}$$

$$T_4 = \frac{T_2 + T_3 + 200}{4}$$

Let's consider $T_2 = T_3 = T_4 = 200$ as the initial guesses for this problem. The guessed value is the average of the temperatures on the boundaries of the plate. These calculations of this example are performed in Excel. Fig. E2.5 demonstrates the calculation steps in Excel. It can be seen in this figure that the Gauss-Seidel method converges to the same values as other methods. The desired convergence criterion determines the number of iterations needed to reach the final results.

2.8 JACOBI METHOD

The Jacobi iterative method is similar to the Gauss-Seidel method with the exception that the newly calculated variables are not replaced until the end of each iteration is reached. In this section we develop the Jacobi method in matrix form.

The matrix of coefficients A can be written as:

$$A = (A - D) + D \tag{2.114}$$

where D is a diagonal matrix whose elements are those of the main diagonal of matrix A. Therefore the matrix $(A - D)$ is similar to A with the difference that its main diagonal elements are equal to zero. Replacing Eq. (2.114) into Eq. (2.13) and rearranging results in:

$$Dx = c - (A - D)x \tag{2.115}$$

from where the vector x can be evaluated:

$$\begin{aligned} x &= D^{-1}c - D^{-1}(A - D)x \\ &= D^{-1}c - (D^{-1}A - I)x \end{aligned} \tag{2.116}$$

In an iterative procedure Eq. (2.116) should be written as:

$$x^{(k)} = D^{-1}c - (D^{-1}A - I)x^{(k-1)} \tag{2.117}$$

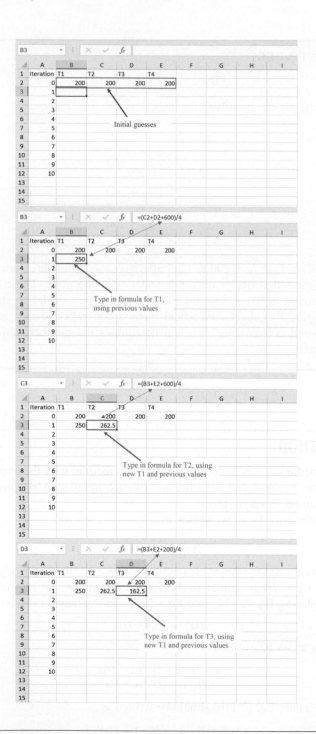

FIG. E2.5

Calculations of the Gauss-Seidel method in Excel.

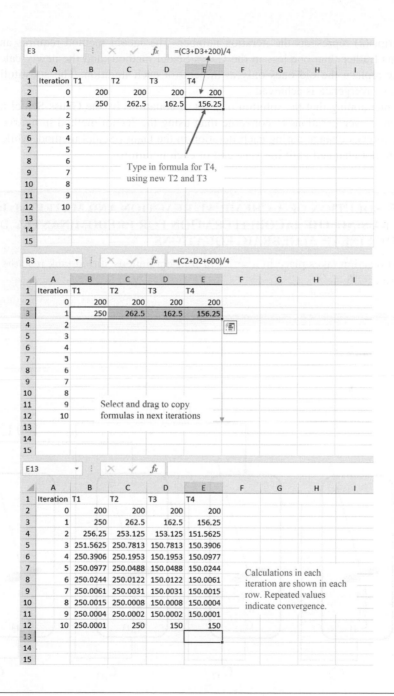

FIG. E2.5

Continued.

where superscript (k) represents the iteration number. The Jacobi method requires an initial guess of all unknowns (rather than one less in the Gauss-Seidel method) and the newly calculated values of the vector x replace the old ones only at the end of each iteration. The substitution procedure continues until convergence is achieved.

It is worth mentioning that the solution of a set of equations by the Gauss-Seidel method in formula form needs fewer iterations to converge than using the Jacobi method in matrix form. This is because the unknowns change during each iteration in the Gauss-Seidel method, while in the Jacobi method they are not changed until the very end of each iteration.

EXAMPLE 2.6 SOLUTION OF A CHEMICAL REACTION AND MATERIAL BALANCE EQUATIONS USING THE JACOBI ITERATION FOR PREDOMINANTLY DIAGONAL SYSTEMS OF LINEAR ALGEBRAIC EQUATIONS

A chemical reaction takes place in a series of four continuous stirred tank reactors arranged as shown in Fig. E2.6.

(Continued)

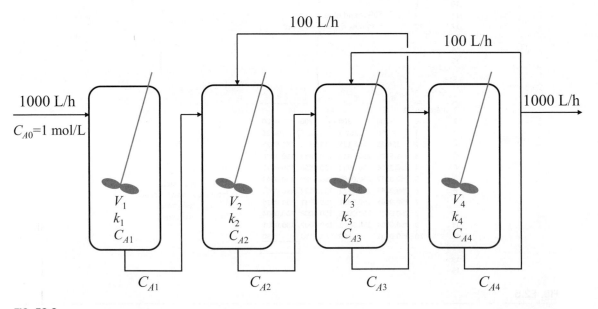

FIG. E2.6

Stirred tank reactors in series.

EXAMPLE 2.6 (CONTINUED)

The chemical reaction is a first-order irreversible reaction of the type:

$$A \xrightarrow{k_i} B$$

The temperatures in each reactor are such that the value of the rate constant k_i is different in each reactor. Also, the volume of each reactor V_i is different. The values of k_i and V_i are given in Table E2.6. The following assumptions can be made regarding this system:

1. The system is at steady state.
2. The reactions are in the liquid phase.
3. There is no change in the volume or density of the liquid.
4. The rate of disappearance of component A in each reactor is given by:

$$R_i = V_i k_i C_{A_i} \, \text{mol}/h$$

Respond to the following questions:

a. Set up the material balance equation for each of the four reactors. What type of equations do you have in this set of material balances?
b. What method do you recommend as the best one to use to solve for the exit concentration (C_{Ai}) from each reactor?
c. Write a MATLAB script to solve this set of equations and find the exit concentration from each reactor.

Method of solution

Part (a): The general unsteady-state material balance for each reactor is:

$$\text{Input} = \text{output} + \text{disappearance by reaction} + \text{accumulation}$$

Because the system is at the steady state, the accumulation term is zero; therefore the material balance simplifies to:

$$\text{Input} = \text{output} + \text{disappearance by reaction}$$

This balance applied to each of the four reactors yields the following set of equations:

$$1000(1) = 1000 \, C_{A_1} + V_1 k_1 C_{A_1}$$
$$1000 \, C_{A_1} + 100 \, C_{A_3} = 1100 \, C_{A_2} + V_2 k_2 C_{A_2}$$
$$1100 \, C_{A_2} + 100 \, C_{A_4} = 1200 \, C_{A_3} + V_3 k_3 C_{A_3}$$
$$1100 \, C_{A_3} = 1100 \, C_{A_4} + V_4 k_4 C_{A_4}$$

($Continued$)

Table E2.6 Reactor data.

Reactor	V_i, L	k_i, h^{-1}	Reactor	V_i, L	k_i, h^{-1}
1	1000	0.1	3	100	0.4
2	1500	0.2	4	500	0.3

EXAMPLE 2.6 (CONTINUED)

Substituting the values of V_i and k_i and rearranging:

$$1100\,C_{A_1} = 1000$$

$$1000\,C_{A_1} - 1400\,C_{A_2} + 100\,C_{A_3} = 0$$

$$1100\,C_{A_2} - 1240\,C_{A_3} + 100\,C_{A_4} = 0$$

$$1100\,C_{A_3} - 1250\,C_{A_4} = 0$$

These are a set of four simultaneous linear algebraic equations. It appears to be a predominantly diagonal system of equations since the coefficients on the diagonal are larger in absolute value than the sum of the absolute values of the other coefficients.

Part (*b*): From the discussion of the Jacobi method (Section 2.8), it would seem that Gauss-Seidel would be the best method of solution for a predominantly diagonal set. However, since calculations in MATLAB are based on matrices, the Jacobi method in matrix form is considerably faster than the Gauss-Seidel method in formula form in the MATLAB workspace.

Part (*c*): The general program, which uses the Jacobi iterative method in matrix form, is described in the next section. An initial guess of unknowns C_{A1} to C_{A4} is needed to start the Jacobi algorithm. Since this system of equations is a predominantly diagonal set, any initial guess for unknowns will yield convergence. However, the initial guess of 0.6 for all four unknowns seems to be an appropriate choice based on the fact that $C_{A0} = 1.0$. Two cases will be run in order to test the ability of the Jacobi method to converge. The first case will use 0.6 as the initial values and the second will use 100 as the starting values.

Program description

The programs and functions developed in this example can be found in: https://www.elsevier. com/books-and-journals/book-companion/9780128229613

The MATLAB function *Jacobi.m* is written to solve a set of linear algebraic equations by the Jacobi iterative method. Inputs to the function are the coefficient matrix, the vector of constants, and the vector of initial guesses for all the unknowns. The default convergence criterion is $|x_i^{(k)} - x_i^{(k-1)}| < 10^{-6}$. However, the user may change this convergence criterion by introducing another value as the fourth input argument into the function.

The next step in the program is building the modified coefficient matrix $(\boldsymbol{D}^{-1}\boldsymbol{A} - \boldsymbol{I})$ and the modified vector of constants $(\boldsymbol{D}^{-1}\boldsymbol{c})$. The function then starts the substitution procedure according to Eq. (2.117), which continues until the convergence criterion is reached for all of the unknowns.

In the program *Example2_6.m* the coefficient matrix and the vector of constants of the set of equations developed in this example are introduced as input data. The program also asks the user to input the convergence criterion and if the user wants to see the results of the calculations at the end of each step. The vector of initial guesses is introduced to the program

(Continued)

EXAMPLE 2.6 (CONTINUED)

in a loop so that the user can redo the calculations with different initial guesses. Then, it calls the function *Jacobi.m* to solve the set of equations. Finally, the program shows the final results of the calculation.

Input and results

>> **Example2_6**
Solution of set of linear algebraic equations by the Jacobi method
Number of equations = 4
Coefficients of Eq. 1 = [1100, 0, 0, 0]
Constant of Eq. 1 = 1000
Coefficients of Eq. 2 = [1000, -1400, 100, 0]
Constant of Eq. 2 = 0
Coefficients of Eq. 3 = [0, 1100, -1240, 100]
Constant of Eq. 3 = 0
Coefficients of Eq. 4 = [0, 0, 1100, -1250]
Constant of Eq. 4 = 0
Convergence criterion = 1e-5
Show step-by-step path to results (0/1)? 1
Vector of initial guess = 0.6*ones(1,4)
Initial guess:
0.6 0.6 0.6 0.6
Iteration no. 1
0.909091 0.471429 0.580645 0.528
Iteration no. 2
0.909091 0.690825 0.460783 0.510968
Iteration no. 3
0.909091 0.682264 0.654036 0.405489
Iteration no. 4
0.909091 0.696068 0.637935 0.575552
Iteration no. 5
0.909091 0.694917 0.663895 0.561383
Iteration no. 6
0.909091 0.696772 0.661732 0.584227
Iteration no. 7
0.909091 0.696617 0.665219 0.582324
Iteration no. 8
0.909091 0.696866 0.664928 0.585393
Iteration no. 9
0.909091 0.696846 0.665397 0.585137
Iteration no. 10

(Continued)

EXAMPLE 2.6 (CONTINUED)

0.909091 0.696879 0.665358 0.585549
Iteration no. 11
0.909091 0.696876 0.665421 0.585515
Iteration no. 12
0.909091 0.696881 0.665416 0.58557
Iteration no. 13
0.909091 0.69688 0.665424 0.585566

Results

CA(1) = 0.9091
CA(2) = 0.6969
CA(3) = 0.6654
CA(4) = 0.5856

Repeat the calculations with another guess (0/1)? 1

Vector of initial guess = 100*ones(1,4)
Initial guess:
100 100 100 100
Iteration no. 1
0.909091 78.5714 96.7742 88
Iteration no. 2
0.909091 7.56179 76.7972 85.1613
Iteration no. 3
0.909091 6.13487 13.5759 67.5816
Iteration no. 4
0.909091 1.61906 10.8923 11.9468
Iteration no. 5
0.909091 1.42738 2.39971 9.58527
Iteration no. 6
0.909091 0.820759 2.03923 2.11175
Iteration no. 7
0.909091 0.79501 0.898394 1.79452
Iteration no. 8
0.909091 0.713522 0.84997 0.790587
Iteration no. 9
0.909091 0.710063 0.69672 0.747973
Iteration no. 10
0.909091 0.699116 0.690215 0.613113
Iteration no. 11
0.909091 0.698652 0.669628 0.607389
Iteration no. 12

(*Continued*)

EXAMPLE 2.6 (CONTINUED)

0.909091 0.697181 0.668755 0.589273
Iteration no. 13
0.909091 0.697119 0.665989 0.588504
Iteration no. 14
0.909091 0.696921 0.665872 0.586071
Iteration no. 15
0.909091 0.696913 0.6655 0.585967
Iteration no. 16
0.909091 0.696886 0.665485 0.58564
Iteration no. 17
0.909091 0.696885 0.665435 0.585626
Iteration no. 18
0.909091 0.696882 0.665433 0.585583
Iteration no. 19
0.909091 0.696882 0.665426 0.585581

Results

CA(1) = 0.9091
CA(2) = 0.6969
CA(3) = 0.6654
CA(4) = 0.5856

Repeat the calculations with another guess (0/1)? 0

Discussion of results

The first case uses the value of 0.6 as the initial guess for the values of the unknowns C_{A1} to C_{A4}. The Jacobi method converges to the solution in 13 iterations. The convergence criterion, which is satisfied by all the unknowns, is 0.000001.

In the second case the value of 100 is used as the initial guess for each of the unknowns C_{A1} to C_A. Convergence to exactly the same answer as in the first case is accomplished in 19 iterations.

2.9 HOMOGENEOUS ALGEBRAIC EQUATIONS AND THE CHARACTERISTIC-VALUE PROBLEM

We mentioned earlier that a homogeneous set of equations:

$$Ax = 0 \tag{2.82}$$

has a nontrivial solution, if and only if the matrix A is singular, that is, if the rank r of A is less than n. The system of equations would consist of r independent equations, r unknowns that can be evaluated independently, and $(n - r)$ unknowns that must be chosen arbitrarily in order to complete the solution. Choosing nonzero values for the $(n - r)$ unknowns transforms the homogeneous set to a nonhomogeneous set of order r. The Gauss and Gauss-Jordan methods, which are applicable to nonhomogeneous systems, can then be used to obtain the complete solution of the problem. In fact, these methods can be used first on the homogeneous system to determine the number of independent equations (or the rank of A) and then applied to the set of r nonhomogeneous independent equations to evaluate the r unknowns. This concept will be demonstrated later in this section in conjunction with the calculation of eigenvectors.

A special class of homogeneous linear algebraic equations arises in the study of vibrating systems, structure analysis, and electric circuit system analysis, and the solution and stability analysis of linear ordinary differential equations (see Chapter 5). This system of equations has the form:

$$Ax = \lambda x \tag{2.118}$$

which can be alternatively expressed as:

$$(A - \lambda I)x = 0 \tag{2.119}$$

where the scalar λ is called an *eigenvalue* (or a *characteristic value*) of matrix A. The vector x is called the *eigenvector* (or *characteristic vector*) corresponding to λ. Matrix I is the identity matrix. The problem often requires the solution of the homogeneous set of equations, represented by Eq. (2.119), to determine the values of λ and x which satisfy this set. In MATLAB $eig(A)$ is a vector containing the eigenvalues of A. The statement $[V, D] = eig(A)$ produces a diagonal matrix D of eigenvalues and a full matrix V whose columns are the corresponding eigenvectors so that $AV = VD$.

Before we proceed with developing methods of solution, let's examine Eq. (2.118) from a geometric perspective. The multiplication of a vector by a matrix is a linear transformation of the original vector to a new vector of a different direction and length. For example, matrix A transforms the vector y to the vector z in the operation

$$Ay = z \tag{2.120}$$

In contrast to this, if x is the eigenvector of A, then the multiplication of the eigenvector x by matrix A yields the same vector x multiplied by a scalar λ, that is, the same vector but of different length:

$$Ax = \lambda x \tag{2.118}$$

It can be stated that for a nonsingular matrix A of order n, there are n characteristic directions in which the operation by A does not change the direction of the vector but only changes its length. More simply stated, matrix A has n eigenvectors and n eigenvalues. The types of eigenvalues that exist for a set of special matrices are listed in Table 2.3.

The homogeneous problem:

$$(A - \lambda I)x = 0 \tag{2.119}$$

possesses nontrivial solutions if the determinant of matrix $(A - \lambda I)$, called the *characteristic matrix* of A, vanishes:

$$|A - \lambda I| = \begin{vmatrix} a_{11} - \lambda & a_{12} & \cdots & a_{1n} \\ a_{21} & a_{22} - \lambda & \cdots & a_{2n} \\ \vdots & \vdots & \ddots & \vdots \\ a_{n1} & a_{n2} & \cdots & a_{nn} - \lambda \end{vmatrix} \tag{2.121}$$

Table 2.3 Eigenvalues of different types of matrices.

Matrix	Eigenvalue
Singular, $|A| = 0$	At least one zero eigenvalue
Nonsingular, $|A| \neq 0$	No zero eigenvalues
Symmetric, $A = A'$	All real eigenvalues
Hermitian	All real eigenvalues
Zero matrix, $A = 0$	All zero eigenvalues
Identity, $A = I$	All unity eigenvalues
Diagonal, $A = D$	Equal to diagonal elements of A
Inverse, A^{-1}	Reciprocals of eigenvalues of A
Transformed, $B = Q^{-1}AQ$	Eigenvalues of B = eigenvalues of A

The determinant can be expanded by minors to yield a polynomial of nth degree:

$$\lambda^n - \alpha_1 \lambda^{n-1} - \alpha_2 \lambda^{n-2} - \cdots - \alpha_n = 0 \tag{2.122}$$

This polynomial, which is called the characteristic equation of matrix A, has n roots which are the eigenvalues of A. These roots may be real distinct, real repeated, or complex depending on matrix A (see Table 2.3). A nonsingular real symmetric matrix of order n has n real nonzero eigenvalues and n linearly independent eigenvectors. The eigenvectors of a real symmetric matrix are orthogonal to each other. The coefficients α_i of the characteristic polynomial are functions of the matrix elements a_{ij}, and must be determined before the polynomial can be used. The well-known Cayley-Hamilton theorem states that a square matrix satisfies its own characteristic equation, that is:

$$A^n - \alpha_1 A^{n-1} - \alpha_2 A^{n-2} - \ldots - \alpha_n I = 0 \tag{2.123}$$

The evaluation of the eigenvalues and eigenvectors of matrices is a complex multistep procedure. Several methods have been developed for this purpose. Some of these apply to symmetric matrices, others to tridiagonal matrices, and a few can be used for general matrices. We can classify these methods into two categories:

1. The methods in this category work with the original matrix A and its characteristic polynomial (Eq. 2.122) to evaluate the coefficients α_i of the polynomial. One such method is the Faddeev-Leverrier procedure, which will be described later. Once the coefficients of the polynomial are known, the methods use root-finding techniques, such as the Newton-Raphson method, to determine the eigenvalues. Finally, the algorithms employ a reduction method, such as Gauss elimination, to calculate the eigenvectors.

2. The methods in this category reduce the original matrix A to tridiagonal form (when A is symmetric) or to Hessenberg form (when A is nonsymmetric) through orthogonal transformations or elementary similarity transformations. They apply successive factorization procedures, such as LR or QR algorithms, to extract the eigenvalues, and, finally, they use a reduction method to calculate the eigenvectors.

In the remaining part of this chapter we will discuss the following methods: (*a*) the Faddeev-Leverrier procedure for calculating the coefficients of the characteristic polynomial, (*b*) the elementary similarity transformation for converting a matrix to Hessenberg form, (*c*) the QR algorithm of successive factorization for the determination of the eigenvalues, and finally, (*d*) the Gauss elimination method applied for the evaluation of the eigenvectors. These methods were chosen for their general applicability to both symmetric and nonsymmetric matrices. For a complete discussion of these and other methods, the reader is referred to Ralston and Rabinowitz. [1]

2.9.1 THE FADDEEV-LEVERRIER METHOD

The Faddeev-Leverrier method [3] calculates the coefficients α_1 to α_n of the characteristic polynomial (Eq. 2.122) by generating a series of matrices A_k whose traces are equal to the coefficients of the polynomial. The starting matrix and first coefficient are:

$$A_1 = A \quad \alpha_1 = trA_1 \tag{2.124}$$

and the subsequent matrices are evaluated from the recursive equations:

$$\left.\begin{array}{l} A_k = A(A_{k-1} - \alpha_{k-1}I) \\ \alpha_k = \dfrac{1}{k}trA_k \end{array}\right\} \quad k = 2, 3, \ldots, n \tag{2.125}$$

In addition to this, the Faddeev-Leverrier method yields the inverse of the matrix A by:

$$A^{-1} = \frac{1}{\alpha_n}(A_{k-1} - \alpha_{k-1}I) \tag{2.126}$$

To elucidate this method, we will determine the coefficients of the characteristic polynomial of the following set of homogeneous equations:

$$\begin{array}{l} (1 - \lambda)x_1 + 2x_2 + x_3 = 0 \\ 3x_1 + (1 - \lambda)x_2 + 2x_3 = 0 \\ 4x_1 + 2x_2 + (1 - \lambda)x_3 = 0 \end{array} \tag{2.127}$$

The characteristic polynomial for this third-order system is:

$$\lambda^3 - \alpha_1\lambda^2 - \alpha_2\lambda - \alpha_3 = 0 \tag{2.128}$$

The matrix A is:

$$A = \begin{bmatrix} 1 & 2 & 1 \\ 3 & 1 & 2 \\ 4 & 2 & 3 \end{bmatrix} \tag{2.129}$$

Application of Eq. (2.124) gives:

$$A_1 = A \; and \; \alpha_1 = trA_1 = 5 \tag{2.130}$$

Application of Eq. (2.125), with $k = 2$, yields:

$$A_2 = A(A_1 - \alpha_1 I)$$

$$= \begin{bmatrix} 1 & 2 & 1 \\ 3 & 1 & 2 \\ 4 & 2 & 3 \end{bmatrix} \left\{ \begin{bmatrix} 1 & 2 & 1 \\ 3 & 1 & 2 \\ 4 & 2 & 3 \end{bmatrix} - \begin{bmatrix} 5 & 0 & 0 \\ 0 & 5 & 0 \\ 0 & 0 & 5 \end{bmatrix} \right\} \tag{2.131}$$

$$= \begin{bmatrix} 6 & -4 & 3 \\ -1 & 6 & 1 \\ 2 & 6 & 2 \end{bmatrix}$$

$$\alpha_2 = \frac{1}{2} tr A_2 = 7 \tag{2.132}$$

Repetition of Eq. (2.125), with $k = 3$, results in:

$$A_3 = A(A_2 - \alpha_2 I)$$

$$= \begin{bmatrix} 1 & 2 & 1 \\ 3 & 1 & 2 \\ 4 & 2 & 3 \end{bmatrix} \left\{ \begin{bmatrix} 6 & -4 & 3 \\ -1 & 6 & 1 \\ 2 & 6 & 2 \end{bmatrix} - \begin{bmatrix} 7 & 0 & 0 \\ 0 & 7 & 0 \\ 0 & 0 & 7 \end{bmatrix} \right\} \tag{2.133}$$

$$= \begin{bmatrix} -1 & 0 & 0 \\ 0 & -1 & 0 \\ 0 & 0 & -1 \end{bmatrix}$$

$$\alpha_3 - \frac{1}{2} tr A_3 = -1 \tag{2.134}$$

Therefore the characteristic polynomial is:

$$\lambda^3 - 5\lambda^2 - 7\lambda + 1 = 0 \tag{2.135}$$

The root-finding techniques described in Chapter 1 may be used to determine the λ's of this polynomial. The eigenvectors corresponding to each eigenvalue may be calculated using the Gauss elimination method. The Faddeev-Leverrier method, the Newton-Raphson method with synthetic division, and the Gauss elimination method constitute a complete algorithm for the evaluation of all the eigenvalues and eigenvectors of this characteristic-value problem. This combination of methods, however, is "fraught with peril" since it is too sensitive to small changes in the coefficients. The use of the QR algorithm, discussed in Section 2.9.3, is preferable.

2.9.2 ELEMENTARY SIMILARITY TRANSFORMATIONS

In Section 2.5.2 we showed that the Gauss elimination method can be represented in matrix form as

$$LA = U \tag{2.59}$$

Matrix A is nonsingular, matrix L is unit lower triangular, and matrix U is upper triangular. The inverse of L is also a unit lower triangular matrix. Postmultiplying both sides of Eq. (2.59) by L^{-1}, we obtain

$$LAL^{-1} = UL^{-1} = B \tag{2.136}$$

This is a similarity transformation of the type described in Section 2.2.2. The transformation coverts matrix A to a *similar* matrix B. The two matrices, A and B, have identical eigenvalues, determinants, and traces.

We therefore conclude that if the Gauss elimination method is extended so that matrix A is post-multiplied by L^{-1}, at each step of the operation, in addition to being premultiplied by L, the resulting matrix B is similar to A. This operation is called the *elementary similarity transformation*.

In the determination of eigenvalues it is desirable to reduce matrix A to a supertriangular matrix of upper Hessenberg form:

$$
H_U = \begin{bmatrix}
h_{11} & h_{12} & h_{13} & \cdots & h_{1n-2} & h_{1n-1} & h_{1n} \\
h_{21} & h_{22} & h_{23} & \cdots & h_{2n-2} & h_{2n-1} & h_{2n} \\
0 & h_{32} & h_{33} & \cdots & h_{3n-2} & h_{3n-1} & h_{3n} \\
\vdots & \vdots & \vdots & \ddots & \vdots & \vdots & \vdots \\
0 & 0 & 0 & \cdots & h_{n-1n-2} & h_{n-1n-1} & h_{n-1n} \\
0 & 0 & 0 & \cdots & 0 & h_{n-1n-2} & h_{nn}
\end{bmatrix}
\tag{2.43}
$$

This can be done by using the $(k+1)$st row to eliminate the elements $(k+2)$ to n of column k. Consequently, the elements of the subdiagonal do not vanish. The transformation matrices that perform this elimination are unit lower triangular of the form shown in Eq. (2.137). The elimination matrix $\overline{L}_1$ that would eliminate the elements of column 1 below the subdiagonal is:

$$
\overline{L}_1 = \begin{bmatrix}
1 & 0 & 0 & 0 & \cdots & 0 \\
0 & 1 & 0 & 0 & \cdots & 0 \\
0 & -\dfrac{h_{31}^{(0)}}{h_{21}^{(0)}} & 1 & 0 & \cdots & 0 \\
0 & -\dfrac{h_{41}^{(0)}}{h_{21}^{(0)}} & 0 & 1 & \cdots & 0 \\
\vdots & \vdots & & \ddots & \ddots & 0 \\
0 & -\dfrac{h_{n1}^{(0)}}{h_{21}^{(0)}} & 0 & 0 & \cdots & 1
\end{bmatrix}
\tag{2.137}
$$

where the superscript (0) indicates that each $\overline{L}_k$ matrix uses the elements $h_{ij}^{(k-1)}$ of the previous transformation step. The reader is encouraged to compare $\overline{L}_1$ with L_1 of Eq. (2.93).

The inverse of $\overline{L}_1$ is given by:

$$
\overline{L}_1^{-1} = \begin{bmatrix}
1 & 0 & 0 & 0 & \cdots & 0 \\
0 & 1 & 0 & 0 & \cdots & 0 \\
0 & \dfrac{h_{31}^{(0)}}{h_{21}^{(0)}} & 1 & 0 & \cdots & 0 \\
0 & \dfrac{h_{41}^{(0)}}{h_{21}^{(0)}} & 0 & 1 & \cdots & 0 \\
\vdots & \vdots & & \ddots & \ddots & 0 \\
0 & \dfrac{h_{n1}^{(0)}}{h_{21}^{(0)}} & 0 & 0 & \cdots & 1
\end{bmatrix}
\tag{2.138}
$$

The complete elementary similarity transformation, which converts matrix A to the upper Hessenberg matrix H, is shown by:

$$\overline{L}_{n-1}\overline{L}_{n-2}\ldots\overline{L}_2\overline{L}_1 A \overline{L}_1^{-1}\overline{L}_2^{-1}\ldots\overline{L}_{n-2}^{-1}\overline{L}_{n-1}^{-1} = H \tag{2.139}$$

Each postmultiplication step by the inverse $\overline{L}_i^{-1}$ preserves the zeros previously obtained in the premultiplication step by L_i [4].

For simplicity in the preceding discussion, the partial pivoting matrices P_{ij} were not applied. However, the use of partial pivoting is strongly recommended to reduce roundoff errors. Premultiplication by P_{ij} interchanges two rows and causes the sign of the determinant to change. Postmultiplication by P_{ij}^{-1} (which is identical to P_{ij}) interchanges the corresponding two columns and causes the sign of the determinant to change again. The premultiplication step must be followed immediately by the postmultiplication step to balance the symmetry of the transformation and to preserve the form of the transformed matrix.

The elementary similarity transformation to produce an upper Hessenberg matrix in formula form is as follows:

Initialization step:

$$h_{ij}^{(0)} = a_{ij} \qquad \begin{cases} i = 1, 2, \ldots, n \\ j = 1, 2, \ldots, n \end{cases} \tag{2.140a}$$

Transformation formula:

$$m_{i,k+1}^{(k)} = \frac{h_{ij}^{(k-1)}}{h_{k+1,k}^{(k-1)}} \qquad \begin{cases} j = n, n-1, \ldots, k \\ i = k+2, \ldots, n \end{cases} \tag{2.140b}$$

Premultiplication step:

$$h_{ij}^{(k-1/2)} = h_{ij}^{(k-1)} - m_{i,k+1}^{(k)} h_{k+1,j}^{(k-1)}, \qquad \begin{cases} j = n, n-1, \ldots, k \\ i = k+2, \ldots, n \end{cases} \begin{cases} k = 1, 2, \ldots, n-2 \\ h_{k+1,k} \neq 0 \end{cases} \tag{2.141}$$

Postmultiplication step:

$$h_{i,k+1}^{(k)} = h_{i,k+1}^{(k-1/2)} + h_{ij}^{(k-1/2)} m_{j,k+1}^{(k)}, \qquad \begin{cases} j = k+2, \ldots, n \\ i = n, n-1, \ldots, 1 \end{cases} \tag{2.142}$$

where the superscript $(k - \frac{1}{2})$ means that only half the complete transformation (i.e., only premultiplication) has been completed at the point.

The QR algorithm, which will be discussed next, uses the upper Hessenberg matrix H to determine its eigenvalues, which are equivalent to the eigenvalues of matrix A.

2.9.3 THE QR ALGORITHM OF SUCCESSIVE FACTORIZATION

The QR algorithm is based on the possible decomposition of a matrix A into a product of two matrices:

$$A = QR \tag{2.143}$$

where Q is orthogonal and R is upper triangular with nonnegative diagonal elements. The decomposition always exists, and when A is nonsingular, the decomposition is unique [1].

This decomposition can be used to form a series of successive matrices A_k, which are *similar* to the original matrix A; therefore their eigenvalues are the same. To do this, let us first define $A_1 = A$ and convert Eq. (2.143) to:

$$A_1 = Q_1 R_1 \tag{2.144}$$

Premultiply each side by Q^{-1} and rearrange to obtain:

$$R_1 = Q_1^{-1} A_1 \tag{2.145}$$

Form a second matrix A_2 from the product of R_1 with Q_1:

$$A_2 = R_1 Q_1 \tag{2.146}$$

and use Eq. (2.145) to eliminate R_1 from Eq. (2.146):

$$A_2 = Q_1^{-1} A_1 Q_1 \tag{2.147}$$

Because Q_1 is an orthogonal matrix, this is an orthogonal transformation of A_1 to A_2; therefore, these two matrices are similar. They have the same eigenvalues. The inverse of an orthogonal matrix is equal to its transpose; thus Eq. (2.147) can also be written as:

$$A_2 = Q_1' A_1 Q_1 \tag{2.148}$$

In the particular case where matrix A is symmetric an orthogonal transformation of A can be found, which yields a diagonal matrix D:

$$D = Q' A Q \tag{2.149}$$

whose diagonal elements are the eigenvalues of A. Our discussion, however, will focus on nonsymmetric matrices that transform into triangular matrices.

The orthogonal matrix Q_1 is determined by finding a series of S'_{ij} orthogonal transformation matrices, each of which eliminates one element, in position ij, below the diagonal of the matrix it is postmultiplying. The complete set of transformations converts matrix A_1 to upper triangular form with nonnegative diagonal elements:

$$S'_{nn-1} \ldots S'_{ij} \ldots S'_{n1} \ldots S'_{31} S'_{21} A_1 = R_1 \tag{2.150}$$

where the counter i increases from $j + 1$ to n, and the counter j increases from 1 to $(n - 1)$.

Each of the S'_{ij} matrices is orthogonal and the product of orthogonal matrices is also orthogonal. Direct comparison of Eq. (2.150) with Eq. (2.145) reveals that Q_1^{-1} is equal to the product of the S'_{ij} matrices:

$$Q_1^{-1} = S'_{nn-1} \ldots S'_{ij} \ldots S'_{31} S'_{21} \tag{2.151}$$

Because the transpose of an orthogonal matrix is equal to its inverse, it follows that:

$$Q_1' = S'_{nn-1} \ldots S'_{ij} \ldots S'_{31} S'_{21} \tag{2.152}$$

and

$$Q_1 = S_{21} S_{31} \ldots S_{ij} \ldots S_{nn-1} \tag{2.153}$$

Therefore Eq. (2.147) can be rewritten in terms of the S_{ij} matrices:

$$A_2 = S'_{nn-1}\ldots S'_{ij}\ldots S'_{31}S'_{21}A_1S_{21}S_{31}\ldots S_{ij}\ldots S_{nn-1} \tag{2.154}$$

As an example of the orthogonal transformation matrices S_{ij}, we give the S_{pq} matrix for a (6×6)-order system, with $p = 6$ and $q = 3$:

$$S_{63} = \begin{bmatrix} 1 & 0 & 0 & 0 & 0 & 0 \\ 0 & 1 & 0 & 0 & 0 & 0 \\ 0 & 0 & s_{33} & 0 & 0 & s_{36} \\ 0 & 0 & 0 & 1 & 0 & 0 \\ 0 & 0 & 0 & 0 & 1 & 0 \\ 0 & 0 & s_{63} & 0 & 0 & s_{63} \end{bmatrix} \tag{2.155}$$

where the diagonal elements of this matrix are specified as:

$$s_{pp} = s_{qq} = \cos\theta \tag{2.156}$$

$$s_{ii} = 1 \text{ for } i \neq p \text{ or } q \tag{2.157}$$

and the off-diagonal elements as:

$$s_{pq} = -s_{qp} = \sin\theta \tag{2.158}$$

$$s_{ij} = 0 \text{ everywhere else} \tag{2.159}$$

Premultiplication of matrix A by S'_{pq} eliminates the element pq and causes rotation of axes in the (p, q) plane. The S_{ij} matrices clearly satisfy the orthogonality requirement that

$$S'_{ij}S_{ij} = I \tag{2.160}$$

The reader is encouraged to verify this equality.

The angle of axis rotation θ, in Eqs. (2.156) and (2.158), is chosen so that by element pq of the matrix being transformed vanishes. It has been shown by Givens [5] that it is not necessary to actually calculate the value of θ itself. The trigonometric terms $\cos\theta$ and $\sin\theta$ can be obtained from the values of the elements of the matrix being transformed. Givens has determined that the elements of the matrix S_{pq} are calculated as follows:

Diagonal elements:

$$s_{pp}^{(k)} = s_{qq}^{(k)} = \frac{a_{qq}^{(k-1)}}{\sqrt{\left(a_{qq}^{(k-1)}\right)^2 + \left(a_{pq}^{(k-1)}\right)^2}} \tag{2.161}$$

$$s_{ii}^{(k)} = 1 \text{ for } i \neq p \text{ or } q \tag{2.162}$$

Off-diagonal elements:

$$s_{pq}^{(k)} = -s_{qp}^{(k)} = \frac{a_{pq}^{(k-1)}}{\sqrt{\left(a_{qq}^{(k-1)}\right)^2 + \left(a_{pq}^{(k-1)}\right)^2}} \tag{2.163}$$

$$s_{ij}^{(k)} = 0 \text{ everywhere else} \tag{2.164}$$

The superscripts $(k - 1)$ have been used in the above equations to remind the reader that the elements $a_{pq}^{(k-1)}$ and $a_{qq}^{(k-1)}$ are those of the matrix from the previous transformation step and not those of the original matrix.

Givens' method of plane rotations can reduce a nonsymmetric matrix to upper triangular form and a symmetric matrix to tridiagonal form. However, a large number of computations are required. It is computationally more efficient to apply first the elementary similarity transformation to reduce the matrix to upper Hessenberg form, as we described in Section 2.9.2 and then use plane rotations to reduce it to triangular form. In the rest of this section we will assume that matrix A has already been reduced to upper Hessenberg form, H_1, and we will show how the QR algorithm further reduces the matrix to obtain its eigenvalues.

If the eigenvalues of matrix H_1 are λ, then the eigenvalues of the matrix $(H_1 - \gamma_1 I)$ are $(\lambda - \gamma_1)$, where γ_1 is called the *shift factor*. The orthogonal transformation applied to A_1 above can also be applied to the shifted matrix $(H_1 - \gamma_1 I)$ as follows:

Decompose the matrix $(H_1 - \gamma_1 I)$ into Q_1 and R_1 matrices:

$$H_1 - \gamma_1 I = Q_1 R_1 \tag{2.165}$$

Rearrange the above equation to obtain R_1:

$$R_1 = Q_1^{-1}(H_1 - \gamma_1 I) \tag{2.166}$$

Form a new matrix $(H_2 - \gamma_1 I)$ from the product of R_1 and Q_1:

$$H_2 - \gamma_1 I = R_1 Q_1 \tag{2.167}$$

Eliminate R_1 using Eq. (2.166):

$$H_2 - \gamma_1 I = Q_1^{-1}(H_1 - \gamma_1 I)Q_1 \tag{2.168}$$

Solve for H_2:

$$H_2 = Q_1^{-1}(H_1 - \gamma_1 I)Q_1 + \gamma_1 I \tag{2.169}$$

It has been shown [1] that if the shift factor γ_1 is chosen to be a good estimate of one of the eigenvalues and that if the magnitudes of the eigenvalues are:

$$|\lambda_1| > |\lambda_2| > \ldots |\lambda_n| \tag{2.170}$$

then the matrix H_k will converge to a triangular form with the elements $h_{n,n-1} \to 0$ and $h_{nn} \to \lambda_n$.

Estimation of the shift factor γ_1 is relatively easy when the matrix has been reduced to upper Hessenberg form:

$$H_1 = \begin{bmatrix} h_{11} & h_{12} & h_{13} & \cdots & h_{1n-2} & h_{1n-1} & h_{1n} \\ h_{21} & h_{22} & h_{23} & \cdots & h_{2n-2} & h_{2n-1} & h_{2n} \\ 0 & h_{32} & h_{33} & \cdots & h_{3n-2} & h_{3n-1} & h_{3n} \\ \vdots & \vdots & \vdots & \ddots & \vdots & \vdots & \vdots \\ 0 & 0 & 0 & \cdots & h_{n-1n-2} & h_{n-1n-1} & h_{n-1n} \\ 0 & 0 & 0 & \cdots & 0 & h_{nn-1} & h_{nn} \end{bmatrix} \tag{2.43}$$

The eigenvalues of the lower (2×2) submatrix:

$$\begin{bmatrix} h_{n-1n-1} & h_{n-1n} \\ h_{nn-1} & h_{nn} \end{bmatrix} \tag{2.171}$$

can be used to determine the shift factor. The two eigenvalues of this matrix are obtained from the quadratic characteristic equation:

$$\gamma^2 - (h_{n-1n-1} + h_{nn})\gamma + (h_{n-1n-1}h_{nn} - h_{nn-1}h_{n-1n}) = 0 \tag{2.172}$$

whose solution is given by the quadratic formula:

$$\gamma_{+,-} = \frac{1}{2}\left[(h_{n-1n-1} + h_{nn}) \pm \sqrt{(h_{n-1n-1} + h_{nn})^2 - 4(h_{n-1n-1}h_{nn} - h_{nn-1}h_{n-1n})} \right] \tag{2.173}$$

The value of γ which is closest to h_{nn} is chosen from the two roots. In the case where the roots are complex conjugates the real part of the root is chosen as the shift factor.

In the QR iteration procedure the subsequent values of the shift factor, $\gamma_2,...,\gamma_k$, are similarly chosen from matrices $H_2,...,H_k$.

The steps of the QR algorithm for calculating the eigenvalues and eigenvectors of a nonsingular nonsymmetric matrix A with real eigenvalues are the following:

1. Use the elementary similarity transformations (Eq. 2.139) to transform matrix A to the upper Hessenberg matrix H_1.
2. Use the lower (2×2) submatrix of H_1 (Eq. 2.171) to estimate the shift factor γ_1 from Eq. (2.172).
3. Construct the shifted matrix $(H_1 - \gamma_1 I)$.
4. Calculate the elements of the transformation matrix S_{21} from the elements of the shifted matrix $(H_1 - \gamma_1 I)$ using Eqs. (2.161)–(2.164).
5. Perform the premultiplication $S'_{21}(H_1 - \gamma_1 I)$, which eliminates the elements in position $(2, 1)$ of the matrix $(H_1 - \gamma_1 I)$.
6. Repeat steps 4 and 5, calculating the transformation matrix S_{pq} and eliminating one element on subdiagonal in each set of steps. The application of steps 4 and 5 for $(n - 1)$ times, with the counter q increasing from 1 to $(n - 1)$ and the counter p set at $(q + 1)$, will convert the Hessenberg matrix H_1 to a triangular matrix R_1:

$$S'_{nn-1}...S'_{31}S'_{21}H_1 = R_1 \tag{2.174}$$

7. Perform the postmultiplication of R_1 by S_{pq} to obtain the transformed shifted matrix $(H_2 - \gamma_1 I)$:

$$H_2 - \gamma_1 I = R_1 S_{21} S_{32}...S_{nn-1} \tag{2.175}$$

8. Solve Eq. (2.175) for the transformed Hessenberg matrix H_2:

$$H_2 = R_1 S_{21} S_{32}...S_{nn-1} + \gamma_1 I \tag{2.176}$$

9. Use H_2 as the new Hessenberg matrix and repeat steps 2 to 8 until $|h_{nn-1}| \le \varepsilon$, where ε is a small convergence criterion. At this point, the element h_{nn} will give one eigenvalue λ_n.

10. Deflate the H_k matrix to order $(n - 1)$ by eliminating the nth row and nth column, and repeat steps 2 to 10 until all the eigenvalues are calculated.

11. Apply the Gauss elimination method with complete pivoting to the matrix $(A - \lambda I)$ to evaluate the eigenvectors corresponding to each eigenvalue. Several different possibilities exist when the eigenvalues are real:

    a. Distinct nonzero eigenvalues: Matrix A is nonsingular and matrix $(A - \lambda I)$ is singular of rank $(n - 1)$. Application of the Gauss elimination method with complete pivoting on the matrix $(A - \lambda I)$ triangularizes the matrix and causes the last row to contain all zero values because the rank is $(n - 1)$. Assume the value of the nth element of the eigenvector to be equal to unity and reduce the problem to finding the remaining $(n - 1)$ elements.

    b. One zero eigenvalue: Matrix A is singular of rank $(n - 1)$ and matrix $(A - \lambda I)$ is singular of rank $(n - 1)$. Application of the Gauss elimination method proceeds as in a. One element of each eigenvector will be found to be a zero element.

    c. One pair of repeated eigenvalues: Matrix A is nonsingular and matrix $(A - \lambda I)$ is of rank $(n - 2)$. Application of the Gauss elimination method with complete pivoting on the matrix $(A - \lambda I)$ triangularizes the matrix and causes the last two rows to contain all zero values because the rank is $(n - 2)$. Assume the values of the last two elements in the eigenvector to be equal to unity and reduce the problem to finding the remaining $(n - 2)$ elements.

The QR algorithm described in this section applies well to both symmetric and nonsymmetric matrices with real eigenvalues. A more general method, called the double QR algorithm, which can evaluate complex eigenvalues, is described by Ralston and Rabinowitz [1].

2.10 USING BUILT-IN MATLAB® AND EXCEL FUNCTIONS

There are different ways in MATLAB to solve a set of linear equations. For example, let's consider the set of equations derived in Example 2.1. This set can be introduced to MATLAB as follows:

$$
\begin{aligned}
&>> A = [4\ \text{-}1\ \text{-}1\ 0; \\
&\text{-}1\ 4\ 0\ \text{-}1; \\
&\text{-}1\ 0\ 4\ \text{-}1; \\
&0\ \text{-}1\ \text{-}1\ 4]; \\
&>> c = [600;\ 600;\ 200;\ 200];
\end{aligned}
$$

This equation can be solved by using one of the following commands:

$$
\begin{aligned}
&>> T = A\ c \\
&>> T = \text{inv}(A) * c \\
&>> T = A^{(\text{-}1)} * c
\end{aligned}
$$

Note that for a large set of equations, $A \backslash c$ can be faster than the other two commands. Also, the determinant of the square matrix A can be calculated in MATLAB by $det(A)$.

The MATLAB function $lu(A)$ performs the LU decomposition on matrix A. The command

$$>> [L, U] = \text{lu}(A)$$

returns a lower triangular matrix L and an upper triangular matrix U such that $A = LU$.

The set of linear algebraic equations can also be solved in Excel. Fig. 2.5 illustrates the steps you need to follow for solving the set of equations derived in Example 2.1. Note that in Excel, the function *minverse* is used for calculating the inverse of a square matrix and the function *mmult* is used to perform matrix multiplication.

2.11 SUMMARY

The aim of this chapter is to present various methods of solving a set of linear algebraic equations. Since these methods can be presented and performed more effectively in matrix form, a review of important matrix and vector operations is discussed first. Various methods of solving a set of linear algebraic equations described in this chapter include Cramer's rule, Gauss elimination, Gauss-Jordan reduction, Gauss-Seidel, and Jacobi. Formula and matrix forms of these methods are given in Sections 2.4 to 2.8 and some illustrative examples are presented for applying these methods in MATLAB and Excel.

Cramer's rule is computationally expensive and is not feasible to apply it to sets of more than three equations. The Gauss elimination method, in addition to solving the set of equations, can be used for calculating the determinant of a matrix. Also, the Gauss-Jordan method can be employed to determine the inverse of a square matrix.

Gauss-Seidel and Jacobi methods are iterative methods for solving predominantly diagonal systems of equations. The Gauss-Jordan method is twice faster as the Jacobi method. At the end of the chapter, built-in MATLAB and Excel functions that can be used for solving sets of linear algebraic equations are discussed.

PROBLEMS

2.1. Solve the following set of equations by using both the Gauss-Seidel method and the Jacobi method.

$$y + x = 3$$
$$-y + 2x = 0$$

In both cases start the iteration from (0, 0). Compare the results of both methods at the end of each iteration.

2.2. Calculate a solution to the following set of linear algebraic equations:

$$4x_1 + x_2 + 2x_3 = 4$$
$$5x_1 + 14x_2 + 6x_3 = 8$$
$$9x_1 + 10x_2 + 20x_3 = 12$$

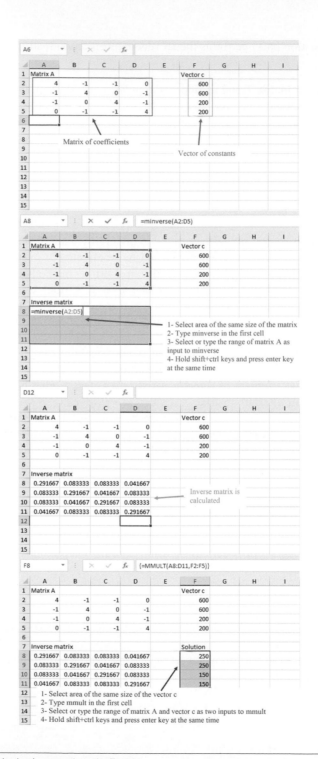

FIG. 2.5

Solving a set of linear algebraic equations in Excel.

(a) Use the Gauss elimination method.

(b) Use the Gauss-Jordan reduction method.

(c) Use the Gauss-Seidel substitution method.

(d) Once you have obtained the answer, verify that it satisfies all three of the equations.

2.3. When a pure sample of gas is bombarded by low-energy electrons in a mass spectrometer, the galvanometer shows peak heights that correspond to individual m/e (mass-to-charge) ratios for the resulting mixture of ions. For the ith peak produced by a pure sample j, one can then assign a sensitivity S_{ij} (peak height per μm of Hg sample pressure). These coefficients are unique for each type of gas.

Distribution of peak heights may also be obtained for an n-component gas mixture that is to be analyzed for the partial pressures $p_1, p_2,..., p_n$ of each of its constituents. The height h_i of a certain peak is a linear combination of the products of the individual sensitivities and partial pressures:

$$\sum_{j=1}^{n} s_{ij}p_j = h_i$$

In general, more than n peaks may be available. However, if the n most distinct ones are chosen, we have $i = 1, 2,..., n$, so that the individual partial pressures are given by the solution of n simultaneous linear equations.

The sensitivities given in Table P2.3 were reported by Carnahan et al. [6] in connection with the analysis of a hydrogen gas mixture. Write a program that will accept values for the sensitivities, $S_{11},..., S_{nn}$, and the peak heights, $h_1,..., h_n$, and compute values for the individual partial pressures, $p_1,..., p_n$.

A particular gas mixture produced the following peak heights: $h_1 = 17.1$, $h_2 = 65.1$, $h_3 = 186.0$, $h_4 = 82.7$, $h_5 = 84.2$, $h_6 = 63.7$, and $h_7 = 119.7$. The measured total pressure of the mixture was 38.78 μm of Hg, which can be compared with the sum of the computed partial pressures.

2.4. A three-component separation system is shown in Fig. P2.4. Feed flow rate and molar composition of feed and outlet streams are shown in the figure. Set up the appropriate

Table P2.3 Sensitivities of a gas mixture.

Peak index, i	m/e	Component index, j						
		1 Hydrogen	2 Methane	3 Ethylene	4 Ethane	5 Propylene	6 Propane	7 n-Pentane
1	2	16.87	0.165	0.2019	0.317	0.234	0.182	0.110
2	16	0	27.70	0.862	0.062	0.073	0.131	0.120
3	26	0	0	22.35	13.05	4.42	6.001	3.043
4	30	0	0	0	11.28	0	1.11	0.371
5	40	0	0	0	0	9.85	1.1684	2.108
6	44	0	0	0	0	0	15.98	2.107
7	72	0	0	0	0	0	0	4.670

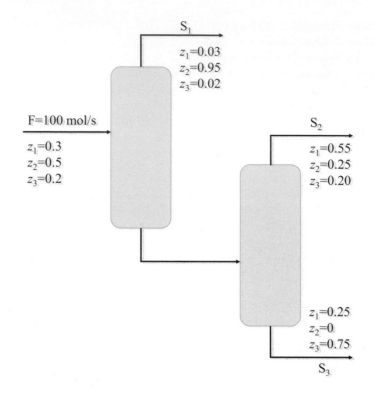

FIG. P2.4

Three-component separation system.

mass balance for the total mass flow rate and mass flow rate of species. Determine the output flow rates by solving the resulting set of linear algebraic equations.

2.5. In the study of chemical reactions Aris [7] developed a technique of writing simultaneous chemical reactions in the form of linear algebraic equations. For example, the following two simultaneous chemical equations:

$$C_2H_6 \rightleftarrows C_2H_4 + H_2$$
$$2C_2H_6 \rightleftarrows C_2H_4 + 2CH_4$$

can be rearranged in the form:

$$C_2H_4 + H_2 - C_2H_6 = 0$$
$$C_2H_4 + 2CH_4 - 2C_2H_6 = 0$$

If we identify A_1 with C_2H_4, A_2 with H_2, A_3 with CH_4, and A_4 with C_2H_6, the set of equations become:

$$A_1 + A_2 - A_4 = 0$$
$$A_1 + 2A_3 - 2A_4 = 0$$

This can be generalized to a system of R reactions between S chemical species by the set of equations represented by:

$$\sum_{j=1}^{S} \alpha_{ij} A_j = 0 \quad i = 1, 2, \ldots, R$$

where α_{ij} are the stoichiometric coefficients of each species A_j in each reaction i.

Aris demonstrated that the number of independent chemical reactions in a set of R reactions is equal to the rank of the matrix of stoichiometric coefficients α_{ij}. Using Aris' method, and the techniques developed in this chapter, determine the number of independent chemical reactions in the following reaction system:

$$
\begin{array}{rcl}
4NH_3 + 5O_2 & \rightleftarrows & 4NO + 6H_2O \\
4NH_3 + 3O_2 & \rightleftarrows & 2N_2 + 6H_2O \\
4NH_3 + 6NO & \rightleftarrows & 5N_2 + 6H_2O \\
2NO + O_2 & \rightleftarrows & 2NO_2 \\
2NO & \rightleftarrows & N_2 + O_2 \\
N_2 + 2O_2 & \rightleftarrows & 2NO_2
\end{array}
$$

2.6. The multistage distillation tower shown in Fig. P2.6 is equipped with a total condenser and a partial boiler. This tower will be used for the separation of a multicomponent mixture. Assume that for this particular mixture, the tower contains the equivalent of 10 equilibrium stages including the reboiler; that is, $N = 10$ and $j = 11$.

The feed to the column has a flow rate $F = 1000$ mol/h. It is a saturated liquid and it enters the column on equilibrium stage 5 ($j = 6$). It contains five components ($n = 5$) whose mole fractions are:

$$z_1 = 0.06 \qquad z_2 = 0.17 \qquad z_3 = 0.22 \qquad z_4 = 0.20 \qquad z_5 = 0.35$$

It is desired to recover a distillate product at a rate of 500 mol/h.

Develop all the material balances for component i for all 10 equilibrium stages and the condenser. For this problem, make the following assumptions:

The external reflux ratio is:

$$\frac{L_1}{D} = 2.5$$

Constant molal overflow occurs in each section of the tower.

The initial guesses of the temperatures corresponding to the equilibrium stages are $T_2 = 140°F$, $T_3 = 150°F$, $T_4 = 160°F$, $T_5 = 170°F$, $T_6 = 180°F$, $T_7 = 190°F$, $T_8 = 200°F$, $T_9 = 210°F$, $T_{10} = 220°F$, and $T_{11} = 230°F$.

The equilibrium constant K_{ji} can be approximated by the following equation:

$$K_{ji} = \alpha_i + \beta_i T_j + \gamma_i T_j^2$$

where the temperatures are in degrees Fahrenheit and the coefficients for each individual component are listed in Table P2.6 [8].

Solve the resulting set of equations to determine the following:

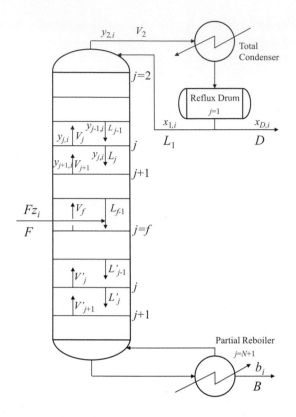

FIG. P2.6

Multistage distillation tower.

Table P2.6 Coefficients of equilibrium constant polynomial.			
Component i	α_i	β_i	γ_i
1	0.70	0.30×10^{-2}	0.65×10^{-4}
2	2.21	1.95×10^{-2}	0.90×10^{-4}
3	1.50	-1.60×10^{-2}	0.80×10^{-4}
4	0.86	-0.97×10^{-2}	0.46×10^{-4}
5	0.71	-0.87×10^{-2}	0.42×10^{-4}

(a) The molar flow rates of all vapor and liquid streams in the tower.

(b) The mole fraction of each component in the vapor and liquid streams.

Note that the mole fractions in each stage do not add up to unity, because the above solution is only a single step in the solution of the multicomponent distillation problem. Assumptions 2 and 3 are only initial guesses which must be subsequently corrected from energy balances and bubble point calculations.

2.7. The system of highly coupled chemical reactions shown in Fig. P2.7 takes place in a batch reactor. The conditions of temperature and pressure in the reactor are such that the kinetic rate constants attain the following values:

$$k_{21} = 0.2 \quad k_{31} = 0.1 \quad k_{32} = 0.1 \quad k_{34} = 0.1 \quad k_{54} = 0.05 \quad k_{64} = 0.2 \quad k_{65} = 0.1$$

$$k_{12} = 0.1 \quad k_{13} = 0.05 \quad k_{23} = 0.05 \quad k_{43} = 0.2 \quad k_{45} = 0.1 \quad k_{46} = 0.2 \quad k_{56} = 0.1$$

If the chemical reaction starts with the following initial concentrations:

$$A_0 = 1.0 \text{ mol}/L \quad D_0 = 0 \quad B_0 = 0 \quad E_0 = 1.0 \text{ mol}/L \quad C_0 = 0 \quad F_0 = 0$$

Calculate the steady-state concentration of all components. Assume that all reactions are of first order.

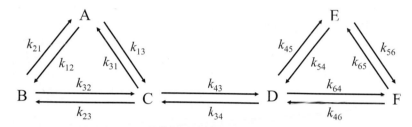

FIG. P2.7

System of coupled chemical reactions.

2.8. Calculate the steady-state concentration of all components for each of the following reaction schemes which take place in a batch reactor. Mechanisms and rate constants are adopted from Levenspiel [9].

(a) Oxidation of sodium sulfide to sodium thiosulfate:

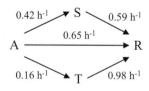

$$A = Na_2S, \ R = Na_2S_2O_3, \ S = Na_2SO_3, \ T = Na_2S_2, \ C_{A0} = 0.185 \text{ mol}/L$$

(b) Progressive chlorination of o- and p-dichlorobenzene (all reactions are + Cl_2):

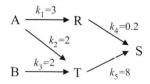

A = *o*-Dichlorobenzene, B = *p*-Dichlorobenzene, R = 1, 2, 3-Trichlorobenzene

T = 1, 2, 4-Trichlorobenzene, S = 1, 2, 3, 4-Tetrachlorobenzene, $C_{A0} = 2$, $C_{B0} = 1$

(c) Oxidation of naphthalene to phthalic anhydride:

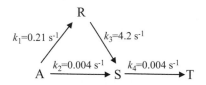

A = naphthalene, R = naphthoquinone, S = phthalic anhydride, T = oxidation products, $C_{A0} = 1$

2.9. A countercurrent stage extraction process is depicted in Fig. P2.9. In such systems a stream containing a weight fraction y_{in} of a chemical enters at a mass flow rate of F_a. Simultaneously, a solvent carrying a weight fraction x_{in} of the same chemical enters from the other side at a flow rate of F_b. Thus, for stage i, a mass balance can be represented as:

$$F_a y_{i-1} + F_b x_{i+1} = F_a y_i + F_b x_i \tag{1}$$

At each stage, an equilibrium is assumed to be established between y_i and x_i as in:

$$K = \frac{x_i}{y_i} \tag{2}$$

where K is called the distribution coefficient. Eq. (2) can be solved for x_i and substituted into Eq. (1) to yield:

$$y_{i-1} - \left(1 + K\frac{F_b}{F_b}\right)y_i + \left(K\frac{F_b}{F_a}\right)y_{i+1} = 0 \tag{3}$$

If $F_a = 500$ kg/h, $y_{in} = 0.12$, $F_b = 800$ kg/h, $x_{in} = 0$, and $K = 0.6$, determine the values of y_{out} and x_{out} if a five-stage extractor is used. Note that Eq. (3) must be modified to account for the inflow weight fractions when applied to the first and last stages.

2.10. A pump delivers a unit flow (Q_1) of a liquid into the piping network illustrated in Fig. P2.10 [9]. Every pipe section has the same length and diameter. The mass and

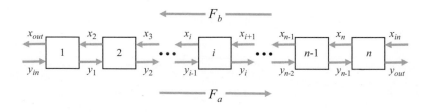

FIG. P2.9

Multistage extraction process.

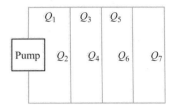

FIG. P2.10

Piping network.

mechanical energy balance can be simplified to obtain the flows in every segment. Solve the following system of equations to obtain the flow in every stream:

$$Q_3 + 2Q_4 - 2Q_2 = 0 \qquad Q_1 = Q_2 + Q_3 \qquad Q_5 + 2Q_6 - 2Q_4 = 0$$

$$Q_3 = Q_4 + Q_5 \qquad 3Q_7 - 2Q_6 = 0 \qquad Q_5 = Q_6 + Q_7$$

2.11. The Lower Colorado River consists of a series of four reservoirs as shown in Fig. P2.11.

Mass balances for chlorides can be written for each reservoir, and the following set of simultaneous linear algebraic equations are obtained [10]:

$$\begin{bmatrix} 13.422 & 0 & 0 & 0 \\ -13.422 & 12.252 & 0 & 0 \\ 0 & -12.252 & 12.377 & 0 \\ 0 & 0 & -12.377 & 11.797 \end{bmatrix} \begin{bmatrix} c_1 \\ c_2 \\ c_3 \\ c_4 \end{bmatrix} = \begin{bmatrix} 750.5 \\ 300 \\ 102 \\ 30 \end{bmatrix}$$

where the right-hand-side vector consists of the loadings of chloride to each of the four lakes and c_1, c_2, c_3, and c_4 are the resulting chloride concentrations for Lakes Powell, Mead, Mohave, and Havasu, respectively.

(a) Use the matrix inverse to solve for the concentrations in each of the four lakes.

(b) How much must the loading to Lake Powell be reduced for the chloride concentration of Lake Havasu to be 75?

(c) Using the column-sum norm, compute the condition number and how many suspect digits would be generated by solving this system.

2.12. A linear mathematical model which has three independent variables, X_1, X_2, and X_3, may be written as:

$$Y = b_1X_1 + b_2X_2 + b_3X_3$$

where b_1, b_2, and b_3 are parameter constants to be determined from experimental observations. It can be shown that the vector of parameters b may be calculated from:

$$b = (X'X)^{-1}X'Y$$

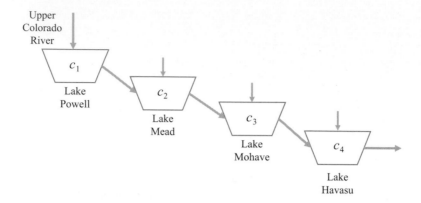

FIG. P2.11

Reservoirs in the Lower Colorado River.

Table P2.12 Experimental observations.			
X_1	X_2	X_3	Y
1	0.2	5.0	1.0
2	0.6	4.1	5.0
3	0.7	3.0	7.0
4	1.0	2.0	10.0
5	1.5	1.2	12.5
6	2.0	0.5	15.0

where X is the matrix that contains the vectors of independent variable observations, X_1, X_2, and X_3, as columns:

$$X = [X_1 \ X_2 \ X_3]$$

and Y is the vector of dependent variable observations (see Chapter 8).

Using the experimental observations shown in Table P2.12 determine the values of the parameters b_1, b_2, and b_3 for this linear model.

REFERENCES

[1] Ralston, A.; Rabinowitz, P.; First, A. *Course in Numerical Analysis*, 2nd ed.; Dover Publications: Mineola, NY, 2001.

[2] Constantinides, A.; Mostoufi, N. *Numerical Methods for Chemical Engineers with MATLAB Applications;* Prentice Hall: Upper Saddle River, NJ, 1999.

[3] Faddeev, D. K.; Faddeeva, U. N. *Computational Methods in Linear Algebra;* Freeman: San Francisco, 1963.

[4]   Johnson, L. W.; Riess, R. D. *Numerical Analysis*, 2nd ed.; Addison-Wesley, Reading: MA, 1982.

[5]   Givens, M. Computation of Plane Unitary Rotation Transforming a General Matrix to Triangular Form. *J. Soc. Ind. Appl. Math.* **1958,** *6*, 26−50.

[6]   Carnahan, B.; Luther, H. A.; Wilkes, J. O. *Applied Numerical Methods;* Wiley: New York, NY, 1969.

[7]   Aris, R. *Introduction to the Analysis of Chemical Reactors;* Prentice-Hall: Englewood Cliffs, NJ, 1965.

[8]   Chang, H. Y.; Over, I. E. *Selected Numerical Methods and Computer Programs for Chemical Engineers;* Sterling Swift: Manchaca, TX, 1981.

[9]   Levenspiel, O. *Chemical Reaction Engineering*, 3rd ed.; John Wiley & Sons: New York, 1999.

[10] Chapra, S. *Applied Numerical Methods with MATLAB for Engineers and Scientists*, 2nd ed.; McGraw-Hill: New York, 2006.

FINITE DIFFERENCE METHODS AND INTERPOLATION

3

CHAPTER OUTLINE

MOTIVATION

Differentiation, integration, and the solution of ordinary and partial differential equations are based on the concept of finite differences. Therefore, the purpose of this chapter is to develop the systematic terminology used in the calculus of finite differences and to derive the relationships between finite differences and differential operators, which are needed in the numerical solution of ordinary and partial differential equations. The calculus of finite differences enables the user to take a differential equation and integrate it numerically by calculating the values of the function at a discrete (finite) number of points, or, conversely, if a set of finite values is available, such as experimental data, these may be differentiated, or integrated, using the calculus of finite differences. Another very useful application of the calculus of finite differences is in the derivation of interpolation/extrapolation formulas, the so-called

Applied Numerical Methods for Chemical Engineers. DOI: https://doi.org/10.1016/B978-0-12-822961-3.00003-0

interpolating polynomials, which can be used to represent experimental data when the actual functionality of these data is not known. A very common example of the application of interpolation is in the extraction of physical properties of water from the steam tables. Interpolating polynomials are also used to estimate the numerical derivative and integral of the tabulated data.

3.1 INTRODUCTION

The most commonly encountered mathematical models in engineering and science are in the form of differential equations. The dynamics of physical systems that have one independent variable can be modeled by ordinary differential equations, while systems with two or more independent variables require the use of partial differential equations. Several types of ordinary differential equations, and a few partial differential equations, render themselves to analytical (closed form) solutions. These methods have been developed thoroughly in differential calculus. However, the great majority of differential equations, especially the nonlinear ones and those which involve large sets of simultaneous differential equations, do not have analytical solutions but require the application of numerical techniques for their solution.

Several numerical methods for differentiation, integration, and the solution of ordinary and partial differential equations are discussed in Chapters 4 to 7 of this book. These methods are based on the concept of *finite differences*. Therefore, the purpose of this chapter is to develop the systematic terminology used in the *calculus of finite differences* and to derive the relationships between finite differences and differential operators, which are needed in the numerical solution of ordinary and partial differential equations.

The calculus of finite differences may be characterized as a "two-way street" that enables the user to take a differential equation and integrate it numerically by calculating the values of the function at a discrete (finite) number of points. Or, conversely, if a set of finite values is available, such as experimental data, these may be differentiated or integrated using the calculus of finite differences. It should be pointed out, however, that numerical differentiation is inherently less accurate than numerical integration.

Another very useful application of the calculus of finite differences is in the derivation of interpolation/extrapolation formulas, the so-called *interpolating polynomials*, which can be used to represent experimental data when the actual functionality of these data is not known. A very common example of the application of interpolation is in the extraction of physical properties of water from the steam tables. Interpolating polynomials are also used to estimate the numerical derivative and integral of the tabulated data (see Chapter 4). The discussion of several interpolating polynomials is given in Sections 3.6 to 3.8.

3.2 SYMBOLIC OPERATORS

In differential calculus, the definition of the derivative is given as:

$$\frac{df(x)}{dx}\bigg|_{x_0} = f'(x_0) = \lim_{x \to x_0} \frac{f(x) - f(x_0)}{x - x_0} \tag{3.1}$$

In the calculus of finite differences, the value of $x - x_0$ does not approach zero but remains a finite quantity. If we represent this quantity by h:

$$h = x - x_0 \tag{3.2}$$

then the derivative may be *approximated* by:

$$f'(x_0) \simeq \frac{f(x_0 + h) - f(x_0)}{h} \tag{3.3}$$

Under certain circumstances, there is a point, ξ, in the interval (a, b) for which the derivative can be calculated *exactly* from Eq. (3.3). This is confirmed by the mean-value theorem of differential calculus:

Mean-value theorem: Let $f(x)$ be continuous in the range $a \leq x \leq b$ and differentiable in the range $a < x < b$; then there exists at least one ξ, $a < \xi < b$, for which:

$$f'(\xi) = \frac{f(b) - f(a)}{b - a} \tag{3.4}$$

This theorem forms the basis for both differential calculus and finite difference calculus.

A function $f(x)$, which is continuous and differentiable in the interval $[x_0, x]$, can be represented by a Taylor series:

$$f(x) = f(x_0) + (x - x_0)f'(x_0) + \frac{(x - x_0)^2 f''(x_0)}{2!} + \frac{(x - x_0)^3 f'''(x_0)}{3!} + \cdots + \frac{(x - x_0)^n f^{(n)}(x_0)}{n!} + R_n(x) \tag{3.5}$$

where $R_n(x)$ is called the remainder. This term lumps together the remaining terms in the infinite series from $(n + 1)$ to infinity; it, therefore, represents the *truncation* error, when the function is evaluated using the terms up to, and including, the nth-order term of the infinite series.

The mean-value theorem can be used to show that there exists a point ξ in the interval (x_0, x) so that the remainder term given by:

$$R_n(x) = \frac{(x - x_0)^{n+1} f^{(n+1)}(\xi)}{(n+1)!} \tag{3.6}$$

The value of ξ is an unknown function of x; therefore, it is impossible to evaluate the remainder, or truncation error, term exactly. The remainder is a term of order $(n + 1)$, as it is a function of $(x - x_0)^{n+1}$ and of the $(n + 1)$st derivative. For this reason, in our discussion of truncation errors, we will always specify the order of the remainder term, and will usually abbreviate it using the notation $O(h^{n+1})$.

The calculus of finite differences is used in conjunction with a series of discrete values, which can be either experimental data, such as:

$$y_{i-3} \quad y_{i-2} \quad y_{i-1} \quad y_i \quad y_{i+1} \quad y_{i+2} \quad y_{i+3}$$

or discrete values of a continuous function $y(x)$:

$$y(x - 3h) \quad y(x - 2h) \quad y(x - h) \quad y(x) \quad y(x + h) \quad y(x + 2h) \quad y(x + 3h)$$

or, equivalently, values of a function $f(x)$:

$$f(x-3h) \quad f(x-2h) \quad f(x-h) \quad f(x) \quad f(x+h) \quad f(x+2h) \quad f(x+3h)$$

In all these cases, the values of the dependent variable, y or f, are those corresponding to *equally spaced* values of the independent variable x. This concept is demonstrated in Fig. 3.1 for a smooth function $y(x)$.

A set of *linear symbolic operators* drawn from differential calculus and finite difference calculus will be defined in conjunction with the foregoing series of discrete values. These definitions will then be used to derive the interrelationships between the operators. The linear symbolic operators are:

D = differential operator
I = integral operator
E = shift operator
Δ = forward difference operator
∇ = backward difference operator
δ = central difference operator
μ = averager operator

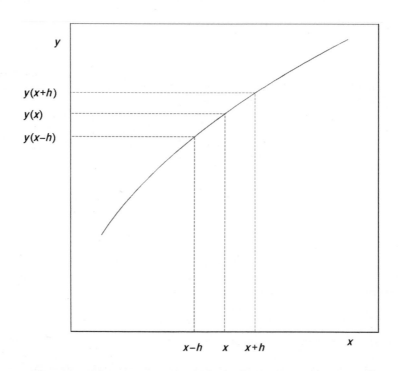

FIG. 3.1

Values of function $y(x)$ at equally spaced points of the independent variable x.

All these operators may be treated as algebraic variables because they satisfy the distributive, commutative, and associative laws of algebra.

The first two operators are well known from differential calculus. The *differential operator D* has the following effect when applied to the function $y(x)$:

$$Dy(x) = \frac{dy(x)}{dx} = y'(x) \tag{3.7}$$

while the integral operator is:

$$Iy(x) = \int_x^{x+h} y(x)dx \tag{3.8}$$

The integral operator is equivalent to the inverse of the differential operator:

$$I = D^{-1} \tag{3.9}$$

The *shift operator* causes the function to shift to the next successive value of the independent variable:

$$Ey(x) = y(x+h) \tag{3.10}$$

The inverse of the shift operator, E^{-1} causes the function to shift in the negative direction of the independent variable:

$$E^{-1}y(x) = y(x-h) \tag{3.11}$$

Higher powers of the shift operator are defined as:

$$E^n y(x) = y(x+nh) \tag{3.12}$$

The shift operator can be expressed in terms of the differential operator by expanding the function $y(x+h)$ into a Taylor series about x:

$$y(x+h) = y(x) + \frac{h}{1!}y'(x) + \frac{h^2}{2!}y''(x) + \frac{h^3}{3!}y'''(x) + \cdots \tag{3.13}$$

Using the differential operator D to indicate the derivatives of y, we obtain:

$$y(x+h) = y(x) + \frac{h}{1!}Dy(x) + \frac{h^2}{2!}D^2 y(x) + \frac{h^3}{3!}D^3 y(x) + \cdots \tag{3.14}$$

Factoring out the term $y(x)$ from the right-hand side of Eq. (3.14),

$$y(x+h) = \left(1 + \frac{h}{1!}D + \frac{h^2}{2!}D^2 + \frac{h^3}{3!}D^3 + \cdots\right)y(x) \tag{3.15}$$

The terms in the parentheses are equivalent to the series expansion:

$$e^{hD} = 1 + \frac{h}{1!}D + \frac{h^2}{2!}D^2 + \frac{h^3}{3!}D^3 + \cdots \tag{3.16}$$

Therefore, Eq. (3.15) can be written as:

$$y(x+h) = e^{hD}y(x) \tag{3.17}$$

Comparing Eq. (3.10) with (3.17), we conclude that the shift operator can be expressed in terms of the differential operator by the relation:

$$E = e^{hD} \tag{3.18}$$

Similarly, the inverse of the shift operator can be related to the differential operator by expanding the function $y(x - h)$ into a Taylor series about x:

$$y(x - h) = y(x) - \frac{h}{1!}y'(x) + \frac{h^2}{2!}y''(x) - \frac{h^3}{3!}y'''(x) + \cdots \tag{3.19}$$

Replacing the derivatives with the differential operators and rearranging, we obtain:

$$y(x - h) = \left(1 - \frac{h}{1!}D + \frac{h^2}{2!}D^2 - \frac{h^3}{3!}D^3 + \cdots\right)y(x) \tag{3.20}$$

The terms in the parentheses are equivalent to the series expansion:

$$e^{-hD} = 1 - \frac{h}{1!}D + \frac{h^2}{2!}D^2 - \frac{h^3}{3!}D^3 + \cdots \tag{3.21}$$

Therefore, Eq. (3.19) can be written as:

$$y(x - h) = e^{-hD}y(x) \tag{3.22}$$

It follows from a comparison of Eq. (3.11) with Eq. (3.22) that:

$$E^{-1} = e^{-hD} \tag{3.23}$$

With these introductory concepts in mind, let us proceed to develop the backward, forward, and central difference operators and the relationships between these and the differential operators.

3.3 BACKWARD FINITE DIFFERENCES

Consider the set of values:

$$y_{i-3} \quad y_{i-2} \quad y_{i-1} \quad y_i \quad y_{i+1} \quad y_{i+2} \quad y_{i+3}$$

or the equivalent set:

$$y(x - 3h) \quad y(x - 2h) \quad y(x - h) \quad y(x) \quad y(x + h) \quad y(x + 2h) \quad y(x + 3h)$$

The *first backward difference* of y at i (or x) is defined as:

$$\nabla y_i = y_i - y_{i-1}$$

or

$$\nabla y(x) = y(x) - y(x - h) \tag{3.24}$$

The *second backward difference* of y at i (or x) is defined as:

$$\nabla^2 y_i = \nabla(\nabla y_i) = \nabla(y_i - y_{i-1}) = \nabla y_i - \nabla y_{i-1}$$
$$= (y_i - y_{i-1}) - (y_{i-1} - y_{i-2})$$
$$\nabla^2 y_i = y_i - 2y_{i-1} + y_{i-2} \tag{3.25}$$

or

$$\nabla^2 y(x) = y(x) - 2y(x-h) + y(x-2h)$$

The *third backward difference* of y at i is defined as:

$$\nabla^3 y_i = \nabla(\nabla^2 y_i) = \nabla(y_i - 2y_{i-1} + y_{i-2})$$
$$= \nabla y_i - 2\nabla y_{i-1} + \nabla y_{i-2}$$
$$= (y_i - y_{i-1}) - 2(y_{i-1} - y_{i-2}) + (y_{i-2} - y_{i-3}) \tag{3.26}$$
$$\nabla^3 y_i = y_i - 3y_{i-1} + 3y_{i-2} - y_{i-3}$$

Higher-order backward differences are similarly derived:

$$\nabla^4 y_i = y_i - 4y_{i-1} + 6y_{i-2} - 4y_{i-3} + y_{i-4} \tag{3.27}$$

$$\nabla^5 y_i = y_i - 5y_{i-1} + 10y_{i-2} - 10y_{i-3} + 5y_{i-4} - y_{i-5} \tag{3.28}$$

The coefficients of the terms in each of the foregoing finite differences correspond to those of the binomial expansion $(a - b)^n$, where n is the order of the finite difference. Therefore, the general formula of the nth-order backward finite difference can be expressed as:

$$\nabla^n y_l = \sum_{m=0}^{n} (-1)^m \frac{n!}{(n-m)!m!} y_{i-m} \tag{3.29}$$

It should also be noted that the sum of the coefficients of the binomial expansion is always equal to zero. This can be used as a check to ensure that higher-order differences have been expanded correctly.

The relationship between backward difference operators and differential operators can now be established. Combine Eqs. (3.22) and (3.24) to obtain:

$$\nabla y(x) = y(x) - y(x-h) = y(x) - e^{-hD}y(x) = (1 - e^{-hD})y(x) \tag{3.30}$$

which shows that the backward difference operator is given by:

$$\nabla = 1 - e^{-hD} \tag{3.31}$$

Using the infinite series expression of e^{-hD} (Eq. 3.21), Eq. (3.31) becomes:

$$\nabla = hD - \frac{h^2 D^2}{2} + \frac{h^3 D^3}{6} - \cdots \tag{3.32}$$

The higher-order backward difference operator, ∇^2, ∇^3,..., can be obtained by raising the first backward difference operator to higher powers:

$$\nabla^2 = (1 - e^{-hD})^2 = (1 - 2e^{-hD} + e^{-2hD}) \tag{3.33}$$

$$\nabla^3 = (1 - e^{-hD})^3 = (1 - 3e^{-hD} + 3e^{-2hD} - e^{-3hD}) \tag{3.34}$$

$$\vdots$$

$$\nabla^n = (1 - e^{-hD})^n \tag{3.35}$$

Expansion of the exponential terms and rearrangement yields the following equations for the second and third backward difference operators:

$$\nabla^2 = h^2 D^2 - h^3 D^3 + \frac{7}{12} h^4 D^4 - \cdots \tag{3.36}$$

$$\nabla^3 = h^3 D^3 - \frac{3}{2} h^4 D^4 + \frac{5}{4} h^5 D^5 - \cdots \tag{3.37}$$

Eqs. (3.32), (3.36), and (3.37) express the backward difference operators in terms of infinite series of differential operators. To complete the set of relationships, equations that express the differential operators in terms of backward difference operators will also be derived. To do so, first rearrange Eq. (3.31) to solve for e^{-hD}:

$$e^{-hD} = 1 - \nabla \tag{3.38}$$

Take the natural logarithm of both sides of this equation:

$$\ln e^{-hD} = -hD = \ln(1 - \nabla) \tag{3.39}$$

Use the infinite series expansion:

$$\ln(1 - \nabla) = -\nabla - \frac{\nabla^2}{2} - \frac{\nabla^3}{3} - \frac{\nabla^4}{4} - \frac{\nabla^5}{5} - \cdots \tag{3.40}$$

Combine Eq. (3.39) with Eq. (3.40) to obtain:

$$hD = \nabla + \frac{\nabla^2}{2} + \frac{\nabla^3}{3} + \frac{\nabla^4}{4} + \frac{\nabla^5}{5} + \cdots \tag{3.41}$$

The higher-order differential operators can be obtained by simply raising both sides of Eq. (3.41) to higher powers:

$$h^2 D^2 = \nabla^2 + \nabla^3 + \frac{11}{12} \nabla^4 + \frac{5}{6} \nabla^5 + \cdots \tag{3.42}$$

$$h^3 D^3 = \nabla^3 + \frac{3}{2} \nabla^4 + \frac{7}{4} \nabla^5 + \cdots \tag{3.43}$$

$$\vdots$$

$$h^n D^n = \left(\nabla + \frac{\nabla^2}{2} + \frac{\nabla^3}{3} + \frac{\nabla^4}{4} + \frac{\nabla^5}{5} + \cdots \right)^n \tag{3.44}$$

The complete set of relationships between backward difference operators and differential operators is summarized in Table 3.1.

EXAMPLE 3.1 : TABLE OF BACKWARD DIFFERENCES

Form the table of backward differences for the data points given in Table E3.1.

Method of solution

The steps for generating the table of backward differences in Excel are shown in Fig. E3.1.

Table 3.1 Backward finite differences.

Backward difference operators	Differential operators
$\nabla = hD - \frac{h^2D^2}{2} + \frac{h^3D^3}{6} - \cdots$	$hD = \nabla + \frac{\nabla^2}{2} + \frac{\nabla^3}{3} + \frac{\nabla^4}{4} + \cdots$
$\nabla^2 = h^2D^2 - h^3D^3 + \frac{7}{12}h^4D^4 - \cdots$	$h^2D^2 = \nabla^2 + \nabla^3 + \frac{11}{12}\nabla^4 + \frac{5}{6}\nabla^5 + \cdots$
$\nabla^3 = h^3D^3 - \frac{3}{2}h^4D^4 + \frac{5}{4}h^5D^5 - \cdots$	$h^3D^3 = \nabla^3 + \frac{3}{2}\nabla^4 + \frac{7}{4}\nabla^5 + \cdots$
$\nabla^n = \left(1 - e^{-hD}\right)^n$	$h^nD^n = \left(\nabla + \frac{\nabla^2}{2} + \frac{\nabla^3}{3} + \frac{\nabla^4}{4} + \frac{\nabla^5}{5} + \cdots\right)^n$

Table E3.1 Numerical data.

x	20	25	30	35	40	45	50
y	0.6071	0.9452	1.3523	1.8210	2.3394	2.8898	3.4474

3.4 FORWARD FINITE DIFFERENCES

The development of forward finite differences follows a course parallel to that used in the development of backward differences.

Consider the set of values:

$$y_{i-3} \quad y_{i-2} \quad y_{i-1} \quad y_i \quad y_{i+1} \quad y_{i+2} \quad y_{i+3}$$

or the equivalent set:

$$y(x-3h) \quad y(x-2h) \quad y(x-h) \quad y(x) \quad y(x+h) \quad y(x+2h) \quad y(x+3h)$$

The *first forward difference* of y at i (or x) is defined as:

$$\Delta y_i = y_{i+1} - y_i$$
$$\text{or}$$
$$\Delta y(x) = y(x+h) - y(x) \tag{3.45}$$

The *second forward difference* of y at i (or x) is defined as:

$$\Delta^2 y_i = \Delta(\Delta y_i) = \Delta(y_{i+1} - y_i) = \Delta y_{i+1} - \Delta y_i$$
$$= (y_{i+2} - y_{i+1}) - (y_{i+1} - y_i)$$
$$\Delta^2 y_i = y_{i+2} - 2y_{i+1} + y_i \tag{3.46}$$
$$\text{or}$$
$$\Delta^2 y(x) = y(x+2h) - 2y(x+h) + y(x)$$

The *third forward difference* of y at i is defined as:

$$\Delta^3 y_i = \Delta\left(\Delta^2 y_i\right) = \Delta(y_{i+2} - 2y_{i+1} + y_i)$$
$$= \Delta y_{i+2} - 2\Delta y_{i+1} + \Delta y_i$$
$$= (y_{i+3} - y_{i+2}) - 2(y_{i+2} - y_{i+1}) + (y_{i+1} - y_i)$$
$$\Delta^3 y_i = y_{i+3} - 3y_{i+2} + 3y_{i+1} - y_i \tag{3.47}$$

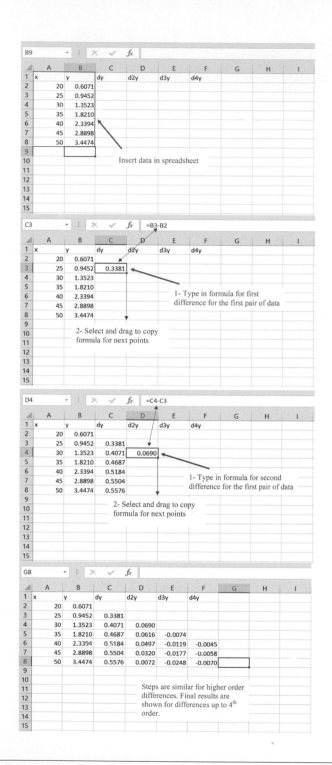

FIG. E3.1

Generating table of backward differences in Excel.

Higher-order forward differences are similarly derived:

$$\Delta^4 y_i = y_{i+4} - 4y_{i+3} + 6y_{i+2} - 4y_{i+1} + y_i \tag{3.48}$$

$$\Delta^5 y_i = y_{i+5} - 5y_{i+4} + 10y_{i+3} - 10y_{i+2} + 5y_{i+1} - y_i \tag{3.49}$$

In similarity to the backward finite differences, the forward finite differences also have coefficients that correspond to those of the binomial expansion $(a - b)^n$. Therefore, the general formula of the nth-order forward finite difference can be expressed as:

$$\Delta^n y_i = \sum_{m=0}^{n} (-1)^m \frac{n!}{(n-m)!m!} y_{i+n-m} \tag{3.50}$$

In MATLAB®, the function *diff(y)* returns forward finite differences of y. Values of nth-order forward finite difference may be obtained from *diff(y, n)*.

The relationship between forward difference operators and differential operators can now be developed. Combine Eqs. (3.45) and (3.17) to obtain:

$$\Delta y(x) = y(x + h) - y(x) = e^{hD} y(x) - y(x) = \left(e^{hD} - 1\right) y(x) \tag{3.51}$$

which shows that the forward difference operator is given by:

$$\Delta = e^{hD} - 1 \tag{3.52}$$

Using the infinite series expression of e^{hD} (Eq. 3.16), Eq. (3.52) becomes:

$$\Delta = hD + \frac{h^2 D^2}{2} + \frac{h^3 D^3}{6} + \cdots \tag{3.53}$$

The higher-order forward difference operator, Δ^2, Δ^3,..., can be obtained by raising the first forward difference operator to higher powers:

$$\Delta^2 = \left(e^{hD} - 1\right)^2 = \left(e^{2hD} - 2e^{hD} + 1\right) \tag{3.54}$$

$$\Delta^3 = \left(e^{hD} - 1\right)^3 = \left(e^{3hD} - 3e^{2hD} + 3e^{hD} - 1\right) \tag{3.55}$$

$$\vdots$$

$$\Delta^n = \left(e^{hD} - 1\right)^n \tag{3.56}$$

Expansion of the exponential terms and rearrangement yields the following equations for the second and third forward difference operators:

$$\Delta^2 = h^2 D^2 + h^3 D^3 + \frac{7}{12} h^4 D^4 + \cdots \tag{3.57}$$

$$\Delta^3 = h^3 D^3 + \frac{3}{2} h^4 D^4 + \frac{5}{4} h^5 D^5 + \cdots \tag{3.58}$$

Eqs. (3.53), (3.57), and (3.58) express the forward difference operators in terms of infinite series of differential operators. To complete the set of relationships, equations that express the differential operators in terms of forward difference operators will also be derived. To do this, first rearrange Eq. (3.52) to solve for e^{hD}:

$$e^{hD} = 1 + \Delta \tag{3.59}$$

Take the natural logarithm of both sides of this equation:

$$\ln e^{hD} = hD = \ln(1 + \Delta) \tag{3.60}$$

Use the infinite series expansion:

$$\ln(1 + \Delta) = \Delta - \frac{\Delta^2}{2} + \frac{\Delta^3}{3} - \frac{\Delta^4}{4} + \frac{\Delta^5}{5} - \cdots \tag{3.61}$$

Combine Eq. (3.60) with Eq. (3.61) to obtain:

$$hD = \Delta - \frac{\Delta^2}{2} + \frac{\Delta^3}{3} - \frac{\Delta^4}{4} + \frac{\Delta^5}{5} - \cdots \tag{3.62}$$

The higher-order differential operators can be obtained by simply raising both sides of Eq. (3.62) to higher powers:

$$h^2 D^2 = \Delta^2 - \Delta^3 + \frac{11}{12}\Delta^4 - \frac{5}{6}\Delta^5 + \cdots \tag{3.63}$$

$$h^3 D^3 = \Delta^3 - \frac{3}{2}\Delta^4 + \frac{7}{4}\Delta^5 - \cdots \tag{3.64}$$

$$\vdots$$

$$h^n D^n = \left(\Delta - \frac{\Delta^2}{2} + \frac{\Delta^3}{3} - \frac{\Delta^4}{4} + \frac{\Delta^5}{5} - \cdots \right)^n \tag{3.65}$$

The complete set of relationships between forward difference operators and differential operators is summarized in Table 3.2.

Table 3.2 Forward finite differences.

Forward difference operators	Differential operators
$\Delta = hD + \frac{h^2 D^2}{2} + \frac{h^3 D^3}{6} + \cdots$	$hD = \Delta - \frac{\Delta^2}{2} + \frac{\Delta^3}{3} - \frac{\Delta^4}{4} + \frac{\Delta^5}{5} - \cdots$
$\Delta^2 = h^2 D^2 + h^3 D^3 + \frac{7}{12}h^4 D^4 + \cdots$	$h^2 D^2 = \Delta^2 - \Delta^3 + \frac{11}{12}\Delta^4 - \frac{5}{6}\Delta^5 + \cdots$
$\Delta^3 = h^3 D^3 + \frac{3}{2}h^4 D^4 + \frac{5}{4}h^5 D^5 + \cdots$	$h^3 D^3 = \Delta^3 - \frac{3}{2}\Delta^4 + \frac{7}{4}\Delta^5 - \cdots$
$\Delta^n = \left(e^{hD} - 1 \right)^n$	$h^n D^n = \left(\Delta - \frac{\Delta^2}{2} + \frac{\Delta^3}{3} - \frac{\Delta^4}{4} + \frac{\Delta^5}{5} - \cdots \right)^n$

EXAMPLE 3.2 TABLE OF FORWARD DIFFERENCES

Form the table of forward differences for the data points given in Table E3.1 (Example 3.1).

Method of solution

The steps for generating the table of forward differences in Excel are shown in Fig. E3.2.

3.5 CENTRAL FINITE DIFFERENCES

As their name implies, central finite differences are *centered* at the pivot position and are evaluated using the values of the function to the right and the left of the pivot position, but located only $h/2$ distance from it.

Consider the series of values used in the previous two sections, but with the additional values at the midpoints of the intervals:

$$y_{i-2} \quad y_{i-1\frac{1}{2}} \quad y_{i-1} \quad y_{i-\frac{1}{2}} \quad y_i \quad y_{i+\frac{1}{2}} \quad y_{i+1} \quad y_{i+1\frac{1}{2}} \quad y_{i+2}$$

or the equivalent set:

$$y(x-2h) \quad y(x-1\tfrac{1}{2}h) \quad y(x-h) \quad y(x-\tfrac{1}{2}h) \quad y(x) \quad y(x+\tfrac{1}{2}h) \quad y(x+h) \quad y(x+1\tfrac{1}{2}h) \quad y(x+2h)$$

The *first central difference* of y at i (or x) is defined as:

$$\delta y_i = y_{i+\frac{1}{2}} - y_{i-\frac{1}{2}}$$

or

$$\delta y(x) = y(x + \tfrac{1}{2}h) - y(x - \tfrac{1}{2}h) \tag{3.66}$$

The *second central difference* of y at i (or x) is defined as:

$$\begin{aligned}
\delta^2 y_i &= \delta(\delta y_i) = \delta\left(y_{i+\frac{1}{2}} - y_{i-\frac{1}{2}}\right) = \delta y_{i+\frac{1}{2}} - \delta y_{i-\frac{1}{2}} \\
&= (y_{i+1} - y_i) - (y_i - y_{i-1}) \\
\delta^2 y_i &= y_{i+1} - 2y_i + y_{i-1} \\
&\text{or} \\
\delta^2 y(x) &= y(x+h) - 2y(x) + y(x-h)
\end{aligned} \tag{3.67}$$

The *third central difference* of y at i is defined as:

$$\begin{aligned}
\delta^3 y_i &= \delta(\delta^2 y_i) = \delta(y_{i+1} - 2y_i + y_{i-1}) \\
&= \delta y_{i+1} - 2\delta y_i + \delta y_{i-1} \\
&= \left(y_{i+1\frac{1}{2}} - y_{i+\frac{1}{2}}\right) - 2\left(y_{i+\frac{1}{2}} - y_{i-\frac{1}{2}}\right) + \left(y_{i-\frac{1}{2}} - y_{i-1\frac{1}{2}}\right) \\
&= y_{i+1\frac{1}{2}} - 3y_{i+\frac{1}{2}} + 3y_{i-\frac{1}{2}} - y_{i-1\frac{1}{2}}
\end{aligned} \tag{3.68}$$

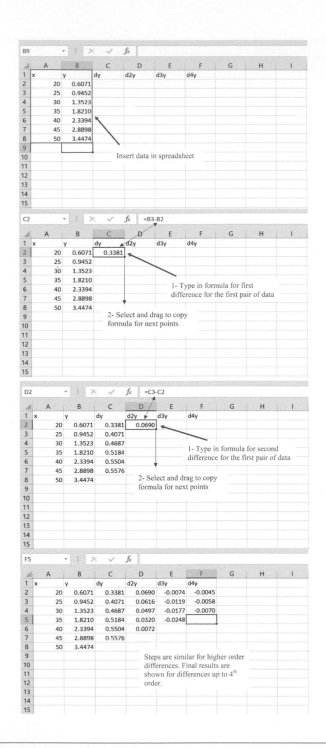

FIG. E3.2

Generating a table of forward differences in Excel.

Higher-order central differences are similarly derived:

$$\delta^4 y_i = y_{i+2} - 4y_{i+1} + 6y_i - 4y_{i-1} + y_{i-2} \tag{3.69}$$

$$\delta^5 y_i = y_{i+2\frac{1}{2}} - 5y_{i+1\frac{1}{2}} + 10y_{i+\frac{1}{2}} - 10y_{i-\frac{1}{2}} + 5y_{i-1\frac{1}{2}} - y_{i-2\frac{1}{2}} \tag{3.70}$$

Consistent with the other finite differences, the central finite differences also have coefficients that correspond to those of the binomial expansion $(a - b)^n$. Therefore, the general formula of the nth-order central finite difference can be expressed as:

$$\delta^n y_i = \sum_{m=0}^{n} (-1)^m \frac{n!}{(n-m)!m!} y_{i-m+n/2} \tag{3.71}$$

It should be noted that the *odd*-order central differences involve values of the function at the midpoint of the intervals, while the even-order central differences involve values at the full intervals. To fully use odd- and even-order central differences we need a set of values of the function y that includes twice as many points as that used in either backward or forward differences. This situation is rather uneconomical, especially in the case where these values must be obtained experimentally. To alleviate this difficulty, we make use of the *averager operator* μ, which is defined as:

$$\mu = \frac{1}{2}\left[E^{1/2} + E^{-1/2}\right] \tag{3.72}$$

The averager operator shifts its operand by a half interval to the right of the pivot and by a half interval to the left of the pivot, evaluates it at these two positions, and averages the two values.

Application of the averager on the odd central differences gives the *first averaged central difference* as follows:

$$\begin{aligned}
\mu\delta y_i &= \frac{1}{2}\left(E^{1/2}\delta y_i + E^{-1/2}\delta y_i\right) \\
&= \frac{1}{2}\left(\delta y_{i+\frac{1}{2}} + \delta y_{i-\frac{1}{2}}\right) \\
&= \frac{1}{2}[(y_{i+1} - y_i) + (y_i - y_{i-1})] \\
&= \frac{1}{2}(y_{i+1} - y_{i-1})
\end{aligned} \tag{3.73}$$

The third averaged central difference is given by:

$$\begin{aligned}
\mu\delta^3 y_i &= \frac{1}{2}\left(E^{1/2}\delta^3 y_i + E^{-1/2}\delta^3 y_i\right) \\
&= \frac{1}{2}\left(\delta^3 y_{i+\frac{1}{2}} + \delta^3 y_{i-\frac{1}{2}}\right) \\
&= \frac{1}{2}[(y_{i+2} - 3y_{i+1} + 3y_i - y_{i-1}) + (y_{i+1} - 3y_i + 3y_{i-1} - y_{i-2})] \\
&= \frac{1}{2}(y_{i+2} - 2y_{i+1} + 2y_{i-1} - y_{i-2})
\end{aligned} \tag{3.74}$$

As expected, the effect of the averager is to remove the midpoint values of the function y from the odd central differences.

It will be shown in Chapter 4 that central differences are more accurate than either backward or forward differences when used to evaluate the derivatives of functions.

The relationships between central difference operators and differential operators can now be developed. Eq. (3.73), representing the first averaged central difference, is combined with Eqs. (3.17) and (3.22) to yield:

$$
\begin{aligned}
\mu\delta y(x) &= \frac{1}{2}[y(x+h) - y(x-h)] \\
&= \frac{1}{2}\left[e^{hD}y(x) - e^{-hD}y(x)\right] \\
&= \frac{1}{2}\left(e^{hD} - e^{-hD}\right)y(x)
\end{aligned}
\tag{3.75}
$$

which shows that the first averaged central difference operator is given by:

$$
\mu\delta = \frac{1}{2}\left(e^{hD} - e^{-hD}\right) = \sinh hD
\tag{3.76}
$$

Using the infinite series expansions of e^{hD} and e^{-hD}, or equivalently the infinite series expansion of the hyperbolic sine:

$$
\sinh hD = hD + \frac{(hD)^3}{3!} + \frac{(hD)^5}{5!} + \frac{(hD)^7}{7!} + \cdots
\tag{3.77}
$$

Eq. (3.76) becomes:

$$
\mu\delta = hD + \frac{h^3 D^3}{6} + \frac{h^5 D^5}{120} + \frac{h^7 D^7}{5040} + \cdots
\tag{3.78}
$$

Similarly, using Eq. (3.67) for the second central difference, and combining it with Eqs. (3.17) and (3.22), we obtain:

$$
\begin{aligned}
\delta^2 y(x) &= y(x+h) - 2y(x) + y(x-h) \\
&= e^{hD}y(x) - 2y(x) + e^{-hD}y(x) \\
&= \left(e^{hD} - 2 + e^{-hD}\right)y(x)
\end{aligned}
\tag{3.79}
$$

which shows that the second central difference operator is equivalent to:

$$
\delta^2 = e^{hD} + e^{-hD} - 2 = 2(\cosh hD - 1) = E + E^{-1} - 2
\tag{3.80}
$$

Expanding the exponentials into their infinite series, or equivalently the infinite series expansion of the hyperbolic cosine in Eq. (3.80), we obtain:

$$
\delta^2 = h^2 D^2 + \frac{h^4 D^4}{12} + \frac{h^6 D^6}{360} + \frac{h^8 D^8}{20160} + \cdots
\tag{3.81}
$$

The higher-order averaged odd central difference operators are obtained by taking products of Eqs. (3.78) and (3.81). The higher-order even central differences are formulated by taking powers of Eq. (3.81). The third and fourth central operators, thus obtained, are as follows:

$$\mu\delta^3 = h^3D^3 + \frac{h^5D^5}{4} + \frac{h^7D^7}{40} + \cdots \tag{3.82}$$

$$\delta^4 = h^4D^4 + \frac{h^6D^6}{6} + \frac{h^8D^8}{80} + \cdots \tag{3.83}$$

To develop the inverse relationships, that is, equations for the differential operators in terms of the central difference operators, we must first derive an algebraic relationship between μ and δ. To do this, we start with Eqs. (3.72) and (3.80). Squaring both sides of Eq. (3.72), we obtain:

$$\mu^2 = \frac{1}{4}\left(E + E^{-1} + 2\right) \tag{3.84}$$

Rearranging Eq. (3.80), we get:

$$\delta^2 + 2 = E + E^{-1} \tag{3.85}$$

Combining Eqs. (3.84) and (3.85), and rearranging, we arrive at the desired relationship:

$$\mu^2 = \frac{\delta^2}{4} + 1 \tag{3.86}$$

Now, take the inverse of Eq. (3.76):

$$hD = \sinh^{-1}\mu\delta \tag{3.87}$$

The infinite series expansion of the inverse hyperbolic sine is:

$$\sinh^{-1}\mu\delta = \mu\delta - \frac{(\mu\delta)^3}{6} + \frac{3(\mu\delta)^5}{40} - \cdots \tag{3.88}$$

Therefore, Eq. (3.87) expands to:

$$hD = \mu\delta - \frac{\mu^3\delta^3}{6} + \frac{3\mu^5\delta^5}{40} - \cdots \tag{3.89}$$

The even powers of μ are eliminated from Eq. (3.89) by using Eq. (3.86) to obtain the first differential operator in terms of central difference operators:

$$hD = \mu\delta - \frac{\mu\delta^3}{6} + \frac{\mu\delta^5}{30} - \cdots \tag{3.90}$$

Higher-order differential operators are obtained by raising Eq. (3.90) to the appropriate power and using Eq. (3.86) to eliminate the even powers of μ. The second, third, and fourth differential operators obtained by this way are as follows:

$$h^2D^2 = \delta^2 - \frac{\delta^4}{12} + \frac{\delta^6}{90} - \cdots \tag{3.91}$$

$$h^3D^3 = \mu\delta^3 - \frac{\mu\delta^5}{4} + \frac{7\mu\delta^7}{120} - \cdots \tag{3.92}$$

Table 3.3 Central finite differences.

Central difference operators	Differential operators
$\mu\delta = hD + \frac{h^3 D^3}{6} + \frac{h^5 D^5}{120} + \frac{h^7 D^7}{5040} + \cdots$	$hD = \mu\delta - \frac{\mu\delta^3}{6} + \frac{\mu\delta^5}{30} - \frac{\mu\delta^7}{140} \cdots$
$\delta^2 = h^2 D^2 + \frac{h^4 D^4}{12} + \frac{h^6 D^6}{360} + \frac{h^8 D^8}{20160} + \cdots$	$h^2 D^2 = \delta^2 - \frac{\delta^4}{12} + \frac{\delta^6}{90} - \cdots$
$\mu\delta^3 = h^3 D^3 + \frac{h^5 D^5}{4} + \frac{h^7 D^7}{40} + \cdots$	$h^3 D^3 = \mu\delta^3 - \frac{\mu\delta^5}{4} + \frac{7\mu\delta^7}{120} - \cdots$
$\delta^4 = h^4 D^4 + \frac{h^6 D^6}{6} + \frac{h^8 D^8}{80} + \cdots$	$h^4 D^4 = \delta^4 - \frac{\delta^6}{6} + \frac{7\delta^8}{240} - \cdots$

$$h^4 D^4 = \delta^4 - \frac{\delta^6}{6} + \frac{7\delta^8}{240} - \cdots \tag{3.93}$$

The complete set of relationships between central difference operators and differential operators is summarized in Table 3.3. These relationships will be used in Chapter 4 to develop a set of formulas expressing the derivatives in terms of central finite differences. These formulas will have higher accuracy than those developed using backward and forward finite differences.

EXAMPLE 3.3 TABLE OF CENTRAL DIFFERENCES

Form the table of central differences for the data points given in Table E3.1 (Example 3.1).

Method of solution

The steps for generating the table of central differences in Excel are shown in Fig. E3.3. Because the odd-order central differences involve values at the midpoint of the intervals, the data points are inserted with empty rows in between.

3.6 INTERPOLATING POLYNOMIALS

Engineers and scientists often face the task of interpreting and correlating experimental observations, which are usually in the form of discrete data and are called upon to either integrate or differentiate these data numerically or graphically. This task is facilitated by the use of interpolation/extrapolation formulas. The calculus of finite differences enables us to develop *interpolating polynomials* that can represent experimental data when the actual functionality of these data is not well known. But, even more significantly, these polynomials can be used to approximate functions that are difficult to integrate or differentiate, thus making the task somewhat easier, albeit approximate.

Let us assume that values of functions $f(x)$ are known at a set of $(n + 1)$ values of the independent variables x:

$$
\begin{array}{ll}
x_0 & f(x_0) \\
x_1 & f(x_1) \\
x_2 & f(x_2) \\
x_3 & f(x_3) \\
\vdots & \vdots \\
x_n & f(x_n)
\end{array}
$$

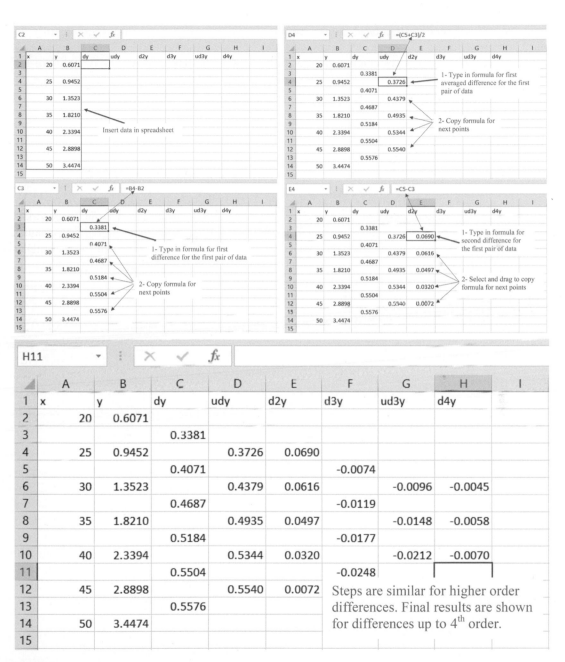

FIG. E3.3

Generating a table of central differences in Excel.

These values are called the *base points* of the function. They are shown graphically in Fig. 3.2a.

The general objective in developing interpolating polynomials is to choose a polynomial of the form:

$$P_n(x) = a_0 + a_1 x + a_2 x^2 + a_3 x^3 + \cdots + a_n x^n \qquad (3.94)$$

so that this equation fits exactly the base points of the function and connects these points with a smooth curve, as shown in Fig. 3.2b. This polynomial can then be used to approximate the function at any value of the independent variable x between the base points.

For the given set of $(n + 1)$ known base points, the polynomial must satisfy the equation:

$$P_n(x_i) = f(x_i) \qquad i = 0, 1, 2, \ldots, n \qquad (3.95)$$

Substitution of the known values of $[x_i, f(x_i)]$ in Eq. (3.94) yields a set of $(n + 1)$ simultaneous linear algebraic equations whose unknowns are the coefficients $a_0, \ldots, a_n$ of the polynomial equation. The solution of this set of linear algebraic equations may be obtained using one of the algorithms discussed in Chapter 2. However, this solution results in an ill-conditioned linear system; therefore, other methods have been favored in the development of interpolating polynomials.

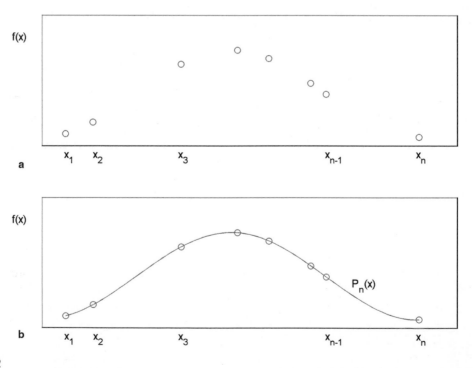

FIG. 3.2

(a) Unequally spaced base points of the function $f(x)$. (b) Unequally spaced base points with interpolating polynomial.

3.7 INTERPOLATION OF EQUALLY SPACED POINTS

In this section, we will develop two interpolation methods for equally spaced data: (1) the *Gregory-Newton formulas*, which are based on forward and backward differences, and (2) *Stirling's interpolation formula* based on central differences.

3.7.1 GREGORY-NEWTON INTERPOLATION

First, we consider a set of known values of the function $f(x)$ at *equally spaced* values of x:

$$
\begin{array}{ll}
x - 3h & f(x - 3h) \\
x - 2h & f(x - 2h) \\
x - h & f(x - h) \\
x & f(x) \\
x + h & f(x + h) \\
x + 2h & f(x + 2h) \\
x + 3h & f(x + 3h)
\end{array}
$$

These points are represented graphically in Fig. 3.3.

The *Gregory-Newton forward interpolation formula* can be derived using the forward finite difference relations derived in Sections 3.2 and 3.4. Eq. (3.17), written for the function f,

$$f(x + h) = e^{hD}f(x) \tag{3.96}$$

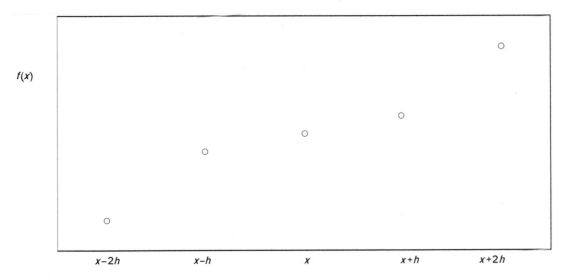

FIG. 3.3

Equally spaced base points for interpolating polynomial.

relates the value of the function at one interval forward of the pivot point x to the value of the function at the pivot point. Applying this equation for n intervals forward, that is, replacing h with nh, we obtain:

$$f(x + nh) = e^{nhD}f(x) \tag{3.97}$$

or equivalently:

$$f(x + nh) = \left(e^{hD}\right)^n f(x) \tag{3.98}$$

We note from Eq. (3.59) that:

$$e^{hD} = 1 + \Delta \tag{3.59}$$

Combining Eqs. (3.98) and (3.59) we obtain:

$$f(x + nh) = (1 + \Delta)^n f(x) \tag{3.99}$$

The term $(1 + \Delta)^n$ can be expanded using the *binomial series:*

$$(1 + \Delta)^n = 1 + n\Delta + \frac{n(n-1)}{2!}\Delta^2 + \frac{n(n-1)(n-2)}{3!}\Delta^3 + \frac{n(n-1)(n-2)(n-3)}{4!}\Delta^4 + \cdots \tag{3.100}$$

Therefore, Eq. (3.99) becomes:

$$f(x + nh) = f(x) + n\Delta f(x) + \frac{n(n-1)}{2!}\Delta^2 f(x) + \frac{n(n-1)(n-2)}{3!}\Delta^3 f(x) + \frac{n(n-1)(n-2)(n-3)}{4!}\Delta^4 f(x) + \cdots \tag{3.101}$$

When n is a positive integer, the binomial series has $(n + 1)$ terms; therefore, Eq. (3.101) is a polynomial of degree n. If $(n + 1)$ base-point values of the function f are known, this polynomial fits all $(n + 1)$ points exactly. Assume that these $(n + 1)$ base-points are $[x_0, f(x_0)]$, $[x_1, f(x_1)]$,..., $[x_n, f(x_n)]$, where $[x_0, f(x_0)]$ is the pivot point and x_i is defined as:

$$x_i = x_0 + ih \tag{3.102}$$

We can now designate the distance of the point of interest from the pivot point as $(x - x_0)$. The value of n is no longer an integer and is replaced by:

$$n = \frac{x - x_0}{h} \tag{3.103}$$

These substitutions convert Eq. (3.101) to:

$$f(x) = f(x_0) + \frac{(x - x_0)}{h}\Delta f(x_0) + \frac{(x - x_0)(x - x_1)}{2!h^2}\Delta^2 f(x_0) + \frac{(x - x_0)(x - x_1)(x - x_2)}{3!h^3}\Delta^3 f(x_0)$$
$$+ \frac{(x - x_0)(x - x_1)(x - x_2)(x - x_3)}{4!h^4}\Delta^4 f(x_0) + \cdots \tag{3.104}$$

This is the *Gregory-Newton forward interpolation formula*. The general formula of the foregoing series is:

$$f(x) = f(x_0) + \sum_{k=1}^{n} \left(\prod_{m=0}^{k-1} (x - x_m) \right) \frac{\Delta^k f(x_0)}{k! h^k} \tag{3.105}$$

In a similar derivation, using backward differences, the *Gregory-Newton backward interpolation formula* is derived as:

$$f(x) = f(x_0) + \frac{(x - x_0)}{h} \nabla f(x_0) + \frac{(x - x_0)(x - x_{-1})}{2! h^2} \nabla^2 f(x_0) + \frac{(x - x_0)(x - x_{-1})(x - x_{-2})}{3! h^3} \nabla^3 f(x_0)$$
$$+ \frac{(x - x_0)(x - x_{-1})(x - x_{-2})(x - x_{-3})}{4! h^4} \nabla^4 f(x_0) + \cdots \tag{3.106}$$

The general formula of this series is:

$$f(x) = f(x_0) + \sum_{k=1}^{n} \left(\prod_{m=0}^{k-1} (x - x_{-m}) \right) \frac{\nabla^k f(x_0)}{k! h^k} \tag{3.107}$$

It was stated earlier that the binomial series (Eq. 3.100) has a finite number of terms, $(n + 1)$ when n is a positive integer. However, in the Gregory-Newton interpolation formulas, n is not usually an integer; therefore, these polynomials have an infinite number of terms. It is known from algebra that if $|\Delta| \leq 1$, then the binomial series for $(1 + \Delta)^n$ converges to the value of $(1 + \Delta)^n$ as the number of terms becomes larger and larger. This implies that finite differences must be small. This is true for a flat, smooth function, or, alternatively, if the known base points are close together, that is, if h is small. Of course, the number of terms that can be used in each formula depends on the highest order of finite differences that can be evaluated from the available known data. It is common sense that for evenly spaced data, the accuracy of interpolation is higher for a large number of data points that are closely spaced together.

For a given set of data points, the accuracy of interpolation can be further enhanced by choosing the pivot point as close to the point of interest as possible so that $x < h$. If this is satisfied, then the series should use as many terms as possible, that is, the number of finite differences in the equation should be maximized. The order of error of the formula applied in each case is equivalent to the order of the finite difference contained in the first truncated term of the series. The flowchart for Gregory-Newton forward interpolation is shown in Fig. 3.4. Note that the flowchart for backward or central interpolation is similar to this figure with the difference in the formula for interpolation polynomial.

EXAMPLE 3.4 GREGORY-NEWTON METHOD FOR INTERPOLATION OF EQUALLY SPACED DATA

An exothermic, relatively slow reaction takes place in a reactor under your supervision. Yesterday, after you left the plant, the temperature of the reactor went out of control, for a yet unknown reason, until the operator put it under control by changing the cooling water flow rate. Your supervisor has asked you to prepare a report regarding this incident. As the first

(Continued)

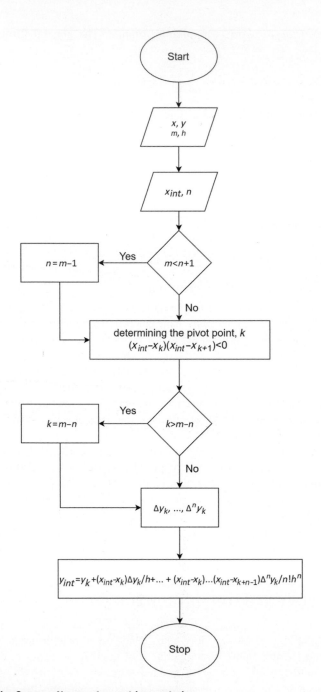

FIG. 3.4 Flowchart of the Gregory-Newton forward interpolation.

In this flowchart, x and y are the vectors of independent and function variables of base points, m is the number of base points, n is the order of interpolation polynomial and x_{int} is the independent variable at which the interpolation is done.

EXAMPLE 3.4 (CONTINUED)

step, you must know when the reactor reached its maximum temperature and what was the value of this maximum temperature. A computer was recording the temperature of the reactor at 1-hour intervals. These time-temperature data are given in Table E3.4. Write a general MATLAB function for n-order one-dimensional interpolation by Gregory-Newton's forward interpolation formula to solve this problem.

Method of solution

The function uses the general formula of the Gregory-Newton forward interpolation (Eq. 3.104) to perform the n-order interpolation. The input to the function specifying the number of base points must be at least $(n + 1)$.

Program description

The programs and functions developed in this example can be found in: https://www.elsevier.com/books-and-journals/book-companion/9780128229613

The MATLAB function *GregoryNewton.m* is developed to perform the Gregory-Newton forward interpolation. The first and second input arguments are the coordinates of the base points. The third input argument is the vector of the independent variable at which the interpolation of the dependent variable is required. The fourth input, n, is the order of interpolation. If no value is introduced to the function through the fourth argument, the function does linear interpolation. For obtaining the results of the higher-order interpolation, this value should be entered as the fourth input argument.

In the beginning, the function checks the inputs. The vectors of coordinates of base points have to be of the same size. The function also checks to see if the vector of the independent variable is monotonic, otherwise, the function terminates calculations. The order of interpolation cannot be more than the intervals (number of base points minus one). In this case, the function displays a warning and continues with the maximum possible order of interpolation. The function then performs the interpolation according to Eq. (3.105).

The main program *Example3_4.m* is written to solve the problem of Example 3.4. It asks the user to input the vector of time (independent variable), the vector of the temperature of the reactor (dependent variable), and the order of interpolation. The program applies the function

(Continued)

Table E3.4 Temperature profile of the reactor.

Time (p.m.)	Temperature (°C)	Time (p.m.)	Temperature (°C)
4	70	9	93
5	71	10	81
6	75	11	68
7	83	12	70
8	92		

EXAMPLE 3.4 (CONTINUED)

GregoryNewton.m to interpolate the temperature between the recorded temperatures and finds its maximum. The user can repeat the calculations with another order of interpolation.

Input and Results

>> **Example3_4**
Vector of time = [4, 5, 6, 7, 8, 9, 10, 11, 12]
Vector of temperature = [70, 71, 75, 83, 92, 93, 81, 68, 70]
Order of interpolation = 2
Maximum temperature of 94.2 C reached at 8.61.
Repeat the calculation (1/0): 0

Discussion of results

Graphical results are shown in Fig. E3.4. As can be seen from this plot and also from the numerical results, the reactor has reached the maximum temperature of 94.2°C at 8:37 p.m. The reader can repeat the calculations with other values for the order of interpolation.

To find the interpolating polynomials, generate the table of forward differences similar to what has been done in Example 3.2. Let us show the temperature with T and time with t. The first two forward differences at $t = 4$ p.m. are $\Delta T = 1$ and $\Delta^2 T = 3$. Then, using Eq. (3.104), the second-degree interpolation polynomial for the first two segments (the first three points) can be written as:

$$T(t) = T(4) + \frac{(t-4)}{1}\Delta T(4) + \frac{(t-4)(t-5)}{2!1^2}\Delta^2 T(4)$$

$$= 70 + (t-4) + \frac{3}{2}(t-4)(t-5)$$

Similar formulas can be presented for other segments and with other orders of the polynomial.

3.7.2 STIRLING'S INTERPOLATION

Stirling's interpolation formula is based on central differences. Its derivation is similar to that of the Gregory-Newton formulas and can be arrived at by using either the symbolic operator relations or the Taylor series expansion of the function. We will use the latter and expand the function $f(x + nh)$ in a Taylor series around x:

$$f(x + nh) = f(x) + \frac{nh}{1!}f'(x) + \frac{n^2 h^2}{2!}f''(x) + \frac{n^3 h^3}{3!}f'''(x) + \cdots \tag{3.108}$$

We replace the derivatives of $f(x)$ with the differential operators to obtain:

$$f(x + nh) = f(x) + \frac{nh}{1!}Df(x) + \frac{n^2 h^2}{2!}D^2 f(x) + \frac{n^3 h^3}{3!}D^3 f(x) + \cdots \tag{3.109}$$

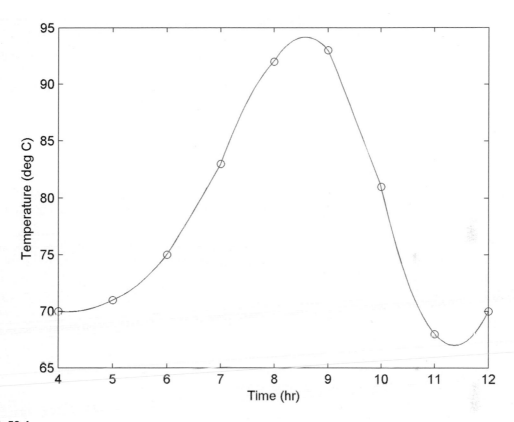

FIG. E3.4

Interpolation of equally spaced points.

The odd-order differential operators in Eq. (3.109) are replaced by averaged central differences and the even-order differential operators by central differences, all taken from Table 3.3. Substituting these into Eq. (3.109) and regrouping of terms yield the formula:

$$f(x + nh) = f(x) + n\mu\delta f(x) + \frac{n^2}{2!}\delta^2 f(x) + \frac{n(n^2 - 1)h^2}{3!}\mu\delta^3 f(x) + \frac{n^2(n^2 - 1)}{4!}\delta^4 f(x) + \cdots \quad (3.110)$$

By applying Eq. (3.103) into Eq. (3.110), we obtain the final form of *Stirling's interpolation formula*:

$$f(x) = f(x_0) + \frac{(x - x_0)}{h}\mu\delta f(x_0) + \frac{(x - x_0)^2}{2!h^2}\delta^2 f(x_0) + \frac{(x - x_{-1})(x - x_0)(x - x_1)}{3!h^3}\mu\delta^3 f(x_0)$$
$$+ \frac{(x - x_{-1})(x - x_0)^2(x - x_1)}{4!h^4}\delta^4 f(x_0) + \cdots \quad (3.111)$$

The general formula for determining the higher-order terms containing odd differences, in the foregoing series, is:

$$\left[\frac{1}{k!h^k} \prod_{m=-(k-1)/2}^{(k-1)/2} (x - x_m) \right] \mu \delta^k f(x_0) \qquad (3.112)$$

where $k = 1, 3,...$, and the formula for terms with even differences is:

$$\left[\frac{(x - x_0)}{k!h^k} \prod_{m=-(k-2)/2}^{(k-2)/2} (x - x_m) \right] \delta^k f(x_0) \qquad (3.113)$$

where $k = 2, 4,...$

Other forms of Stirling's interpolation formula exist, which make use of base points spaced at half intervals (i.e., at $h/2$). Our choice of using *averaged* central differences to replace the odd differential operators eliminated the need for having base points located at the midpoints.

EXAMPLE 3.5 STIRLING'S METHOD FOR INTERPOLATION OF EQUALLY SPACED DATA

Using the data given in Table E3.4, derive the second-order interpolation polynomial for the starting segments of the data set.

Method of solution

At the pivot point $t = 5$ p.m., the first averaged central difference and the second central difference are $\mu \delta T = 2.5$ and $\delta^2 T = 3$, respectively. Then, using Eq. (3.111), the second-degree interpolation polynomial for the first two segments (the first three points) can be written as:

$$T(t) = T(5) + \frac{(t-5)}{1} \mu \delta T(5) + \frac{(t-5)^2}{2!1^2} \delta^2 T(5)$$
$$= 71 + 2.5(t-5) + 1.5(t-5)^2$$

Similar formulas can be presented at other pivot points and with other orders of the polynomial.

3.8 INTERPOLATION OF UNEQUALLY SPACED POINTS

In this section, we will develop two interpolation methods for unequally spaced data: the *Lagrange polynomials* and *spline interpolation*.

3.8.1 LAGRANGE POLYNOMIALS

Consider a set of unequally spaced base points, such as those shown in Fig. 3.2A. Define the polynomial:

$$P_n(x) = \sum_{k=0}^{n} p_k(x) f(x_k) \qquad (3.114)$$

which is the sum of the *weighted* values of the function at all $(n + 1)$ base points. The weights $p_k(x)$ are nth-degree polynomial functions corresponding to each base point. Eq. (3.114) is actually a linear combination of nth-degree polynomials; therefore, $P_n(x)$ is also an nth-degree polynomial.

For the interpolating polynomial to fit the function exactly at all the base points, each particular weighting polynomial $p_k(x)$ must be chosen so that it has the value of unity when $x = x_k$, and the value of zero at all other base points, that is,

$$p_k(x) = \begin{cases} 0 & i \neq k \\ 1 & i = k \end{cases} \tag{3.115}$$

The *Lagrange polynomials*, which have the form:

$$p_k(x) = C_k \prod_{\substack{i=0 \\ i \neq k}}^{n} (x - x_i) \tag{3.116}$$

satisfy the first part of the condition (3.115) because there will be a term $(x_i - x_i)$ in the product series of Eq. (3.116) whenever $x = x_i$. The constant C_k is evaluated to make the Lagrange polynomial satisfy the second part of the condition (3.115):

$$C_k = \frac{1}{\displaystyle\prod_{\substack{i=0 \\ i \neq k}}^{n} (x_k - x_i)} \tag{3.117}$$

Combination of Eqs. (3.116) and (3.117) gives the Lagrange polynomials:

$$p_k(x) = \prod_{\substack{i=0 \\ i \neq k}}^{n} \left(\frac{x - x_i}{x_k - x_i} \right) \tag{3.118}$$

The interpolating polynomial $P_n(x)$ has a remainder term that can be obtained from Eq. (3.6):

$$R_n(x) = \prod_{i=0}^{n} (x - x_i) \frac{f^{(n+1)}(\xi)}{(n+1)!} \quad x_0 < \xi < x_n \tag{3.119}$$

The flowchart for interpolation by Lagrange polynomials is shown in Fig. 3.5.

EXAMPLE 3.6 LAGRANGE POLYNOMIAL FOR INTERPOLATION OF UNEQUALLY SPACED DATA

Repeat Example 3.4 with the reduced data set given in Table E3.6. Do not write a program this time, but derive the appropriate Lagrange polynomial and analytically determine when the maximum temperature has reached.

Method of solution

It can be seen in Table E3.6 that we encounter unequally spaced data. Therefore, the Lagrange polynomials can be used for interpolation. Because there are four points in this

(Continued)

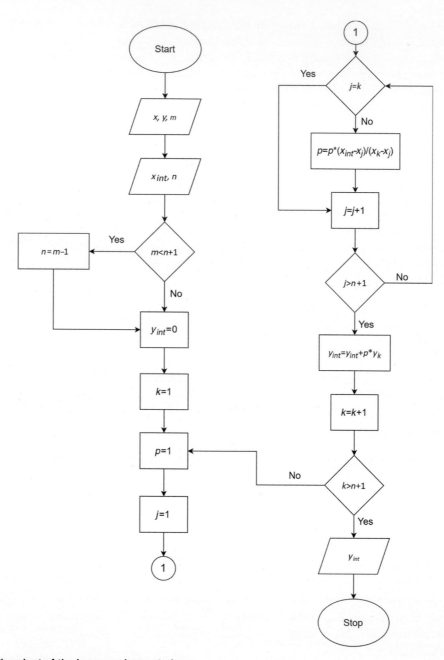

FIG. 3.5 Flowchart of the Lagrange interpolation.

In this flowchart, x and y are the vectors of independent and function variables of base points, m is the number of base points, n is the order of interpolation polynomial and x_{int} is the independent variable at which the interpolation is done.

Table E3.6 Temperature profile of the reactor.	
Time (p.m.)	**Temperature ($^\circ$C)**
4	70
7	83
9	93
12	70

EXAMPLE 3.6 (CONTINUED)

table, we can use a third-degree polynomial to cover the whole domain. The third-degree Lagrange polynomial for this case becomes:

$$
\begin{aligned}
T(t) &= \frac{(t - t_2)(t - t_3)(t - t_4)}{(t_1 - t_2)(t_1 - t_3)(t_1 - t_4)} T(t_1) + \frac{(t - t_1)(t - t_3)(t - t_4)}{(t_2 - t_1)(t_2 - t_3)(t_2 - t_4)} T(t_2) \\
&+ \frac{(t - t_1)(t - t_2)(t - t_4)}{(t_3 - t_1)(t_3 - t_2)(t_3 - t_4)} T(t_3) + \frac{(t - t_1)(t - t_2)(t - t_3)}{(t_4 - t_1)(t_4 - t_2)(t_4 - t_3)} T(t_4) \\
&= \frac{(t - 7)(t - 9)(t - 12)}{(4 - 7)(4 - 9)(4 - 12)} \times 70 + \frac{(t - 4)(t - 9)(t - 12)}{(7 - 4)(7 - 9)(7 - 12)} \times 83 \\
&+ \frac{(t - 4)(t - 7)(t - 12)}{(9 - 4)(9 - 7)(9 - 12)} \times 93 + \frac{(t - 4)(t - 7)(t - 9)}{(12 - 4)(12 - 7)(12 - 9)} \times 70
\end{aligned}
$$

The maximum of this function occurs at $t = 9.41$ hours when the temperature is estimated to be 93.4°C.

3.8.2 SPLINE INTERPOLATION

When we deal with a large number of data points, high-degree interpolating polynomials are likely to fluctuate between base points instead of passing smoothly through them. This situation is illustrated in Fig. 3.6a. Although the interpolating polynomial passes through all the base points, it is not able to predict the value of the function satisfactorily in between these points. To avoid such an undesired behavior of the high-degree interpolating polynomial, a series of lower-degree interpolating polynomials may be used to connect a smaller number of base points. This set of interpolating polynomials are called *spline functions*. Fig. 3.6b shows the result of such interpolation using third-degree (or cubic) splines. Compared to the higher-order interpolation illustrated in Fig. 3.6a, third-degree splines shown in Fig. 3.6b provide a much more acceptable approximation.

The most common spline used in engineering problems is the *cubic spline*. In this method, a cubic polynomial is used to approximate the curve between each two adjacent base points. Because there would be an infinite number of third-degree polynomials passing through each pair of points, additional constraints are necessary to make the spline unique. Therefore, it is set that all the polynomials should have equal first and second derivatives at the base points. These conditions imply that the slope and the curvature of the spline polynomials are continuous across the base points.

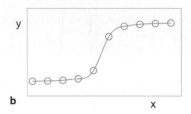

FIG. 3.6

(a) Fluctuation of high-degree interpolating polynomials between base points. (b) Cubic spline interpolation.

The cubic spline of the interval $[x_{i-1}, x_i]$ has the following general form:

$$P_i(x) = a_i x^3 + b_i x^2 + c_i x + d_i \tag{3.120}$$

There are four unknown coefficients in Eq. (3.120) and n such polynomials for the whole range of data points $[x_0, x_n]$. Therefore, there are $4n$ unknown coefficients and we need $4n$ equations to evaluate these coefficients. The required equations come from the following conditions:

a. Each spline passes from the base points of the edge of its interval ($2n$ equations).
b. The first derivative of the splines is continuous across the interior base points ($n-1$ equations).
c. The second derivative of the splines is continuous across the interior base points ($n-1$ equation).
d. The second derivative of the end splines is zero at the end base points (two equations). This is called the *natural* condition. Another commonly used condition is to set the third derivative of the end splines equal to the third derivative of the neighboring splines. The latter is called the *not-a-knot* condition.

Simultaneous solution of the foregoing $4n$ linear algebraic equations results in the determination of all cubic interpolating polynomials. However, for programming purposes, there is an alternative method of determination of the coefficients, which needs a simultaneous solution of only $(n-1)$ algebraic equations. This method is described in detail in this section.

The second derivative of the Eq. (3.120) is a line:

$$P_i''(x) = y'' = 6a_i x + 2b_i \tag{3.121}$$

From Eq. (3.121) it can be concluded that the second derivative of the interpolating polynomial at any point in the interval $[x_{i-1}, x_i]$ can be given by the first-order Lagrange interpolation formula:

$$y'' = \frac{x - x_i}{x_{i-1} - x_i} y''_{i-1} + \frac{x - x_{i-1}}{x_i - x_{i-1}} y''_i \tag{3.122}$$

The expression for the spline can be obtained by twice integrating Eq. (3.122):

$$y = \frac{(x - x_i)^3}{6(x_{i-1} - x_i)} y''_{i-1} + \frac{(x - x_{i-1})^3}{6(x_i - x_{i-1})} y''_i + C_1 x + C_2 \tag{3.123}$$

where the constants C_1 and C_2 in Eq. (3.123) are evaluated from the following boundary conditions:

$$\begin{aligned} y(x_{i-1}) &= y_{i-1} \\ y(x_i) &= y_i \end{aligned} \tag{3.124}$$

By evaluating the constants C_1 and C_2 from the conditions (3.124), substituting them into Eq. (3.123), and further rearrangement, we find the following cubic equation:

$$P_i(x) = y = \frac{1}{6}\left[\frac{(x-x_i)^3}{x_{i-1}-x_i} - (x_{i-1}-x_i)(x-x_i)\right]y''_{i-1} + \frac{1}{6}\left[\frac{(x-x_{i-1})^3}{x_i-x_{i-1}} - (x_i-x_{i-1})(x-x_{i-1})\right]y''_i$$
$$+ \left(\frac{x-x_i}{x_{i-1}-x_i}\right)y_{i-1} + \left(\frac{x-x_{i-1}}{x_i-x_{i-1}}\right)y_i \tag{3.125}$$

Note that Eqs. (3.120) and (3.125) are equivalent and the relations between their coefficients are given by the following equation:

$$a_i = \frac{1}{6}\left(\frac{y''_{i-1}-y''_i}{x_{i-1}-x_i}\right)$$
$$b_i = \frac{1}{2}\left(\frac{x_{i-1}y''_i - x_iy''_{i-1}}{x_{i-1}-x_i}\right)$$
$$c_i = \frac{1}{2}\left(\frac{x_i^2y''_{i-1}-x_{i-1}^2y''_i}{x_{i-1}-x_i}\right) + \frac{1}{6}(x_{i-1}-x_i)(y''_i - y''_{i-1}) + \frac{y_{i-1}-y_i}{x_{i-1}-x_i} \tag{3.126}$$
$$d_i = \frac{1}{6}\left(\frac{x_{i-1}^3y''_i - x_i^3y''_{i-1}}{x_{i-1}-x_i}\right) + \frac{1}{6}(x_{i-1}-x_i)(x_iy''_{i-1}-x_{i-1}y''_i) + \frac{x_{i-1}y_i - x_iy_{i-1}}{x_{i-1}-x_i}$$

Although Eq. (3.125) is a more complicated expression than Eq. (3.120), it contains only two unknowns, namely y''_{i-1} and y''_i. To determine the y'' values, we apply the condition of continuity of the first derivative of splines at the interior base points, that is,

$$y'_{i+1} = y'_i \tag{3.127}$$

Differentiating Eq. (3.125) and applying the resulting expression in the condition (3.127), followed by rearranging of the terms, results in:

$$(x_i - x_{i-1})y''_{i-1} + 2(x_{i+1}-x_{i-1})y''_i + (x_{i+1}-x_i)y''_{i+1} = 6\frac{y_{i+1}-y_i}{x_{i+1}-x_i} - 6\frac{y_i-y_{i-1}}{x_i-x_{i-1}} \tag{3.128}$$

where $i = 1, 2,..., n - 1$ and $y''_0 = y''_n = 0$ (natural spline).

Eq. (3.128) represents an $(n - 1)$-order tridiagonal set of simultaneous equations that in matrix form becomes

$$\begin{bmatrix} 2(x_2-x_0) & (x_2-x_1) & 0 & 0 & \cdots & & 0 & 0 \\ (x_2-x_1) & 2(x_3-x_1) & (x_3-x_2) & 0 & \cdots & & 0 & 0 \\ 0 & (x_3-x_2) & 2(x_4-x_2) & (x_4-x_3) & \cdots & & 0 & 0 \\ \cdots & \cdots & \cdots & \cdots & \cdots & \cdots & \cdots & \cdots \\ 0 & 0 & & \cdots & 0 & (x_{n-2}-x_{n-3}) & 2(x_{n-1}-x_{n-3}) & (x_{n-1}-x_{n-2}) \\ 0 & 0 & & \cdots & 0 & 0 & (x_{n-1}-x_{n-2}) & 2(x_n-x_{n-2}) \end{bmatrix}$$

$$\times \begin{bmatrix} y''_1 \\ y''_2 \\ y''_3 \\ \vdots \\ y''_{n-2} \\ y''_{n-1} \end{bmatrix} = 6\begin{bmatrix} \frac{y_2-y_1}{x_2-x_1} - \frac{y_1-y_0}{x_1-x_0} \\ \frac{y_3-y_2}{x_3-x_2} - \frac{y_2-y_1}{x_2-x_1} \\ \frac{y_4-y_3}{x_4-x_3} - \frac{y_3-y_2}{x_3-x_2} \\ \vdots \\ \frac{y_{n-1}-y_{n-2}}{x_{n-1}-x_{n-2}} - \frac{y_{n-2}-y_{n-3}}{x_{n-2}-x_{n-3}} \\ \frac{y_n-y_{n-1}}{x_n-x_{n-1}} - \frac{y_{n-1}-y_{n-2}}{x_{n-1}-x_{n-2}} \end{bmatrix} \tag{3.129}$$

After calculating the values of the second derivatives at each base point, Eq. (3.125) can be used for interpolating the value of the function in every interval.

EXAMPLE 3.7 LAGRANGE POLYNOMIALS AND CUBIC SPLINES FOR INTERPOLATION OF UNEQUALLY SPACED DATA

The pressure drop of a basket-type filter is measured at different flow rates as shown in Table E3.7. Write a program to estimate the pressure drop of the filter at any flow rate within the experimental range. This program should call general MATLAB functions for interpolating unequally spaced data using Lagrange polynomials and cubic splines.

Method of solution

The Lagrange interpolation is done based on Eqs. (3.114) and (3.118). The order of interpolation is an input to the function. The cubic spline interpolation is done based on Eq. (3.125). The values of the second derivatives at base points, assuming a natural spline, are calculated from Eq. (3.129).

Program description

The programs and functions developed in this example can be found in: https://www.elsevier.com/books-and-journals/book-companion/9780128229613

The general MATLAB function *Lagrange.m* performs the nth-order Lagrange interpolation. This function consists of the following three parts:

In the beginning, it checks the inputs and sets the order of interpolation if necessary. If not introduced to the function, the interpolation is done by the first-order Lagrange polynomial (linear interpolation).

In the second part of the function, locations of all the points at which the values of the function are to be evaluated are found between the base points. Because matrix operations are much faster than element-by-element operations in MATLAB, the required number of independent and dependent variables are arranged in two interim matrices at each location. These matrices are used in the interpolation section for doing the interpolation in vector form.

The last part of the function is interpolation itself. In this section, $p_k(x)$ subpolynomials are calculated according to Eq. (3.118). The terms of summation (3.114) are then calculated, and finally, the function value is determined based on Eq. (3.114). To be time efficient, all these calculations are done in vector form and at all the required points simultaneously.

The MATLAB function *NaturalSPLINE.m* also consists of three parts. The first and second parts are more or less similar to those of *Lagrange.m*. However, instead of forming the interim matrices, the interpolation locations are kept in a vector.

In the last section of the function, the matrix of coefficients and the vector of constants are built according to Eq. (3.129) and the values of the second derivatives at the base points are evaluated. The interpolation is then performed, in the vector form, based on Eq. (3.125).

(Continued)

Table E3.7 Pressure drop of a basket-type filter.

Flow rate (L/s)	Pressure drop (kPa)	Flow rate (L/s)	Pressure drop (kPa)
0	0	32.56	1.781
10.80	0.299	36.76	2.432
16.03	0.576	39.88	2.846
22.91	1.036	43.68	3.304

EXAMPLE 3.7 (CONTINUED)

Input and results

>> Example3_7
Vector of flow rates = [0, 10.80, 16.03, 22.91, 28.24, 32.56, 36.76, 39.88, 43.68]
Vector of pressure drops = [0, 0.299, 0.576, 1.036, 1.383, 1.781, 2.432, 2.846, 3.304]
Order of the Lagrange interpolation = 3

Discussion of results

The order of the Lagrange interpolation is chosen to be three for comparison of the results with that of the cubic spline which is also third-order interpolation. Fig. E3.7 shows the results of the calculations. There is no essential difference between the two methods. The cubic spline, however, passes smoothly through the base points, as expected. Because the Lagrange interpolation is performed in the subsets of four base points with no restriction related to their neighboring base points, it can be seen that the slope of the resulting curve is not continuous through most of the base points.

3.9 USING BUILT-IN MATLAB® FUNCTIONS

MATLAB has several functions for interpolation. The command

$$>> y_i = interp1\,(x, y, x_i)$$

takes the values of the independent variable x and the dependent variable y (base points) and does the one-dimensional interpolation based on x_i to find y_i. The default method of interpolation is linear. However, the user can choose the method of interpolation in the fourth input argument from "*nearest*" (nearest neighbor interpolation), "*linear*" (linear interpolation), "*spline*" (cubic spline interpolation), and "*cubic*" (cubic interpolation). If the vector of the independent variable is not equally spaced, the function *interp1q* may be used instead. It is faster than *interp1* because it does not check the input arguments. MATLAB also has the function *spline* to perform one-dimensional interpolation by cubic splines, using the *not-a-knot* method. It can also return coefficients of piecewise polynomials if required. Different MATLAB functions for interpolation are listed in Table 3.4.

The functions *interp2*, *interp3*, and *interpn* perform two-, three-, and n-dimensional interpolation, respectively.

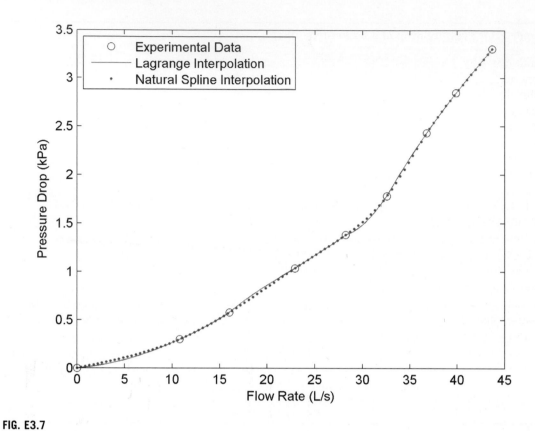

FIG. E3.7

Lagrange polynomials and cubic splines.

Table 3.4 MATLAB functions for interpolating one-dimensional data.	
Function	**Description**
interp1	General one dimensional interpolation
griddedInterpolant	Gridded data interpolation
pchip	Piecewise cubic Hermite interpolation polynomial
spline	Cubic spline interpolation
ppval	Evaluate piecewise polynomial
mkpp	Make piecewise polynomial
unmkpp	Piecewise polynomial details
padecoef	Padé approximation of time delay
interpft	One dimensional interpolation using FFT method

3.10 SUMMARY

Finite differences are the base of numerical integration, differentiation, interpolation, and also the numerical solution of ordinary and partial differential equations. In this chapter, various types of finite differences are first introduced and discussed. Related operators, including D (differential operator), I (integral operator), E (shift operator), Δ (forward difference operator), ∇ (backward difference operator), δ (central difference operator), and μ (averager operator) are defined and their properties, as well as their relationships, are established. These relationships are used later in this chapter for interpolation, as well as in Chapter 4 for differentiation and integration, in Chapters 5 and 6 for the solution of ordinary differential equations and in Chapter 7 for solving partial differential equations.

Interpolation is used when the trend of a function is to be estimated based on discrete data. Polynomials of various degrees are used for interpolation. For equally spaced data, the Gregory-Newton backward, the Gregory-Newton forward, and Stirling's formulas (which are developed based on backward differences, forward differences, and central differences, respectively) can be used.

For unequally spaced data, the Lagrange polynomial of the desired degree can be employed. However, when dealing with a large number of data points, a high-degree polynomial is likely to fluctuate between the base points instead of passing smoothly through them. In this case, a series of lower-degree interpolating polynomials, called spline functions, may be used to connect a smaller number of base points. Third-degree splines usually provide an acceptable approximation.

At the end of the chapter, built-in MATLAB functions which perform interpolation are described.

PROBLEMS

3.1. Show that all the interpolating formulas discussed here reduce to the same formula when a first-order interpolation is used.

3.2. Derive the Gregory-Newton backward interpolation formula.

3.3. Using the experimental data in Table P3.3
   (a) Develop the forward difference table.
   (b) Develop the backward difference table.

Table P3.3 Data of penicillin fermentation.

Time (hours)	Penicillin concentration (units/mL)	Time (hours)	Penicillin concentration (units/mL)
0	0	120	9430
20	106	140	10,950
40	1600	160	10,280
60	3000	180	9620
80	5810	200	9400
100	8600		

(c) Apply the Gregory-Newton interpolation formulas to evaluate the concentration at $x = 10, 50, 90, 130, 170,$ and 190 hours.

3.4. Write a MATLAB function that uses the Gregory-Newton backward interpolation formula to evaluate the function $f(x)$ from a set of $(n + 1)$ equally spaced input values. Write the function in a general fashion so that n can be any positive integer. Also, write a MATLAB script that reads the data and shows how this MATLAB function fits the data. Use the experimental data of Table P3.3 to verify the program, and evaluate the function at $x = 10, 50, 90, 130, 170,$ and 190.

3.5. Using the experimental data of Problem 3.3
   (a) Develop the central difference table.
   (b) Apply Stirling's interpolation formula to evaluate the function at $x = 10, 50, 90, 130, 170,$ and 190.

3.6. Write a MATLAB function that uses Stirling's interpolation formula to evaluate the function $f(x)$ from a set of $(n + 1)$ equally spaced input values. Write the function in a general fashion so that n can be any positive integer. Also, write a MATLAB script that reads the data and shows how this MATLAB function fits the data. Use the experimental data of Table P3.3 to verify the program, and evaluate the function at $x = 10, 50, 90, 130, 170,$ and 190.

3.7. With the set of unequally spaced data points in Table P3.7, use Lagrange polynomials and spline interpolation to evaluate the function at $x = 2, 4, 5, 8, 9,$ and 11.

3.8. Vapor pressure of lithium chloride is given in Table P3.8 [1]. Use these data to present the vapor pressure of lithium chloride in the following tables:
   (a) From 800°C to 1350°C at 50°C increments.
   (b) From 10 kPa to 100 kPa at 10 kPa increments.

Table P3.7 Unequally spaced data points.

x	$f(x)$	x	$f(x)$
1	7	10	8.2
3	3.5	12	9.0
6	3.2	13	9.2
7	3.9		

Table P3.8 Vapor pressure of lithium chloride.

Pressure (mm Hg)	Temperature (°C)	Pressure (mm Hg)	Temperature (°C)
1	783	60	1081
5	883	100	1129
10	932	200	1203
20	987	400	1290
40	1045	760	1382

3.9. The zeta potential of particles in a suspension is an indication of the sign and the density of the surface charge of the particles. The isoelectric point (i.e. p.) refers to the pH where zeta-potential is zero. Use data from Rashchi et al. [2], shown in Table P3.9, to determine the isoelectric points of silica in the presence of 10^{-4} M $Pb(NO_3)_2$.

3.10. Consider the base points given in Table P3.10. Evaluate y for x values from 0 to 10 by interpolation at each 0.1 interval. In a chart, show the base points by a symbol and the interpolated values by a curve.

3.11. Solubility of oxygen in the 10 g/L salt in water solution as a function of temperature is given in Table P3.11. Determine the solubility of oxygen in this solution at 22.4°C by the following methods and compare the results:
  (a) Gregory-Newton forward formula
  (b) Gregory-Newton backward formula
  (c) Stirling formula

Table P3.9 Zeta-potential of silica in the presence of 10^{-4} M $Pb(NO_3)_2$ as a function of pH.

pH	Zeta potential (mV)	pH	Zeta potential (mV)
1.74	−5.30	6.00	−33.2
2.72	−10.8	6.53	−15.7
3.72	−21.8	6.70	−10.0
4.09	−32.0	7.29	13.7
4.32	−35.8	8.06	32.2
4.70	−36.9	10.02	24.0
5.00	−36.7	11.12	6.90
5.55	−37.7	12.15	−30.0

Table P3.10 Numerical data.

x	y	x	y
0	2.00	6	0.95
1	1.84	7	1.09
2	1.45	8	1.12
3	1.05	9	1.05
4	0.81	10	0.94
5	0.81		

Table P3.11 Solubility of oxygen in the salt solution.

Temperature (°C)	5	10	15	20	25	30
Solubility (ppm)	11.6	10.3	9.1	8.2	7.4	6.8

(d) Lagrange polynomials (second and third degree)

(e) Cubic spline

3.12. Experimental data for emittance e of tungsten as a function of temperature T are summarized in Table P3.12 [3].

It was found that the equation:

$$e(T) = 0.0242 \left(\frac{T}{303.16} \right)^{1.27591}$$

correlates the data accurately to three digits. Find the Newton-Cotes interpolating polynomial for these data. Compare the interpolated values with the emittance obtained from the foregoing equation at the points midway between the tabulated temperatures. Plot emissions calculated by interpolation and correlation as a function of temperature in the interval 300 K to 2000 K.

3.13. The following data given in Table P3.13 show the values of the thermal conductivity and viscosity of helium gas at various temperatures.

In each case, determine the simplest interpolating polynomial that is likely to predict k and μ in the specific ranges of temperature.

3.14. Use the portion of steam given in Table P3.14 for saturated steam:

(a) The corresponding enthalpy of vaporization ΔH_{vap} for a density of 100 kg/m^3 with linear interpolation

Table P3.12 Emittance of tungsten as a function of temperature.

T (K)	e	T (K)	e
300	0.024	1200	0.140
400	0.035	1300	0.155
500	0.046	1400	0.170
600	0.056	1500	0.186
700	0.067	1600	0.202
800	0.083	1700	0.219
900	0.097	1800	0.235
1000	0.111	1900	0.252
1100	0.125	2000	0.269

Table P3.13 Experimental data.

T (K)	k (mW/mK)	T (°C)	$\mu \times 10^6$ (kg/m · s)
100	75.5	−129	12.55
200	119.3	−73	15.66
300	156.7	−18	18.17
400	190.6	93	23.05
500	222.3	204	27.50
600	252.4	316	31.13

Table P3.14 Properties of saturated steam at 200 MPa.

ρ (kg/m³)	96.7273	126.115	170.497
ΔH_{vap} (kJ/kg)	1000.50	818.531	585.133

    (b) The same corresponding enthalpy of vaporization using quadratic interpolation

    (c) The density corresponding to the enthalpy of vaporization of 900 kJ/kg using inverse interpolation.

3.15. The data shown in Table P3.15 for the density of nitrogen gas versus temperature come from a table that was measured with high precision. Use all possible interpolating polynomials to estimate the density at a temperature of 325 K. Which polynomial provides the best estimate?

Table P3.15 Density of nitrogen.

T (K)	200	250	300	350	400	450
Density (kg/m³)	1.708	1.367	1.139	0.967	0.854	0.759

3.16. A chemical experiment produces the data shown in Table P3.16. Find an appropriate interpolating polynomial. Plot the interpolation polynomial and compare it with data points.

Table P3.16 Experimental data.

Temperature (°C)	0	5	10	15	20	25
Concentration (ppm)	12.5	10.6	9.3	8.3	7.1	6.5

3.17. Vapor pressure, P^{sat}, of water as a function of temperature, T, is given in Table P3.17. Find the interpolating polynomial of these data and estimate saturation pressure at 5°C, 45°C, and 95°C. Compare your results with the known values of the pressure: 0.008721 bar, 0.095848 bar, 0.84528 bar, respectively.

Table P3.17 Vapor pressure of water.

T (K)	0	10	20	30	40	60	80	100
P^{sat} (bar)	0.0061	0.0123	0.0234	0.0424	0.0738	0.1992	0.4736	1.0133

3.18. The world historical and predicted population from 1600 and 2050 is given in Table P3.18.

    (a) Find a natural spline that interpolates data in this table.

    (b) What does this spline show for the population in 1975?

Table P3.18 World population (millions).

Year	Population	Year	Population
1600	660	1950	2521
1700	710	1999	6008
1750	791	2008	6707
1800	978	2010	6896
1850	1262	2012	7052
1900	1650	2050	9725

REFERENCES

[1] Green, D. W.; Southard, M. Z. *Perry's Chemical Engineers' Handbook*, 9th ed.; McGraw-Hill: New York, 2018.

[2] Rashchi, F.; Xu, Z.; Finch, J. A. Adsorption of Silica in Pb- and Ca-SO_4-CO_3 Systems. *Colloids and Surfaces A: Physicochemical and Engineering Aspects* **1998,** *132*, 159−171.

[3] Kharab, A.; Guenther, R. B. *An Introduction to Numerical Methods: A MATLAB Approach*, 3rd ed.; Chapman and Hall/CRC: Boca Raton, FL, 2011.

DIFFERENTIATION AND INTEGRATION

4

CHAPTER OUTLINE

MOTIVATION

The solution of many engineering problems requires the calculation of the derivative of a function at a known point or integration of the derivative over a known range of the independent variable.

Applied Numerical Methods for Chemical Engineers. DOI: https://doi.org/10.1016/B978-0-12-822961-3.00004-2

The simplest example of such problems is root-finding by the Newton-Raphson method, which needs calculation of the derivative of the function in each iteration (see Section 1.7). Although in some cases the analytical derivative of the function may be derived, it is more convenient to obtain it numerically if the function is complicated and/or the calculation is done by a computer program. The same applies to integration, for which there is no algebraic expression for the experimental data or analytical integration does not exist for the function. Therefore numerical integration is inevitable.

4.1 INTRODUCTION

In chemical reaction kinetics one of the methods for the determination of the order of a chemical reaction is the method of initial rates. In this method the reaction starts with different initial concentrations of the reactant A, and changes in the concentration of A with time are measured. For each initial concentration, the initial reaction rate can be calculated from the differentiation of concentration with respect to time at the beginning of the reaction:

$$-r_{A_0} = -\frac{dC_A}{dt}\Big|_{t=0} \tag{4.1}$$

If the reaction rate could be expressed by:

$$-r_A = kC_A^n \tag{4.2}$$

then taking the logarithm of both sides of this equation at $t = 0$ results in:

$$\ln(-r_{A_0}) = \ln k + n \ln C_{A_0} \tag{4.3}$$

The reaction order can be obtained by calculation of the slope of the line $\ln(-r_{A0})$ versus $\ln C_{A0}$.

Experimental determination of the rate of drying of a given material can be done by placing the moist material in a tray that is exposed to the drying air stream. A balance indicates the weight of the moist material, which is being recorded at different time intervals, during drying. The drying rate is calculated for each point by:

$$R = -\frac{1}{A}\frac{dW}{dt} \tag{4.4}$$

where R is the drying rate, A is the exposed surface area for drying, W is the mass of the moist material, and t is time.

In the study of the hydrodynamics of multiphase reactors the velocity profiles of solids may be determined experimentally by using the radioactive particle tracking (RPT) velocimetry technique [1]. In this technique a radioactive tracer is being followed for several time intervals, and coordinates of this tracer are evaluated at each time interval. The instantaneous velocity of the tracer then can be calculated from:

$$V_i = \frac{dx_i}{dt} \tag{4.5}$$

where V_i is the velocity of the tracer in direction i, x_i is the ith component of the coordinate of the tracer, and dt is the time increment used at the time of data acquisition. The steady-state velocity profile of solid particles in the reactor is calculated by averaging the instantaneous velocities in small compartments inside the reactor. Once the velocity profile is determined, solid velocity fluctuation is calculated by:

$$V'_i = V_i - <V_i>$$

(4.6)

where V'_i is the velocity fluctuation (a function of time) and $<V_i>$ is the average velocity (a function of position), both in i-direction. Having the aforementioned information, the turbulent eddy diffusivity of solids (D_i) may be obtained from the Lagrangian autocorrelation integral of velocity fluctuations:

$$D_i(t) = \int_0^t <V'_i(t)V'_i(t)> d\tau$$

(4.7)

The height of a cooling tower is calculated from the following equation:

$$z = \frac{G}{MK_Ga P} \int_{H_1}^{H_2} \frac{dH}{H^* - H}$$

(4.8)

where z is the height of the tower, G is the dry air mass flow, M is the molecular weight of air, K_Ga is the overall mass transfer coefficient, P is the pressure, H is the enthalpy of moist air, and H^* is the enthalpy of moist air at saturation. The integral in Eq. (4.8) should be calculated from H_1 at the inlet of the tower to H_2 at its outlet. To calculate this integral, enthalpies between H_1 and H_2 may be read from the psychrometric chart.

The calculation of the volume of a nonisothermal chemical reactor usually needs the use of numerical integration. For example, consider the first-order reaction $A \rightarrow B$ in the liquid phase, taking place in an adiabatic plug flow reactor. Pure A enters the reactor and it is desired to find the conversion X_1 at the outlet. The volume of this reactor is given by:

$$V = \frac{v_0}{k_1} \int_0^{x_1} \frac{dX}{(1 - X) \exp\left[\frac{E_a}{R}\left(\frac{1}{T_1} - \frac{1}{T}\right)\right]}$$

(4.9)

where V is the volume of the reactor, v_0 is the inlet volumetric flow rate of A, k_1 is the rate constant at the temperature T_1, E_a is the activation energy of the reaction, R is the ideal gas constant, T is the temperature of the reactor where the conversion is X, and T_1 is a reference temperature.

We must relate X and T through the energy balance to carry out this integration. For an adiabatic plug flow reactor, assuming constant heat capacities for both A and B results in:

$$T = T_0 + \frac{X(-\Delta H_R)}{C_{p_A} + X(C_{p_B} - C_{p_A})}$$

(4.10)

In this equation ΔH_R is the heat of the reaction, C_{pA} and C_{pB} are the heat capacities of A and B, respectively, and T_0 is a reference temperature.

To calculate the volume of the reactor from Eq. (4.9), the interval $[0, X]$ is divided into small ΔXs first, and from Eq. (4.10), the temperature in each increment can then be evaluated. Knowing both X and T, the function in the denominator of the integral in Eq. (4.9) is calculated. Finally, using a numerical technique for integration, the volume of the reactor can be calculated from Eq. (4.9).

In addition to calculating definite integrals, numerical integration can also be used to solve simple differential equations of the form:

$$y' = \frac{dy}{dx} = f(x) \tag{4.11}$$

The solution to the differential Eq. (4.11), after rearrangement, is given as:

$$y = y(x_0) + \int_{x_0}^{x} f(x)dx \tag{4.12}$$

In this chapter we deal with numerical differentiation in Sections 4.2 to 4.5 and integration in Sections 4.6 to 4.10.

4.2 DIFFERENTIATION BY BACKWARD FINITE DIFFERENCES

The relationships between backward difference operators and differential operators, which are summarized in Table 3.1, enable us to develop a variety of formulas expressing derivatives of functions in terms of backward finite differences, and vice versa. In addition, these formulas may have any degree of accuracy desired, provided that a sufficient number of terms is retained in the manipulation of these infinite series. This concept will be demonstrated in the remainder of this section.

4.2.1 FIRST DERIVATIVE

Rearrange Eq. (3.32) to solve for the differential operator D:

$$D = \frac{1}{h}\nabla + \frac{hD^2}{2} - \frac{h^2D^3}{6} + \cdots \tag{4.13}$$

Apply this operator to the function y at i:

$$Dy_i = \frac{1}{h}\nabla y_i + \frac{hD^2 y_i}{2} - \frac{h^2D^3 y_i}{6} + \cdots \tag{4.14}$$

Truncate the series, retaining only the first term, and show the order of the truncation error:

$$Dy_i = \frac{1}{h}\nabla y_i + O(h) \tag{4.15}$$

Express the differential and backward operators in terms of their respective definitions:

$$\frac{dy_i}{dx} = \frac{1}{h}(y_i - y_{i-1}) + O(h) \tag{4.16}$$

Eq. (4.16) therefore enables us to evaluate the first derivative of y at position i in terms of backward finite differences.

We can derive equations for the first derivative with higher accuracies. Rearrange Eq. (4.32) to solve for hD:

$$hD = \nabla + \frac{h^2 D^2}{2} - \frac{h^3 D^3}{6} + \cdots \tag{4.17}$$

Rearrange Eq. (3.36) to solve for $h^2 D^2$:

$$h^2 D^2 = \nabla^2 + h^3 D^3 - \frac{7}{12} h^4 D^4 + \cdots \tag{4.18}$$

Combine these two equations to eliminate $h^2 D^2$:

$$
\begin{aligned}
hD &= \nabla + \frac{1}{2}\left(\nabla^2 + h^3 D^3 - \frac{7}{12} h^4 D^4 + \cdots\right) - \frac{h^3 D^3}{6} + \cdots \\
&= \nabla + \frac{1}{2}\nabla^2 + \frac{h^3 D^3}{3} - \cdots
\end{aligned}
\tag{4.19}
$$

Divide through by h, and apply this operator to the function y at i:

$$Dy_i = \frac{1}{h}\nabla y_i + \frac{1}{2h}\nabla^2 y_i + \frac{h^2 D^3 y_i}{3} - \cdots \tag{4.20}$$

Truncate the series, retaining only the first *two* terms, and express the operators in terms of their respective definitions:

$$
\begin{aligned}
\frac{dy_i}{dx} &= \frac{1}{h}(y_i - y_{i-1}) + \frac{1}{2h}(y_i - 2y_{i-1} + y_{i-2}) + O(h^2) \\
&= \frac{1}{2h}(3y_i - 4y_{i-1} + y_{i-2}) + O(h^2)
\end{aligned}
\tag{4.21}
$$

The term $O(h^n)$ is used to represent the order of the first term in the truncated portion of the series. When $h < 1$ and the function is smooth and continuous, the first term in the truncated portion of the series is the predominant term. It should be emphasized that for $h < 1$,

$$h > h^2 > h^3 > h^4 > \ldots > h^n$$

Therefore when $h < 1$, formulas with the higher-order error term, $O(h^n)$ have smaller truncation errors; that is, they are more accurate approximations of derivatives.

On the other hand, when $h > 1$,

$$h < h^2 < h^3 < h^4 < \ldots < h^n$$

Therefore formulas with higher-order error terms have larger truncation errors and are less accurate approximations of derivatives.

In this section the first derivative of y is obtained with errors of order h and h^2. For the case where $h < 1$, Eq. (4.21) is a more accurate approximation of the first derivative than Eq. (4.16). To obtain higher accuracy, however, a larger number of terms is involved in the calculation.

It is obvious then that the choice of step size h is very important in determining the accuracy and stability of numerical integration and differentiation. This concept will be discussed in detail in Chapters 5 and 7.

4.2.2 SECOND DERIVATIVE

Rearrange Eq. (3.36) to solve for D^2:

$$D^2 = \frac{1}{h^2}\nabla^2 + hD^3 - \frac{7}{12}h^2D^4 + \cdots \tag{4.22}$$

Apply this operator to the function y at i:

$$D^2y_i = \frac{1}{h^2}\nabla^2y_i + hD^3y_i - \frac{7}{12}h^2D^4y_i + \cdots \tag{4.23}$$

Truncate the series, retaining only the first term, and express the operators in terms of their respective definition:

$$\frac{d^2y_i}{dx^2} = \frac{1}{h^2}(y_i - 2y_{i-1} + y_{i-2}) + O(h) \tag{4.24}$$

This equation evaluates the second derivative of y at position i, in terms of backward finite differences, with an error of order h.

We can derive equations for the second derivative with higher accuracies. Rearrange Eq. (3.36) to solve for h^2D^2:

$$h^2D^2 = \nabla^2 + h^3D^3 - \frac{7}{12}h^4D^4 + \cdots \tag{4.25}$$

Rearrange Eq. (3.37) to solve for h^3D^3:

$$h^3D^3 = \nabla^3 + \frac{3}{2}h^4D^4 - \frac{5}{4}h^5D^5 + \cdots \tag{4.26}$$

Combine these two equations to eliminate h^3D^3:

$$\begin{aligned} h^2D^2 &= \nabla^2 + \left(\nabla^3 + \frac{3}{2}h^4D^4 - \frac{5}{4}h^5D^5 + \cdots\right) - \frac{7}{12}h^4D^4 + \cdots \\ &= \nabla^2 + \nabla^3 + \frac{11}{12}h^4D^4 - \cdots \end{aligned} \tag{4.27}$$

Divide through by h^2 and apply the operator to the function y at i:

$$D^2y_i = \frac{1}{h^2}\nabla^2y_i + \frac{1}{h^2}\nabla^3y_i + \frac{11}{12}h^2D^4y_i - \cdots \tag{4.28}$$

Truncate the series, retaining only the first two terms, and express the operators in terms of their respective definitions:

$$\begin{aligned} \frac{d^2y_i}{dx^2} &= \frac{1}{h^2}(y_i - 2y_{i-1} + y_{i-2}) + \frac{1}{h^2}(y_i - 3y_{i-1} + 3y_{i-2} - y_{i-3}) + O(h^2) \\ &= \frac{1}{h^2}(2y_i - 5y_{i-1} + 4y_{i-2} - y_{i-3}) + O(h^2) \end{aligned} \tag{4.29}$$

It should be noted that this same equation could have been derived using Eq. (4.42) and an equation for ∇^4 (not shown here). This statement applies to all these examples, which can be solved using both sets of equations shown in Table 3.1.

The formulas for the first and second derivatives, developed in the preceding two sections, together with those of the third and fourth derivatives, are summarized in Table 4.1. It can be concluded, from these examples, that any derivative can be expressed in terms of finite differences with any degree of accuracy desired. These formulas may be used to differentiate the function $y(x)$ given a set of values of this function at equally spaced intervals of x, such as a set of experimental data. Conversely, these same formulas may be used in the numerical integration of differential equations, as shown in Chapters 5 to 7.

Table 4.1 Derivatives in terms of backward finite differences.

Error of order h

$$\frac{dy_i}{dx} = \frac{1}{h}(y_i - y_{i-1}) + O(h)$$

$$\frac{d^2y_i}{dx^2} = \frac{1}{h^2}(y_i - 2y_{i-1} + y_{i-2}) + O(h)$$

$$\frac{d^3y_i}{dx^3} = \frac{1}{h^3}(y_i - 3y_{i-1} + 3y_{i-2} - y_{i-3}) + O(h)$$

$$\frac{d^4y_i}{dx^4} = \frac{1}{h^4}(y_i - 4y_{i-1} + 6y_{i-2} - 4y_{i-3} + y_{i-4}) + O(h)$$

Error of order h^2

$$\frac{dy_i}{dx} = \frac{1}{2h}(3y_i - 4y_{i-1} + y_{i-2}) + O(h^2)$$

$$\frac{d^2y_i}{dx^2} = \frac{1}{h^2}(2y_i - 5y_{i-1} + 4y_{i-2} - y_{i-3}) + O(h^2)$$

$$\frac{d^3y_i}{dx^3} = \frac{1}{2h^3}(5y_i - 18y_{i-1} + 24y_{i-2} - 14y_{i-3} + 3y_{i-4}) + O(h^2)$$

$$\frac{d^4y_i}{dx^4} = \frac{1}{h^4}(3y_i - 14y_{i-1} + 26y_{i-2} - 24y_{i-3} + 11y_{i-4} - 2y_{i-5}) + O(h^2)$$

4.3 DIFFERENTIATION BY FORWARD FINITE DIFFERENCES

The relationships between forward difference operators and differential operators, which are summarized in Table 3.2, enable us to develop a variety of formulas expressing derivatives of functions in terms of forward finite differences, and vice versa. As was demonstrated in Section 4.2, these formulas may have any degree of accuracy desired, provided that a sufficient number of terms is retained in the manipulation of these infinite series. A set of expressions, parallel to those of Section 4.2, will be derived using the forward finite differences.

4.3.1 FIRST DERIVATIVE

Rearrange Eq. (3.53) to solve for the differential operator D:

$$D = \frac{1}{h}\Delta - \frac{hD^2}{2} - \frac{h^2D^3}{6} - \cdots \tag{4.30}$$

Apply this operator to the function y at i:

$$Dy_i = \frac{1}{h}\Delta y_i - \frac{hD^2y_i}{2} - \frac{h^2D^3y_i}{6} - \cdots \tag{4.31}$$

Truncate the series, retaining only the first term:

$$Dy_i = \frac{1}{h}\Delta y_i + O(h) \tag{4.32}$$

Express the differential and forward operators in terms of their respective definitions:

$$\frac{dy_i}{dx} = \frac{1}{h}(y_{i+1} - y_i) + O(h) \tag{4.33}$$

Eq. (4.33) enables us to evaluate the first derivative of y at position i in terms of forward finite differences with an error of order h.

We can derive equations for the first derivative with higher accuracies. Rearrange Eq. (3.53) to solve for hD:

$$hD = \Delta - \frac{h^2 D^2}{2} - \frac{h^3 D^3}{6} - \cdots \tag{4.34}$$

Rearranging Eq. (3.57) to solve for $h^2 D^2$:

$$h^2 D^2 = \Delta^2 - h^3 D^3 - \frac{7}{12} h^4 D^4 - \cdots \tag{4.35}$$

Combine these two equations to eliminate $h^2 D^2$:

$$
\begin{aligned}
hD &= \Delta - \frac{1}{2}\left(\Delta^2 - h^3 D^3 - \frac{7}{12} h^4 D^4 - \cdots\right) - \frac{h^3 D^3}{6} - \cdots \\
&= \Delta - \frac{1}{2}\Delta^2 + \frac{h^3 D^3}{3} + \cdots
\end{aligned}
\tag{4.36}
$$

Divide through by h, and apply this operator to the function y at i:

$$Dy_i = \frac{1}{h}\Delta y_i - \frac{1}{2h}\Delta^2 y_i + \frac{h^2 D^3 y_i}{3} + \cdots \tag{4.37}$$

Truncate the series, retaining only the first *two* terms, and express the operators in terms of their respective definitions:

$$
\begin{aligned}
\frac{dy_i}{dx} &= \frac{1}{h}(y_{i+1} - y_i) - \frac{1}{2h}(y_{i+2} - 2y_{i+1} + y_i) + O(h^2) \\
&= \frac{1}{2h}(-y_{i+2} + 4y_{i+1} - 3y_i) + O(h^2)
\end{aligned}
\tag{4.38}
$$

4.3.2 SECOND DERIVATIVE

Rearrange Eq. (3.57) to solve for D^2:

$$D^2 = \frac{1}{h^2}\Delta^2 - hD^3 - \frac{7}{12}h^2 D^4 - \cdots \tag{4.39}$$

Apply this operator to the function y at i:

$$D^2 y_i = \frac{1}{h^2}\Delta^2 y_i - hD^3 y_i - \frac{7}{12}h^2 D^4 y_i - \cdots \tag{4.40}$$

Truncate the series, retaining only the first term, and express the operators in terms of their respective definitions:

$$\frac{d^2 y_i}{dx^2} = \frac{1}{h^2}(y_{i+2} - 2y_{i+1} + y_i) + O(h) \tag{4.41}$$

This equation evaluates the second derivative of y at position i, in terms of forward finite differences, with an error of order h.

We can derive equations for the second derivative with higher accuracies. Rearrange Eq. (3.57) to solve for h^2D^2:

$$h^2D^2 = \Delta^2 - h^3D^3 - \frac{7}{12}h^4D^4 - \cdots \tag{4.42}$$

Rearrange Eq. (3.58) to solve for h^3D^3:

$$h^3D^3 = \Delta^3 - \frac{3}{2}h^4D^4 - \frac{5}{4}h^5D^5 - \cdots \tag{4.43}$$

Combine these two equations to eliminate h^3D^3:

$$\begin{aligned}
h^2D^2 &= \Delta^2 - \left(\Delta^3 - \frac{3}{2}h^4D^4 - \frac{5}{4}h^5D^5 - \cdots\right) - \frac{7}{12}h^4D^4 - \cdots \\
&= \Delta^2 - \Delta^3 + \frac{11}{12}h^4D^4 - \cdots
\end{aligned} \tag{4.44}$$

Divide through by h^2 and apply the operator to the function y at i:

$$D^2y_i = \frac{1}{h^2}\Delta^2y_i - \frac{1}{h^2}\Delta^3y_i + \frac{11}{12}h^2D^4y_i - \cdots \tag{4.45}$$

Truncate the series, retaining only the first two terms, and express the operators in terms of their respective definitions:

$$\frac{d^2y_i}{dx^2} = \frac{1}{h^2}(-y_{i+3} + 4y_{i+2} - 5y_{i+1} + 2y_i) + O(h^2) \tag{4.46}$$

Table 4.2 Derivatives in terms of forward finite differences.

Error of order h

$$\frac{dy_i}{dx} = \frac{1}{h}(y_{i+1} - y_i) + O(h)$$

$$\frac{d^2y_i}{dx^2} = \frac{1}{h^2}(y_{i+2} - 2y_{i+1} + y_i) + O(h)$$

$$\frac{d^3y_i}{dx^3} = \frac{1}{h^3}(y_{i+3} - 3y_{i+2} + 3y_{i+1} - y_i) + O(h)$$

$$\frac{d^4y_i}{dx^4} = \frac{1}{h^4}(y_{i+4} - 4y_{i+3} + 6y_{i+2} - 4y_{i+1} + y_i) + O(h)$$

Error of order h^2

$$\frac{dy_i}{dx} = \frac{1}{2h}(-y_{i+2} + 4y_{i+1} - 3y_i) + O(h^2)$$

$$\frac{d^2y_i}{dx^2} = \frac{1}{h^2}(-y_{i+3} + 4y_{i+2} - 5y_{i+1} + 2y_i) + O(h^2)$$

$$\frac{d^3y_i}{dx^3} = \frac{1}{2h^3}(-3y_{i+4} + 14y_{i+3} - 24y_{i+2} + 18y_{i+1} - 5y_i) + O(h^2)$$

$$\frac{d^4y_i}{dx^4} = \frac{1}{h^4}(-2y_{i+5} + 11y_{i+4} - 24y_{i+3} + 26y_{i+2} - 14y_{i+1} + 3y_i) + O(h^2)$$

The formulas developed in these sections for the first and second derivatives are summarized in Table 4.2, together with those of the third and fourth derivatives.

It should be pointed out that all the finite difference approximations of derivatives obtained in this section and the previous section have coefficients that add up to zero. This is a rule of thumb that applies to all such combinations of finite differences.

From a comparison between Tables 4.1 and 4.2, we conclude that derivatives can be expressed in their backward or forward differences, with formulas that are very similar to each other in the number of terms involved, and in the order of truncation error. The choice between using forward or backward differences will depend on the problem and its boundary conditions. This will be discussed further in Chapters 5 to 7.

4.4 DIFFERENTIATION BY CENTRAL FINITE DIFFERENCES

The relationships between central difference operators and differential operators, which are summarized in Table 3.3, will be used in the following sections to develop a set of formulas expressing the derivatives in terms of central finite differences. These formulas will have higher accuracy than those developed in the previous two sections using backward and forward finite differences.

4.4.1 FIRST DERIVATIVE

Rearrange Eq. (3.78) to solve for D:

$$D = \frac{1}{h}\mu\delta - \frac{h^2 D^3}{6} - \frac{h^4 D^5}{120} - \cdots \tag{4.47}$$

Apply this operator to the function y at i:

$$Dy_i = \frac{1}{h}\mu\delta y_i - \frac{h^2 D^3 y_i}{6} - \frac{h^4 D^5 y_i}{120} - \cdots \tag{4.48}$$

Truncate the series, retaining only the first term:

$$Dy_i = \frac{1}{h}\mu\delta y_i + O(h^2) \tag{4.49}$$

Express the differential and averaged central difference operators in terms of their respective definitions:

$$\frac{dy_i}{dx} = \frac{1}{2h}(y_{i+1} - y_{i-1}) + O(h^2) \tag{4.50}$$

This equation enables us to evaluate the first derivative of y at position i in terms of central finite differences. Comparing this equation with Eq. (4.16) and Eq. (4.33) reveals that the use of central differences increases the accuracy of the formulas for the same number of terms retained.

We can derive equations for the first derivative with higher accuracies. Rearrange Eq. (3.78) to solve for hD:

$$hD = \mu\delta - \frac{h^3D^3}{6} - \frac{h^5D^5}{120} - \cdots \tag{4.51}$$

Rearrange Eq. (3.82) to solve for h^3D^3:

$$h^3D^3 = \mu\delta^3 - \frac{h^5D^5}{4} - \frac{h^7D^7}{40} - \cdots \tag{4.52}$$

Combine these two equations to eliminate h^3D^3:

$$hD = \mu\delta - \frac{1}{6}\left(\mu\delta^3 - \frac{h^5D^5}{4} - \frac{h^7D^7}{40} - \cdots\right) - \frac{h^5D^5}{120} - \cdots$$
$$= \mu\delta - \frac{1}{6}\mu\delta^3 + \frac{h^5D^5}{30} + \cdots \tag{4.53}$$

Divide through by h and apply this operator to the function y at i:

$$Dy_i = \frac{1}{h}\mu\delta y_i - \frac{1}{6h}\mu\delta^3 y_i + \frac{h^4D^5y_i}{30} + \cdots \tag{4.54}$$

Truncate the series, retaining only the first *two* terms, and express the operators in terms of their respective definitions:

$$\frac{dy_i}{dx} = \frac{1}{2h}(y_{i+1} - y_{i-1}) - \frac{1}{12h}(y_{i+2} - 2y_{i+1} + 2y_{i-1} - y_{i-2}) + O(h^4)$$
$$= \frac{1}{12h}(-y_{i+2} + 8y_{i+1} - 8y_{i-1} + y_{i-2}) + O(h^4) \tag{4.55}$$

4.4.2 SECOND DERIVATIVE

Rearrange Eq. (3.81) to solve for D^2:

$$D^2 = \frac{1}{h^2}\delta^2 - \frac{h^2D^4}{12} - \frac{h^4D^6}{360} - \cdots \tag{4.56}$$

Apply this operator to the function y at i:

$$D^2y_i = \frac{1}{h^2}\delta^2 y_i - \frac{h^2D^4y_i}{12} - \frac{h^4D^6y_i}{360} - \cdots \tag{4.57}$$

Truncate the series, retaining only the first term:

$$D^2y_i = \frac{1}{h^2}\delta^2 y_i + O(h^2) \tag{4.58}$$

Express the differential and central difference operators in terms of their respective definitions:

$$\frac{d^2y_i}{dx^2} = \frac{1}{h^2}(y_{i+1} - 2y_i + y_{i-1}) + O(h^2) \tag{4.59}$$

We can derive equations for the second derivative with higher accuracies. Rearrange Eq. (3.81) to solve for $h^2 D^2$:

$$h^2 D^2 = \delta^2 - \frac{h^2 D^4}{12} - \frac{h^6 D^6}{360} - \cdots \tag{4.60}$$

Rearrange Eq. (3.83) to solve for $h^4 D^4$:

$$h^4 D^4 = \delta^4 - \frac{h^6 D^6}{6} - \frac{h^8 D^8}{80} - \cdots \tag{4.61}$$

Combine these two equations to eliminate $h^4 D^4$:

$$\begin{aligned} h^2 D^2 &= \delta^2 - \frac{1}{12}\left(\delta^4 - \frac{h^6 D^6}{6} - \frac{h^8 D^8}{80} - \cdots\right) - \frac{h^6 D^6}{360} - \cdots \\ &= \delta^2 - \frac{1}{12}\delta^4 + \frac{h^6 D^6}{90} - \cdots \end{aligned} \tag{4.62}$$

Divide through by h^2 and apply this operator to function y at i:

$$D^2 y_i = \frac{1}{h^2}\delta^2 y_i - \frac{1}{12h^2}\delta^4 y_i + \frac{h^4 D^6}{90} - \cdots \tag{4.63}$$

Truncate the series, retaining only the first two terms, and express the operators in terms of their respective definitions:

$$\begin{aligned} \frac{d^2 y_i}{dx^2} &= \frac{1}{h^2}(y_{i+1} - 2y_i + y_{i-1}) - \frac{1}{12h^2}(y_{i+2} - 4y_{i+1} + 6y_i - 4y_{i-1} + y_{i-2}) + O(h^4) \\ &= \frac{1}{12h^2}(-y_{i+2} + 16y_{i+1} - 30y_i + 16y_{i-1} - y_{i-2}) + O(h^4) \end{aligned} \tag{4.64}$$

Table 4.3 Derivatives in terms of central finite differences.

Error of order h^2

$$\frac{dy_i}{dx} = \frac{1}{2h}(y_{i+1} - y_{i-1}) + O(h^2)$$

$$\frac{d^2 y_i}{dx^2} = \frac{1}{h^2}(y_{i+1} - 2y_i + y_{i-1}) + O(h^2)$$

$$\frac{d^3 y_i}{dx^3} = \frac{1}{2h^3}(y_{i+2} - 2y_{i+1} + 2y_{i-1} - y_{i-2}) + O(h^2)$$

$$\frac{d^4 y_i}{dx^4} = \frac{1}{h^4}(y_{i+2} - 4y_{i+1} + 6y_i - 4y_{i-1} + y_{i-2}) + O(h^2)$$

Error of order h^4

$$\frac{dy_i}{dx} = \frac{1}{12h}(-y_{i+2} + 8y_{i+1} - 8y_{i-1} + y_{i-2}) + O(h^4)$$

$$\frac{d^2 y_i}{dx^2} = \frac{1}{12h^2}(-y_{i+2} + 16y_{i+1} - 30y_i + 16y_{i-1} - y_{i-2}) + O(h^4)$$

$$\frac{d^3 y_i}{dx^3} = \frac{1}{8h^3}(-y_{i+3} + 8y_{i+2} - 13y_{i+1} + 13y_{i-1} - 8y_{i-2} + y_{i-3}) + O(h^4)$$

$$\frac{d^4 y_i}{dx^4} = \frac{1}{6h^4}(-y_{i+3} + 12y_{i+2} - 39y_{i+1} + 56y_i - 39y_{i-1} + 12y_{i-2} - y_{i-3}) + O(h^4)$$

The formulas derived in Sections 4.4.1 and 4.4.2 for the first and second derivatives are summarized in Table 4.3, along with those for the third and fourth derivatives. Development of formulas with higher accuracy and for the higher derivatives are left as exercises for the reader (see Problems).

EXAMPLE 4.1 DERIVATIVE OF VECTORS OF EQUALLY SPACED POINTS

Calculate the solids volume fraction in a riser of a bench-scale gas-solid fluidized bed whose axial pressure profile is given in Table E4.1. Assume fully developed solids flow in the riser and neglect wall shear and solids stress. The densities of gas and solids phases are 1.2 kg/m^3 and 2650 kg/m^3, respectively. Solve this problem in Excel. Also, write a general MATLAB$^®$ function for solving this problem to calculate first- to fourth-order derivatives of a series of data presented numerically in a matrix whose columns represent vectors of the dependent variable. The user should be able to choose between backward, forward, or central differentiation, as well as the order of the truncation error.

Method of solution

The equations in Tables 4.1, 4.2, and 4.3 are used to differentiate the columns of the matrix **y** with the desired order of truncation error. Differentiation is done based on equally spaced segments of the independent variable.

Writing the momentum balance equation for the two-phase flow, we find that pressure drop in the aforementioned conditions is balanced by the weight of the bed, that is,

$$-\frac{dP}{dz} = \left[\rho_g(1 - \varepsilon_s) + \rho_s\varepsilon_s\right]g \tag{1}$$

where p is the pressure, z is the axial position, ρ_g and ρ_s are the densities of gas and solids, respectively, ε_s is the volume fraction of the solids, and g is the gravitational acceleration.

Eq. (1) can be solved for ε_s:

(Continued)

Table E4.1 Pressure profile along the riser.

No.	Axial position (m)	Pressure (kPa)	No.	Axial position (m)	Pressure (kPa)
1	0.00	27.2	11	8.40	14.9
2	0.84	22.9	12	9.24	13.5
3	1.68	21.2	13	10.08	12.9
4	2.52	20.6	14	10.92	12.3
5	3.36	19.7	15	11.76	11.5
6	4.20	18.5	16	12.60	10.8
7	5.04	17.8	17	13.44	10.2
8	5.88	17.1	18	14.28	9.1
9	6.72	16.4	19	15.12	5.7
10	7.56	15.6	20	15.96	0

EXAMPLE 4.1 (CONTINUED)

$$\varepsilon_s = \frac{(-dp/dz) - \rho_g g}{\left(\rho_s - \rho_g\right)g} \tag{2}$$

The solids volume fraction profile can be calculated from Eq. (2) once the pressure gradient is extracted from the data tabulated in Table E4.1.

Before we solve this problem in Excel, let us compare the pressure gradients evaluated by different derivative formulas, at one point, say point No. 5. Note that $h = 0.84$ m in Table E4.1.

In terms of backward finite difference with the error of order h:

$$\frac{dy_i}{dx} = \frac{1}{h}(y_i - y_{i-1}) + O(h)$$

$$\frac{dp_5}{dz} = \frac{1}{h}(p_5 - p_4) = \frac{19.7 - 20.6}{0.84} = -1.071 \quad \text{kPa/m}$$

In terms of backward finite difference with the error of order h^2:

$$\frac{dy_i}{dx} = \frac{1}{2h}(3y_i - 4y_{i-1} + y_{i-2}) + O(h^2)$$

$$\frac{dp_5}{dx} = \frac{1}{2h}(3p_5 - 4p_4 + p_3) = \frac{3 \times 19.7 - 4 \times 20.6 + 21.1}{2 \times 0.84} = -1.310 \quad kPa/m$$

In terms of forward finite difference with the error of order h:

$$\frac{dy_i}{dx} = \frac{1}{h}(y_{i+1} - y_i) + O(h)$$

$$\frac{dp_5}{dx} = \frac{1}{h}(p_6 - p_5) = \frac{18.5 - 19.7}{0.84} = -1.429 \quad kPa/m$$

In terms of forward finite difference with the error of order h^2:

$$\frac{dy_i}{dx} = \frac{1}{2h}(-y_{i+2} + 4y_{i+1} - 3y_i) + O()$$

$$\frac{dp_5}{dx} = \frac{1}{2h}(-p_7 + 4p_6 - 3p_5) = \frac{-17.8 + 4 \times 18.5 - 3 \times 19.7}{2 \times 0.84} = -1.726 \quad kPa/m$$

In terms of central finite difference with the error of order h^2:

$$\frac{dy_i}{dx} = \frac{1}{2h}(y_{i+1} - y_{i-1}) + O(h^2)$$

(Continued)

EXAMPLE 4.1 (CONTINUED)

$$\frac{dp_5}{dx} = \frac{1}{2h}(p_6 - p_4) = \frac{18.5 - 20.6}{2 \times 0.84} = -1.250 \quad kPa/m$$

In terms of central finite difference with the error of order h^4:

$$\frac{dy_i}{dx} = \frac{1}{12h}(-y_{i+2} + 8y_{i+1} - 8y_{i-1} + y_{i-2}) + O(h^4)$$

$$\frac{dp_5}{dx} = \frac{1}{12h}(-p_7 + 8p_6 - 8p_4 + p_3) = \frac{-17.8 + 8 \times 18.5 - 8 \times 20.6 + 21.2}{12 \times 0.84} = -1.329 \quad kPa/m$$

For solving the problem in Excel, we use the formula of the first derivative in terms of central finite difference with the error of order h^2. This formula cannot be used at the start and endpoints. Instead the forward and backward differences formulas are employed at start and endpoints, respectively, both with the error of order h^2 (for being consistent with the central difference formula). The steps for evaluating the pressure gradient and then the solids volume fraction in Excel are shown in Fig. E4.1a.

Program description

The programs and functions developed in this example can be found in: https://www.elsevier.com/books-and-journals/book-companion/9780128229613

The MATLAB function *deriv.m* is written to calculate first to fourth-order derivatives of a matrix of input data. The first part of the program is initialization, in which the values of h, the order of derivative, method of finite difference used, and order of truncation error are assigned if not entered as input to the function. If only y is given as input to the function, the program calculates the central finite differences of y as the output. The second input argument is the increment of the independent variable. The third one is the order of derivative. A value of -1, 0, or 1 as the fourth input argument results in the calculation of the derivative based on backward, central, or forward finite differences, respectively. The fifth argument is the value of the order of truncation error (1 or 2 for backward and forward differences and 2 or 4 for central differences).

The derivative matrix returned by the function *deriv.m* has the same number of elements as the vector of input data itself. However, it is important to note that, depending on the method of finite difference used, some elements at one or both ends of the derivative vector are evaluated by a different method of differentiation. For example, in first-order differentiation with the forward finite difference method with truncation error $O(h)$, the last element of the returned derivative vector is calculated by backward differences. Another example is the calculation of the second derivative of a vector by the central finite difference method with truncation error $O(h^2)$, where the function evaluates the first two elements of the vector of derivatives by forward differences and the last two elements of the vector of derivatives by backward differences. The reader should pay special attention to the fact that when the

(Continued)

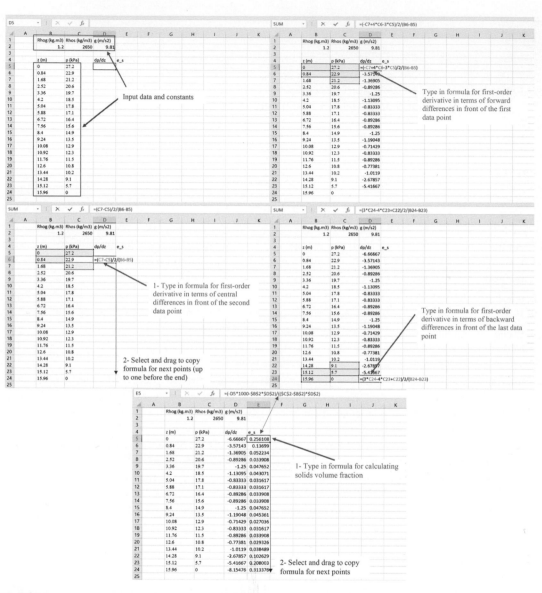

FIG. E4.1a

Calculating solids volume fraction along the riser based on pressure profile in Excel.

EXAMPLE 4.1 (CONTINUED)

function calculates the derivative by the central finite difference method with the truncation error of the order $O(h^4)$, the starting and ending rows of derivative values are calculated by forward and backward finite differences with truncation error of the order $O(h^2)$.

The main program *Example4_1.m* asks the reader to input the data from the keyboard. It then applies the function *deriv.m* to evaluate the pressure gradient and calculates the solids volume fraction from Eq. (2). In the end the program plots the result of the foregoing calculations.

Input and results

>> Example4_1
Vector of pressures (kPa) = [27.2, 22.9, 21.2, 20.6, 19.7, 18.5, 17.8, 17.1, 16.4, 15.6, 14.9, 13.5, 12.9, 12.3, 11.5, 10.8, 10.2, 9.1, 5.7, 0]
Axial distance between pressure probes (m) = 0.84
Density of the gas (kg/m^3) = 1.2
Density of solids (kg/m^3) = 2650

Discussion of results

Fig. E4.1b shows the results graphically. It can be seen from this figure that the solids fraction is higher at the bottom and top sections of the riser, while it is basically constant in the middle section (the fully developed zone). The pressure profile is also shown in the same

(Continued)

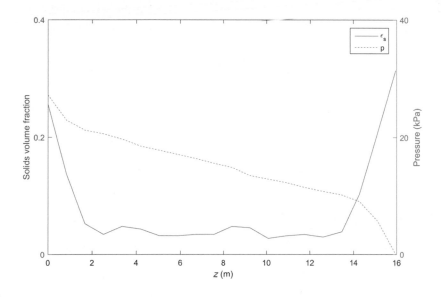

FIG. E4.1b

Solids volume fraction and pressure profiles.

EXAMPLE 4.1 (CONTINUED)

figure and it visually demonstrates how pressure decrease from the bottom toward the top of the riser. The trend of constant solids volume fraction in the majority of the length of the riser confirms the assumption made at the beginning that the measurements were done in the fully developed zone where the solids move with a constant velocity.

4.5 SPLINE DIFFERENTIATION

In some situations tabulated function values are available instead of their algebraic expression, and it is desired to evaluate the derivative of the function at a point (or points) between the tabulated values. A practical method in this situation is to interpolate the base points first and calculate the value of the derivative from differentiating the interpolating polynomial. Among different interpolating polynomials, cubic splines have the advantage of continuity of the first derivative through all base points.

By cubic spline interpolation of the function, its derivative at any point x in the interval $[x_{i-1},$ $x_i]$ can be calculated from differentiating Eq. (3.125):

$$\frac{dy}{dx} = \frac{1}{2}\frac{(x-x_i)^2}{x_{i-1}-x_i}y''_{i-1} + \frac{1}{2}\frac{(x-x_{i-1})^2}{x_i-x_{i-1}}y''_i - \frac{1}{6}(x_{i-1}-x_i)(y''_{i-1} - y''_i) + \frac{y_{i-1}-y_i}{x_{i-1}-x_i} \qquad (4.65)$$

Before calculating the derivative from Eq. (4.65), the values of the second derivative at the base points should be calculated from Eq. (3.129). Note that if a natural spline interpolation is employed, the second derivatives for the first and the last intervals are equal to zero.

The reader can easily modify the MATLAB function *NaturalSpline.m* (see Example 3.7) to calculate at any point the first derivative of a function from a series of tabulated data. It is enough to replace the formula of the interpolation section with the differentiation formula, Eq. (4.65). Also, the MATLAB function *spline.m* is able to give the piecewise polynomial coefficients from which the derivative of the function can be evaluated. A good example of applying such a method can be found in Hanselman and Littlefield [2]. As mentioned before, *spline.m* applies the not-a-knot algorithm for calculating the polynomial coefficients.

4.6 INTEGRATION FORMULAS

In the following sections we develop integration formulas. This operation is represented by:

$$S = \int_{x_0}^{x_n} f(x)dx \qquad (4.66)$$

which is the integral of the function $y = f(x)$, or *integrand*, with respect to the independent variable x, evaluated between the limits $x = x_0$ to $x = x_n$. If the function $f(x)$ is such that it can be

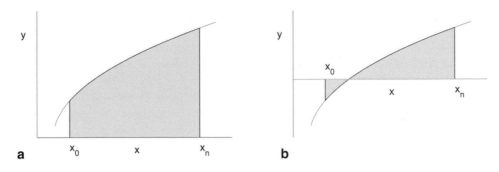

FIG. 4.1

Graphical presentation of the integral (a) positive area only (b) positive and negative areas.

integrated analytically, numerical methods are not needed for this problem. However, in many cases the function $f(x)$ is very complicated, or the function is only a set of tabulated values of x and y, such as experimental data. Under these circumstances, the integral in Eq. (4.66) must be developed numerically. This operation is known as *numerical quadrature*.

It is known from differential calculus that the integral of a function $f(x)$ is equivalent to the area between the function and the x-axis enclosed within the limits of integration, as shown in Fig. 4.1a. Any portion of the area that is below the x-axis is counted as a negative area (Fig. 4.1b). Therefore one way of evaluating the integral in Eq. (4.66) is to plot the function graphically and then simply measure the area enclosed by the function. However, this is a very impractical and inaccurate way of evaluating integrals.

A more accurate and systematic way of evaluating integrals is to perform the integration numerically. In the next two sections we derive Newton-Cotes integration formulas for equally spaced intervals and Gauss quadrature for unequally spaced points.

4.7 NEWTON-COTES FORMULAS OF INTEGRATION

This method is accomplished by first replacing the function $y = f(x)$ with a polynomial approximation, such as the Gregory-Newton forward interpolation formula (Eq. 3.104). In practice, the interval $[x_0, x_n]$ is divided into several segments, each of width h, and the Gregory-Newton forward interpolation formula becomes (note that $x_{i+1} = x_i + h$):

$$y = y_0 + \frac{(x - x_0)}{h}\Delta y_0 + \frac{(x - x_0)(x - x_1)}{2!h^2}\Delta^2 y_0 + \frac{(x - x_0)(x - x_1)(x - x_2)}{3!h^3}\Delta^3 y_0 + \cdots \qquad (4.67)$$

Because this interpolation formula fits the function exactly at a finite number of points $(n + 1)$, we divide the total interval of integration $[x_0, x_n]$ into n segments, each of width h. In the next step by using Eq. (4.67), Eq. (4.66) can be integrated. The upper limits of integration can be chosen to include an increasing set of segments of integration, each of width h. In each case we retain a number of finite differences in the finite series of Eq. (4.67) equal to the number of segments of integration. This operation yields the well-known *Newton-Cotes formulas of integration*. The first three of the Newton-Cotes formulas are also known by the names *trapezoidal rule*, *Simpson's 1/3 rule*, and *Simpson's 3/8 rule*, respectively. These are developed in the next three sections.

4.7.1 THE TRAPEZOIDAL RULE

In developing the first Newton-Cotes formula we use one segment of width h and fit the polynomial through two points (x_0, y_0) and (x_1, y_1) (Fig. 4.2). This is tantamount to fitting a straight line between these points. We retain the first two terms of the Gregory-Newton polynomial (up to, and including, the first forward finite difference), and group together the rest of the terms of the polynomial into a remainder term. Thus the integral equation becomes:

$$S_1 = \int_{x_0}^{x_1} \left[y_0 + \frac{(x - x_0)}{h} \Delta y_0 \right] dx + \int_{x_0}^{x_1} R_n(x) dx \tag{4.68}$$

The first integral on the right-hand side is integrated with respect to x and the first forward difference is replaced with its definition of $\Delta y_0 = y_1 - y_0$, to obtain:

$$S_1 = \frac{h}{2}(y_0 + y_1) + \int_{x_0}^{x_1} R_n(x) dx \tag{4.69}$$

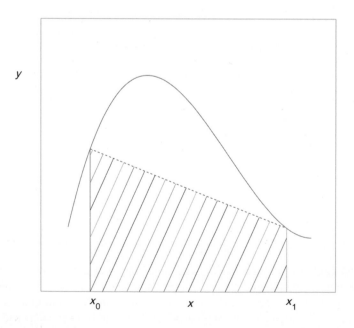

FIG. 4.2

Enlargement segment showing the application of the trapezoidal rule.

The remainder term is evaluated as follows:

$$\int_{x_0}^{x_1} R_n(x)dx = \int_{x_0}^{x_1} \left[\frac{(x-x_0)(x-x_1)}{2!h^2}\Delta^2 y_0 + \frac{(x-x_0)(x-x_1)(x-x_2)}{3!h^3}\Delta^3 y_0 + \cdots \right] dx$$

$$= -\frac{1}{12}h\Delta^2 y_0 + \frac{1}{24}h\Delta^3 y_0 - \cdots \tag{4.70}$$

The forward difference operators, Δ^2, Δ^3, ..., are replaced by their equivalent in terms of differential operators (Eqs. 3.57 and 3.58) and the remainder term becomes:

$$\int_{x_0}^{x_1} R_n(x)dx = -\frac{1}{12}h^3 D^2 y_0 - \frac{1}{24}h^4 D^3 y_0 + \cdots \tag{4.71}$$

The remainder series can be replaced by one term evaluated at ξ_1; therefore:

$$\int_{x_0}^{x_1} R_n(x)dx = -\frac{1}{12}h^3 D^2 f(\xi_1) \tag{4.72}$$

This is a term of order h^3 and is abbreviated by $O(h^3)$. Therefore, Eq. (4.69) can be written as:

$$S_1 = \frac{h}{2}(y_0 + y_1) + O(h^3) \tag{4.73}$$

This equation is known as the *trapezoidal rule* because the term $(h/2)(y_0 + y_1)$ is essentially the formula for calculating the area of a trapezoid. In this case the segment of integration is a trapezoid standing on its side. It was mentioned earlier that fitting a polynomial through only two points is equivalent to fitting a straight line through these points. This causes the shape of the integration segment to be a trapezoid, shown as the shaded area in Fig. 4.2. The area between $y = f(x)$ and the straight line represents the truncation error of the trapezoidal rule. If the function $f(x)$ is actually linear, then the trapezoidal rule calculates the integral exactly, because $D^2 f(\xi_1) = 0$, which causes the remainder term to vanish.

The trapezoidal rule in the form of Eq. (4.73) gives the integral of only one integration segment of width h. To obtain the total integral, Eq. (4.68) must be applied over each of the n segments (with the appropriate limits of integration) to obtain the following series of equations:

$$S_1 = \frac{h}{2}(y_0 + y_1) + O(h^3) \tag{4.73}$$

$$S_2 = \frac{h}{2}(y_1 + y_2) + O(h^3) \tag{4.74}$$

$$\vdots$$

$$S_n = \frac{h}{2}(y_{n-1} + y_n) + O(h^3) \tag{4.75}$$

The addition of all these equations over the total interval gives the *multiple-segment trapezoidal rule*:

$$S = \frac{h}{2}\left(y_0 + 2\sum_{i=1}^{n-1} y_i + y_n \right) + nO(h^3) \tag{4.76}$$

Fig. 4.3 illustrates the flowchart of the multiple-segment trapezoidal rule of integration.

For simplicity, the error term has been shown as $nO(h^3)$ in Eq. (4.76). This is only an approximation because the remainder term includes the second derivative of y evaluated at unknown values

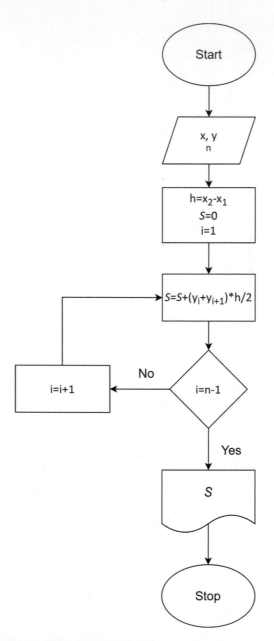

FIG. 4.3 Flowchart of the multiple-segment trapezoidal rule.

In this flowchart x and y are the vectors of independent and function variables of base points, n is the number of points, and S is the integral value.

of ξ_i, each ξ_i being specific for that interval of integration. The absolute value of the error term cannot be calculated, but its relative magnitude can be measured by order of the term. Because n is inversely proportional to h,

$$n = \frac{x_n - x_0}{h} \tag{4.77}$$

the error term for the multiple-segment trapezoidal rule becomes

$$nO(h^3) = \frac{x_n - x_0}{h} O(h^3) \simeq O(h^2) \tag{4.78}$$

That is, the repeated application of the trapezoidal rule over multiple segments has lowered the error term by approximately one order of magnitude. A more rigorous analysis of the truncation error is given in the next chapter.

4.7.2 SIMPSON'S 1/3 RULE

In the derivation of the second Newton-Cotes formula of integration we use two segments of width h (Fig. 4.4) and fit the polynomial through three points, (x_0, y_0), (x_1, y_1), and (x_2, y_2). This is

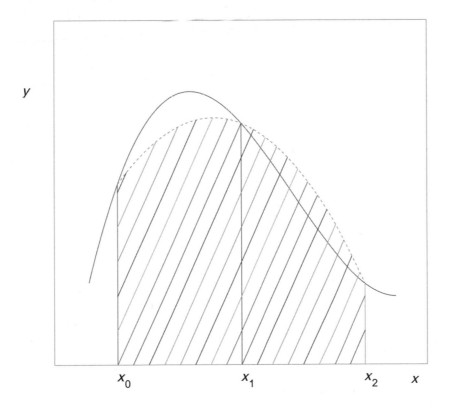

FIG. 4.4

Application of Simpson's 1/3 rule over two segments of integration.

equivalent to fitting a parabola through these points. We retain the first three terms of the Gregory-Newton polynomial (up to, and including, the second forward finite difference) and group together the rest of the terms of the polynomial into the remainder term. The integral equation becomes:

$$S_1 = \int_{x_0}^{x_2} \left[y_0 + \frac{(x - x_0)}{h} \Delta y_0 + \frac{(x - x_0)(x - x_1)}{2! h^2} \Delta^2 y_0 \right] dx + \int_{x_0}^{x_2} R_n(x) dx \qquad (4.79)$$

Integration of Eq. (4.79) and substitution of the relevant finite difference relations simplify this equation to:

$$S_1 = \frac{h}{3}(y_0 + 4y_1 + y_2) - \frac{1}{90} h^5 D^4 f(\xi_1) \qquad (4.80)$$

The error term is of order h^5 and may be abbreviated by $O(h^5)$. We would have expected to obtain an error term of $O(h^4)$ because three terms were retained in the Gregory-Newton polynomial. However, the term containing h^4 in the remainder has a zero coefficient, thus giving this fortuitous result. The final form of the second Newton-Cotes formula, which is better known as *Simpson's 1/3 rule*, is:

$$S_1 = \frac{h}{3}(y_0 + 4y_1 + y_2) + O(h^5) \qquad (4.81)$$

This equation calculates the integral over two segments of integration. Repeated application of Simpson's 1/3 rule over subsequent pairs of segments, and summation of all formulas over the total interval, gives the *multiple-segment Simpson's 1/3 rule*:

$$S = \frac{h}{3}\left(y_0 + 4\sum_{i=1}^{n/2} y_{2i-1} + 2\sum_{i=1}^{n/2-1} y_{2i} + y_n \right) + O(h^4) \qquad (4.82)$$

Fig. 4.5 illustrates the flowchart of multiple-segment Simpson's 1/3 rule of integration.

Because Simpson's 1/3 rule fits *pairs* of segments, the total interval must be subdivided into an *even* number of segments. The first summation term in Eq. (4.82) sums up the odd-subscripted terms and the second summation adds up the even-subscripted terms.

The order of error of the multiple-segment Simpson's 1/3 rule was reduced by one order of magnitude to $O(h^4)$ for the same reason as in Section 4.7.1. Simpson's 1/3 rule is more accurate than the trapezoidal rule but requires additional arithmetic operations.

4.7.3 SIMPSON'S 3/8 RULE

In the derivation of the third Newton-Cotes formula of integration we use three segments of width h (Fig. 4.6) and fit the polynomial through four points, (x_0, y_0), (x_1, y_1), (x_2, y_2), and (x_3, y_3). This, in fact, is equivalent to fitting a cubic equation through the four points. We retain the first four terms of the Gregory-Newton polynomial (up to, and including, the third forward finite difference) and group together the rest of the terms of the polynomial into the remainder term. The integral equation becomes:

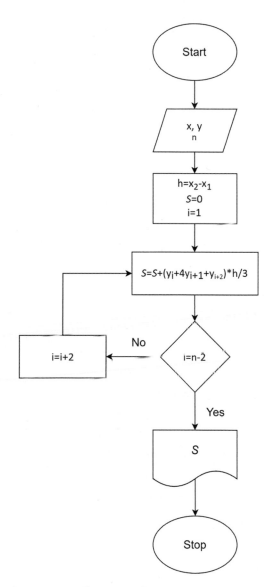

FIG. 4.5 Flowchart of the multiple-segment Simpson's 1/3 rule.

In this flowchart x and y are the vectors of independent and function variables of base points, n is the number of points, and S is the integral value. Note that n should be an odd number (even number of divisions).

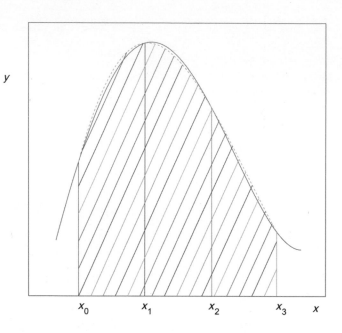

FIG. 4.6

Application of Simpson's 3/8 rule over three segments of integration.

$$S_1 = \int_{x_0}^{x_3} \left[y_0 + \frac{(x - x_0)}{h} \Delta y_0 + \frac{(x - x_0)(x - x_1)}{2!h^2} \Delta^2 y_0 + \frac{(x - x_0)(x - x_1)(x - x_2)}{3!h^3} \Delta^3 y_0 \right] dx$$
$$+ \int_{x_0}^{x_3} R_n(x)dx \tag{4.83}$$

Integration of Eq. (4.83) and substitution of the relevant finite difference relations simplify the equation to:

$$S_1 = \frac{3h}{8}(y_0 + 3y_1 + 3y_2 + y_3) - \frac{3}{80} h^5 D^4 f(\xi_1) \tag{4.84}$$

The error term is of order h^5 and may be abbreviated by $O(h^5)$. The final form of this equation, which is better known as *Simpson's 3/8 rule*, is given by:

$$S_1 = \frac{3h}{8}(y_0 + 3y_1 + 3y_2 + y_3) + O(h^5) \tag{4.85}$$

The *multiple-segment Simpson's 3/8 rule* is obtained by repeated application of Eq. (4.83) over triplets of segments and summation over the total interval of integration:

$$S = \frac{3h}{8} \left(y_0 + 3 \sum_{i=1}^{n/3} (y_{3i-2} + y_{3i-1}) + 2 \sum_{i=1}^{n/3-1} y_{3i} + y_n \right) + O(h^4) \tag{4.86}$$

The flowchart of the multiple-segment Simpson's 3/8 rule of integration is similar to that in Fig. 4.5 with the exception that in this case S should be evaluated from Eq. (4.85) and we should move three segments forward in each step (thus the number of segments should be divisible by 3).

Comparison of the error terms of Simpson's 1/3 rule and Simpson's 3/8 rule shows that they are both of the same order, with the latter being only slightly more accurate. For this reason, Simpson's 1/3 rule is usually preferred because it achieves the same order of accuracy with three points rather than the four points required by the 3/8 rule.

4.7.4 SUMMARY OF NEWTON-COTES INTEGRATION

The three Newton-Cotes formulas of integration derived in the previous sections are summarized in Table 4.4. In the derivation of the Newton-Cotes formulas the function $y = f(x)$ is approximated by the Gregory-Newton polynomial $P_n(x)$ of degree n with remainder $R_n(x)$. The evaluation of the integral is performed:

$$\int_a^b y\,dx = \int_a^b P_n(x)dx + \int_a^b R_n(x)dx \tag{4.87}$$

This results in a formula of the general form:

$$\int_a^b y\,dx = \sum_{i=0}^n w_i y_i + O\left[h^{n+2}, D^{n+1}f(\xi)\right] \tag{4.88}$$

where the x_i are $(n+1)$ equally spaced base points in the interval $[a, b]$. The weights w_i are determined by fitting the $P_n(x)$ polynomial to $(n+1)$ base points. The integral is exact, that is,

$$\int_a^b y\,dx = \sum_{i=0}^n w_i y_i \tag{4.89}$$

for any function $y = f(x)$ which is of polynomial form up to degree n, because the derivative $D^{n+1}f(\xi)$ is zero for polynomials of degree $\leq n$; thus the error term $O[h^{n+2}, D^{n+1}f(\xi)]$ vanishes.

Table 4.4 Summary of the Newton-Cotes numerical integration formulas.	
Rectangle rule[1]	$\int_{x_0}^{x_1} y\,dx = hy_0 - \frac{1}{2}h^2 Df(\xi)$
Trapezoidal rule	$\int_{x_0}^{x_1} y\,dx = \frac{h}{2}(y_0 + y_1) - \frac{1}{12}h^3 D^2 f(\xi)$
Simpson's 1/3 rule	$\int_{x_0}^{x_2} y\,dx = \frac{h}{3}(y_0 + 4y_1 + y_2) - \frac{1}{90}h^5 D^4 f(\xi)$
Simpson's 3/8 rule	$\int_{x_0}^{x_3} y\,dx = \frac{3h}{8}(y_0 + 3y_1 + 3y_2 + y_3) - \frac{3}{80}h^5 D^4 f(\xi)$
General quadrature formula	$\int_{x_0}^{x_n} y\,dx = \sum_{i=0}^n w_i y_i + O\left[h^{n+2}, D^{n+1}f(\xi)\right]$

[1]We did not develop this method, also known as the Riemann sum, in this section. It should be mentioned that the formula for this method can be derived by retaining the first term of the Gregory-Newton polynomial Eq. (4.67) and integrating it from x_0 to x_1.

EXAMPLE 4.2 INTEGRATION FORMULAS: TRAPEZOIDAL AND SIMPSON'S 1/3 RULES

Write a general MATLAB function for integrating experimental data using Simpson's 1/3 rule. Compare the results of this function and the existing MATLAB function *trapz* (trapezoidal rule) for the solution of the following problem:

Two very important quantities in the study of fermentation processes are the carbon dioxide evolution rate and the oxygen uptake rate. These are calculated from experimental analysis of the inlet and exit gases of the fermenter and the flow rates, temperature, and pressure of these gases. The ratio of carbon dioxide evolution rate to oxygen uptake rate yields the respiratory quotient, which is a good indicator of the metabolic activity of the microorganism. In addition, the foregoing rates can be integrated to obtain the total amounts of carbon dioxide produced and oxygen consumed during the fermentation. Table E4.2a shows a set of rates calculated from the fermentation of *Penicillium chrysogenum*, which produces penicillin antibiotics.

Using Simpson's 1/3 rule, calculate the total amounts of carbon dioxide produced and oxygen consumed during this 10-hour period of fermentation. Repeat this calculation using the trapezoidal rule and compare the results obtained from the two methods.

Method of solution

In this problem the carbon dioxide evolution rate data and the oxygen uptake rate data are integrated separately. There are 11 data points (10 intervals) for each rate; therefore we can use either the trapezoidal rule or Simpson's 1/3 rule for this integration. We first use Simpson's 1/3 rule and then repeat using the trapezoidal rule, as the problem specifies.

For a better understanding of the procedure, we show the calculations by the formula in the following (only for carbon dioxide):

Trapezoidal rule:

For the first segment:

$$\int_{140}^{141} ydx = \frac{(141 - 140)}{2}(15.72 + 15.53) = 15.625$$

For the whole interval:

$$\int_{140}^{150} ydx = \frac{1}{2}[15.53 + 2 \times (15.53 + 15.19 + ... + 17.75) + 18.95] = 168.345$$

Simpson's 1/3 rule:

For the first two segments:

$$\int_{140}^{142} ydx = \frac{(141 - 140)}{3}(15.72 + 4 \times 15.53 + 15.19) = 30.9967$$

(Continued)

EXAMPLE 4.2 (CONTINUED)

For the whole interval:

$$\int_{140}^{150} ydx = \frac{1}{3}(15.53 + 4 \times 15.53 + 2 \times 15.19 + \dots + 2 \times 17.60 + 4 \times 17.75 + 18.95) = 168.6633$$

Let us also show how we can do these calculations in Excel. Calculation steps are shown in Fig. E4.2. Only the trapezoidal rule is shown in this figure.

Program description

The programs and functions developed in this example can be found in: https://www.elsevier.com/books-and-journals/book-companion/9780128229613

The MATLAB function *Simpson.m* first tests the input arguments which are the vector of the independent variable (x) and the vector of function values (y). These two vectors should be of the same length. Elements of vector x have to be equally spaced values. Also, the number of elements of these vectors (n) should be odd (even number of intervals). If the vectors contain an even number of elements (odd number of intervals), the function calculates the value of the integral up to the point ($n - 1$) and adds the value of the integral, approximated by trapezoidal rule, for the last interval. The user should pay special attention to this case because the truncation errors for Simpson's 1/3 rule and trapezoidal rule are not of the same order. After checking the aforementioned conditions, the function calculates the value of the integral based on Eq. (4.82). If necessary, the function adds the value of the integral for the last segment according to Eq. (4.75).

The main program *Example4_2.m* asks the user to input the data from the keyboard, calls the functions *trapz* and *Simpson* for integration, and displays the results.

Input and results

```
>> Example4_2
```

(Continued)

Table E4.2a Fermentation data.		
Time of fermentation (h)	**Carbon dioxide evolution rate (g/h)**	**Oxygen uptake rate (g/h)**
140	15.72	15.49
141	15.53	16.16
142	15.19	15.35
143	16.56	15.13
144	16.21	14.2
145	17.39	14.23
146	17.36	14.29
147	17.42	12.74
148	17.6	14.74
149	17.75	13.68
150	18.95	14.51

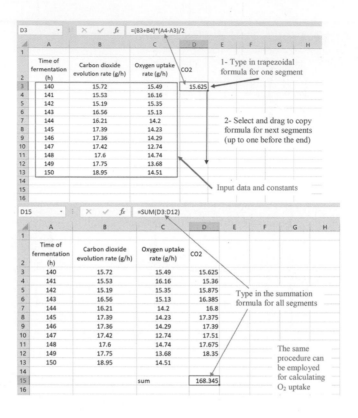

FIG. E4.2

Integration by trapezoidal rule in Excel.

EXAMPLE 4.2 (CONTINUED)

Vector of time = [140:150]

Carbon dioxide evolution rate (g/h) = [15.72, 15.53, 15.19, 16.56, 16.21, 17.39, 17.36, 17.42, 17.60, 17.75, 18.95]

Oxygen uptake rate (g/h) = [15.49, 16.16, 15.35, 15.13, 14.20, 14.23, 14.29, 12.74, 14.74, 13.68, 14.51]

Total carbon dioxide evolution = 168.3450 (evaluated by the trapezoidal rule)

Total carbon dioxide evolution = 168.6633 (evaluated by the Simpson 1/3 rule)

Total oxygen uptake = 145.5200 (evaluated by the trapezoidal rule)

Total oxygen uptake = 144.9733 (evaluated by the Simpson 1/3 rule)

Discussion of results

The integration of the experimental data, using both Simpson's 1/3 rule and the trapezoidal rule, yields the total amounts of carbon dioxide and oxygen shown in Table E4.2b.

Table E4.2b Evaluated oxygen consumption and carbon dioxide production.		
	Simpson's 1/3	**Trapezoidal**
Total CO_2 (g)	168.6633	168.345
Total O_2 (g)	144.9733	145.52

4.8 GAUSS QUADRATURE

In the development of the Newton-Cotes formulas we have assumed that the interval of integration could be divided into segments of equal width. This is usually possible when integrating continuous functions. However, if experimental data are to be integrated, such data may be used with a variable-width segment. It has been suggested by Chapra and Canale [3] that a combination of the trapezoidal rule with Simpson's rules may be feasible for integrating certain sets of unevenly spaced data points.

Gauss quadrature is a powerful method of integration that employs unequally spaced base points. This method uses the Lagrange polynomial to approximate the function and then applies orthogonal polynomials to locate the loci of the base points. If no restrictions are placed on the location of the base points, they may be chosen to be the locations of the roots of certain orthogonal polynomials to achieve higher accuracy than the Newton-Cotes formulas for the same number of base points. This concept is used in the Gauss quadrature method which is discussed in this section.

4.8.1 TWO-POINT GAUSS-LEGENDRE QUADRATURE

To illustrate the approach, we first develop the integration formula for the two-point problem. In the Newton-Cotes method the location of the base points is determined and integration is done based on the values of the function at these base points. This is shown in Fig. 4.2 for the trapezoidal rule, which approximates the integral by taking the area under the straight line connecting the function values at the ends of the integration interval.

Now, consider the case that the restriction of fixed points is withdrawn and we are able to estimate the integral from the area under a straight line which joins any two points on the curve. By choosing these points in proper positions, a straight line that balances the positive and negative errors can be drawn, as illustrated in Fig. 4.7. As a result, we obtain an improved estimate of the integral.

To derive the two-point Gauss quadrature, the function $y = f(x)$ is replaced by a linear polynomial and a remainder,

$$y = \left[\frac{x - x_1}{x_0 - x_1} y_0 + \frac{x - x_0}{x_1 - x_0} y_1 \right] + R(x) \tag{4.90}$$

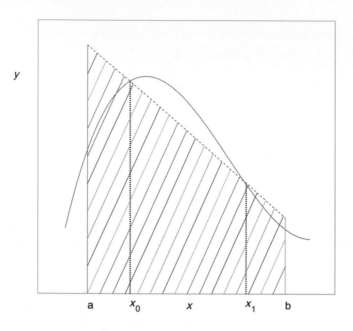

FIG. 4.7

Application of two-point Gauss quadrature to improve integral estimation.

The integral $\int_a^b y\,dx$ is evaluated by:

$$\int_a^b y\,dx = \int_a^b \left[\frac{x - x_1}{x_0 - x_1} y_0 + \frac{x - x_0}{x_1 - x_0} y_1 \right] dx + \int_a^b R(x)\,dx \qquad (4.91)$$

Without loss of generality, the interval $[a, b]$ is changed to $[-1, 1]$. The linear transformation equation for converting between x in the interval $[a, b]$ and z in the interval $[-1, 1]$ is the following:

$$z = \frac{2x - (a + b)}{b - a} \qquad (4.92)$$

Using Eq. (4.92), the transformed integral is given by:

$$\int_a^b y(x)\,dx = \frac{b - a}{2} \int_{-1}^1 Y(z)\,dz \qquad (4.93)$$

and

$$\int_{-1}^1 Y\,dz = w_0 Y_0 + w_1 Y_1 + \int_{-1}^1 R(z)\,dz \qquad (4.94)$$

where the use of Y (instead of y) indicates that the function value at the variable z (rather than x) should be used. The weights w_0 and w_1 are calculated from:

$$w_0 = \int_{-1}^{1} \frac{z - z_1}{z_0 - z_1} dz = \frac{-2z_1}{z_0 - z_1} \tag{4.95}$$

and

$$w_1 = \int_{-1}^{1} \frac{z - z_0}{z_1 - z_0} dz = \frac{-2z_0}{z_1 - z_0} \tag{4.96}$$

Up to this point, the development of this method is equivalent to that of the trapezoidal rule. The Gauss quadrature method goes a step beyond this to make the error term in Eq. (4.94) vanish. To do so, the integral of the error term is expanded in terms of second-degree Legendre polynomial (see Appendix):

$$\int_{-1}^{1} R(z) dz = \frac{3}{2} z^2 - \frac{1}{2} \tag{4.97}$$

The values of z_0 and z_1 are chosen as the root of the second-degree Legendre polynomial, that is, $z_0 = +1/\sqrt{3}$ and $z_1 = -1/\sqrt{3}$. This choice of roots causes the error term to vanish. Therefore Eq. (4.94) becomes:

$$\int_{-1}^{1} Y dz = w_0 Y_0 + w_1 Y_1 = Y_0 + Y_1 \tag{4.98}$$

Calculation of the integral through Eqs. (4.93) and (4.97) implies that instead of evaluating the function at $z_0 = -1$ and $z_1 = 1$ (using function values at base points), which is the case in the trapezoidal rule, function values at $z_0 = -1/\sqrt{3}$ and $z_1 = 1/\sqrt{3}$ should be used in the Gauss quadrature method. This improves the precision of calculation, as illustrated in Fig. 4.7. This is roughly equivalent to the application of the five-point trapezoidal rule.

The Gauss quadrature formula developed in this section is known as the *Gauss-Legendre quadrature* because of the use of the Legendre polynomials. Other orthogonal polynomials, such as Chebyshev, Laguerre, or Hermite, may be used in a similar manner to develop a variety of Gauss quadrature formulas.

4.8.2 HIGHER-POINT GAUSS-LEGENDRE FORMULAS

The function $y = f(x)$ is replaced by the Lagrange polynomial (see Section 3.8.1) and its remainder,

$$y = \sum_{i=0}^{n} L_i y_i + R_n(x) \tag{4.99}$$

where

$$L_i = \prod_{\substack{j=0 \\ j \neq i}}^{n} \frac{x - x_j}{x_i - x_j} \tag{4.100}$$

and

$$R_n(x) = \prod_{i=0}^{n} (x - x_i) \frac{f^{(n+1)}(\xi)}{(n+1)!} \quad a < \xi < b \tag{3.119}$$

The integral $\int_a^b y\,dx$ is evaluated by:

$$\int_a^b y\,dx = \int_a^b \left[\sum_{i=0}^n L_i y_i\right] dx + \int_a^b R_n(x)\,dx \tag{4.101}$$

Converting the interval from $[a, b]$ to $[-1, 1]$ through Eqs. (4.92) and (4.93), the transformed integral is given by:

$$\int_{-1}^1 Y\,dz = \sum_{i=0}^n w_i Y_i + \int_{-1}^1 R_n(z)\,dz \tag{4.102}$$

where the weights w_i are calculated from:

$$w_i = \int_{-1}^1 L_i(z)\,dz = \int_{-1}^1 \prod_{\substack{j=0 \\ j \neq i}}^n \frac{z - z_j}{z_i - z_j}\,dz \tag{4.103}$$

and the error term is given by:

$$\int_{-1}^1 R_n(z)\,dz = \int_{-1}^1 \prod_{i=0}^n (z - z_i)q_n(z)\,dz \tag{4.104}$$

The $q_n(z)$ and $\prod_{i=0}^n (z - z_i)$ are polynomials of degree n and $n + 1$, respectively.

Up to this point, the development of this method is different from that of the Newton-Cotes formulas in only one respect: the use of the Lagrange interpolation formula for unequally spaced points instead of the Gregory-Newton formula. The Gauss quadrature method goes a step beyond this to make the error term (Eq. 4.104) vanish. To do so, the two polynomials in the error term are expanded in terms of Legendre orthogonal polynomials (see the Appendix). The values of z_i are chosen as the roots of the $(n + 1)$st-degree Legendre polynomial. This choice of roots, combined with the orthogonality property of the Legendre polynomials (see Eq. A.4 in the Appendix), causes the error term to vanish. Therefore Eq. (4.102) becomes:

$$\int_{-1}^1 Y\,dz = \sum_{i=0}^n w_i Y_i \tag{4.105}$$

Because the vanishing error term was of degree $(n + 1)$, Eq. (4.105) yields the integral of the function Y *exactly* when Y is a polynomial of degree $(2n + 1)$ or less. In effect, the judicious choice of the $(n + 1)$ base points at the $(n + 1)$ roots of the Legendre polynomial has increased the accuracy of the integration from n to $(2n + 1)$. As usual, however, the increase in accuracy has been obtained at the cost of having to perform a larger number of arithmetic calculations. The error of Gauss-Legendre formulas is given by [4]:

$$\int_a^b R_n(x)\,dx = \frac{2^{2n+3}[(n+1)!]^4}{(2n + 3)[(2n+2)!]^3} f^{(2n+2)}(\xi) \quad a < \xi < b \tag{4.106}$$

The roots z_i of the Legendre polynomials can be evaluated after calculating the coefficients of the polynomial from the formula given in Table A.1 in the Appendix. The values of the weights w_i corresponding to these roots have been calculated for the integration interval $[-1, 1]$. Table 4.5 lists the roots and weights of the Gauss-Legendre quadrature for selected values of n.

Table 4.5 Roots of Legendre polynomials $P_{n+1}(z)$ and the weight factors for the Gauss-Legendre quadrature.

Number of points	Roots (z_i)	Weight factors (w_i)
Two-point formula $(n + 1 = 2)$	± 0.57735026918926	1.000000000000000
Three-point formula $(n + 1 = 3)$	0 ± 0.774596669241483	0.888888888888888 0.555555555555555
Four-point formula $(n + 1 = 4)$	± 0.339981043584856 ± 0.861136311594053	0.652145154862546 0.347854845137454
Five-point formula $(n + 1 = 5)$	0 ± 0.538469310105683 ± 0.906179845938664	0.568888888888889 0.478628670499366 0.236926885056189
Six-point formula $(n + 1 = 6)$	± 0.238619186083197 ± 0.661209386466265 ± 0.932469514203152	0.467913934572691 0.360761573048139 0.171324492379170
10-point formula $(n + 1 = 10)$	± 0.148874338981631 ± 0.433395394129247 ± 0.679409568299024 ± 0.865063366688985 ± 0.973906528517172	0.295524224714753 0.269266719309996 0.219086362515982 0.149451349150581 0.066671344308688
15-point formula $(n + 1 = 15)$	0 ± 0.201194093997435 ± 0.394151347077543 ± 0.570972172608539 ± 0.724417731360170 ± 0.848206583410427 ± 0.937273392400706 ± 0.987992518020485	0.202578241925561 0.198431485327111 0.186161000115562 0.166269205816994 0.139570677926154 0.107159220467172 0.070366047488108 0.030753241996117

EXAMPLE 4.3 INTEGRATION FORMULAS: GAUSS-LEGENDRE QUADRATURE

Write a general MATLAB function for integrating a function using a general Gauss-Legendre quadrature. Apply this function to solve of the following problem:

A liquid film, initially at temperature T_0, is falling (in the z-direction) a vertical solid wall (xz-plane). The wall is maintained at a temperature (T_S) higher than that of the falling film. It is desired to know the temperature profile of the fluid as a function of y and z, near the wall. The partial differential equation which describes the temperature of the liquid for is:

$$\rho C_p \nu_z \frac{\partial T}{\partial z} = k \frac{\partial^2 T}{\partial y^2}$$

where ρ is the density, C_p is the heat capacity, ν_z is the velocity, k is the thermal conductivity, and T is the temperature of the liquid.

(Continued)

EXAMPLE 4.3 (CONTINUED)

The velocity profile of the falling liquid is given by Bird et al. [5]:

$$\nu_z = \frac{\delta^2 \rho g}{2\mu}\left[2\frac{y}{\delta} - \left(\frac{y}{\delta}\right)^2\right]$$

where δ is the thickness of the film, g is gravity acceleration, and μ is the viscosity of the liquid. Therefore near the wall, where $y \ll \delta$, the velocity simplifies to:

$$\nu_z = \frac{\rho g \delta}{\mu}y$$

Putting this velocity profile into the energy balance equation, we get:

$$y\frac{\partial T}{\partial z} = \beta\frac{\partial^2 T}{\partial y^2}$$

in which $\beta = \mu k/\rho^2 C_p g\delta$. For short contact time, we may write the boundary conditions as:

$$\text{At } z = 0, \quad T = T_0 \quad \text{for } y > 0$$
$$\text{At } y = 0, \quad T = T_s \quad \text{for } z > 0$$
$$\text{At } y = \infty, \quad T = T_0 \quad \text{for } z \text{ finite}$$

The analytical solution to this problem is [5]:

$$\Theta = \frac{T - T_0}{T_s - T_0} = 1 - \frac{1}{\Gamma\left(\frac{4}{3}\right)}\int_0^\eta e^{-\eta^3}d\eta$$

where $\eta = y/\sqrt{9\beta z}$ is a dimensionless variable and the gamma function, $\Gamma(x)$ is defined as:

$$\Gamma(x) = \int_0^\infty t^{x-1}e^{-t}dt \quad x > 0$$

Using Gauss-Legendre quadrature, calculate the foregoing temperature profile and plot it against η.

Method of solution

To evaluate the temperature profile (Θ), we first have to integrate the function $e^{-\eta^3}$ for several values of $\eta \geq 0$. The temperature profile itself, then, can be calculated from the equation described earlier.

Program description

The programs and functions developed in this example can be found in: https://www.elsevier.com/books-and-journals/book-companion/9780128229613

The function *GaussLegendre.m* numerically evaluates the integral of a function by n-point Gauss-Legendre quadrature. The program checks the inputs to the function to be sure that they have valid values. If no value is introduced for the integration step, the function sets it to the integration interval. Also, the default value for the number of points of the Gauss-Legendre formula is 2.

(Continued)

EXAMPLE 4.3 (CONTINUED)

The next step in the function is the calculation of the coefficients of the nth degree Legendre polynomial. Once these coefficients are calculated, the program evaluates the roots of the Legendre polynomial (z_1 to z_n) using the MATLAB function *roots*. Then the function calculates the coefficients of the Lagrange polynomial terms (L_1 to L_n) and evaluates the weight factors, w_i, as defined in Eq. (4.103). Finally, using the values of z_i and w_i, the integral is numerically evaluated by Eq. (4.105).

To solve the problem described in this example, the main program *Example4_3.m* is written to calculate the temperature profile for a specific range of the dimensionless number η. The function to be integrated is introduced in the MATLAB function *Ex4_3_func.m*.

Input and results

>> Example4_3
Initial value of Vector of independent variable (eta) = [0: 0.2: 2]
Name of m-file containing the function subject to integration = 'ex4_4_func'

Discussion of results

The temperature profile of the liquid near the wall is calculated by the program *Example4_3. m* for $0 \le \eta \le 2$ and is plotted in Fig. E4.3. We can verify the solution at the boundaries of y and z from Fig. E4.3:

i. The results represented in Fig. E4.3 show that at $\eta = 0$, the temperature of the liquid is identical to that of the plate (i.e., $\Theta = 1$, therefore $T = T_s$). The variable η attains a value of zero at only two situations:

  a. In the liquid next to the wall (at $y = 0$ and at all values of z).

  b. After an infinite distance from the origin of flow (at $z = \infty$ and at all values of y).

    Situation a is consistent with the boundary conditions given in the statement of the problem, while situation b is an expected result, as passing a long-enough distance along the wall, all the liquid will be at the same temperature as the wall.

ii. Fig. E4.3 also shows that at high enough dimensionless number η, the temperature of the liquid is equal to the initial temperature of the liquid, that is,

$$\lim_{\eta \to \infty} \Theta = 0$$

iii. The variable η becomes infinity under the following circumstances:

  a. In the fluid far away from the wall (at $y = \infty$ and at all values of z).

  b. At the origin of the flow (at $z = 0$ and at all values of y).

  Both these situations are specified as boundary conditions of the problem.

4.9 SPLINE INTEGRATION

Another method of integrating unequally spaced data points is to interpolate the data using a suitable interpolation method, such as cubic splines, and then evaluate the integral from the relevant

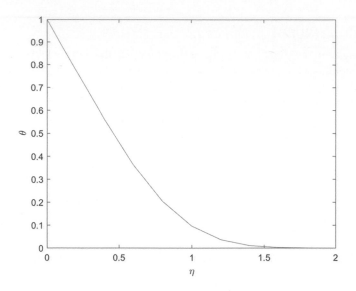

FIG. E4.3

Temperature profile.

polynomial. Therefore the integral of Eq. (4.66) may be calculated by integrating Eq. (3.125) over the interval $[x_{i-1}, x_i]$ and summing up these terms for all the intervals:

$$\int_{x_0}^{x_n} y dx = \sum_{i=1}^{n} \left[\frac{1}{2}(x_i - x_{i-1})(y_i + y_{i-1}) + \frac{1}{36}(x_{i-1} - x_i)^3(y''_{i-1} + y''_i) \right] \tag{4.109}$$

Prior to calculating the integral from Eq. (4.109), the values of the second derivative at the base points should be calculated from Eq. (3.129). Note that if a natural spline interpolation is employed, the second derivatives for the first and the last intervals are equal to zero. Eq. (4.109) is basically an improved trapezoidal formula in which the value of the integral by trapezoidal rule [the first term in the bracket of Eq. (4.109)] is corrected for the curvature of the function [the second term in the bracket of Eq. (4.109)].

The reader can easily modify the MATLAB function *NaturalSpline.m* (see Example 3.7) to calculate the integral of a function from a series of tabulated data. It is enough to replace the formula of the interpolation section with the integration formula, Eq. (4.109). Also, the MATLAB function *spline.m* is able to give the piecewise polynomial coefficients from which the integral of the function can be evaluated. A good example of applying such a method can be found in Hanselman and Littlefield [2]. Remember that *spline.m* applies the not-a-knot algorithm for calculating the polynomial coefficients.

4.10 MULTIPLE INTEGRALS

In this section we discuss the evaluation of double integrals. Evaluation of integrals with more than two dimensions can be obtained in a similar manner. Let us start with a simple case of double integral with constant limits, that is, integration over a rectangle in the xy plane:

$$S = \int_a^b \int_c^d f(x,y)dydx = \int_a^b \left(\int_c^d f(x,y)dy \right)dx \qquad (4.108)$$

The inner integral may be calculated by one of the methods described in Sections 4.7 to 4.9. We use the trapezoidal rule (Eq. 4.76) for simplicity:

$$\int_c^d f(x,y)dy = \frac{k}{2}\left[f(x,c) + 2\sum_{j=1}^{m-1} f(x,y_j) + f(x,d) \right] \qquad (4.109)$$

where m is the number of divisions and k is the integration step in the y-direction and x is considered as constant. Replacing Eq. (4.109) into Eq. (4.108) results in:

$$S = \frac{k}{2}\int_a^b f(x,c)dx + k\sum_{j=1}^{m-1}\int_a^b f(x,y_j)dx + \frac{k}{2}\int_a^b f(x,d)dx \qquad (4.110)$$

Now, we apply the trapezoidal rule to each of the integrals of Eq. (4.110):

$$\int_a^b f(x,y_j)dx = \frac{h}{2}\left[f(a,y_j) + 2\sum_{i=1}^{n-1} f(x_i,y_i) + f(b,y_j) \right] \qquad (4.111)$$

Here, n is the number of divisions and h is the integration step in the x-direction and y_j is considered as constant.

Finally, we combine Eqs. (4.110) and (4.111) to calculate the estimated value of the integral (4.108):

$$
\begin{aligned}
S = \frac{hk}{4}&\left[f(a,c) + 2\sum_{i=1}^{n-1} f(x_i,c) + f(b,c) \right] \\
+ \frac{hk}{2}&\left[\sum_{j=1}^{m-1} f(a,y_j) + 2\sum_{j=1}^{m-1}\sum_{i=1}^{n-1}(x_i,y_j) + \sum_{j=1}^{m-1} f(b,y_j) \right] \\
+ \frac{hk}{4}&\left[f(a,d) + 2\sum_{i=1}^{n-1} f(x_i,d) + f(b,d) \right]
\end{aligned}
\qquad (4.112)
$$

The method described earlier may be slightly modified to be applicable to the double integrals with variable inner limits of the form:

$$S = \int_a^b \int_{c(x)}^{d(x)} f(x,y)dydx \qquad (4.113)$$

Because the length of the integration interval for the inner integral (i.e., $[c, d]$) changes with the value of x, we may either keep the number of divisions constant in the y-direction and let the integration step change with x $[k = k(x)]$ or keep the integration step in the y-direction constant and use a different number of divisions at each x value $[m = m(x)]$. However, to maintain the same order of error throughout the calculation, the second condition (i.e., constant step size) should be employed. Therefore Eq. (4.109) can be written at each position x_i in the following form to account for the variable limits:

$$\int_{c(x_i)}^{d(x_i)} f(x_i,y)dy = \frac{k}{2}\left[f(x_i,c(x_i)) + 2\sum_{j=1}^{m_i-1} f(x_i,y_j) + f(x_i,d(x_i)) \right] \qquad (4.114)$$

where m_i indicates that the number of divisions in the y-direction is a function of x. In practice, at each x value, we may have to change the step size k slightly to obtain an integer value for the number of divisions. Although this does not change the order of magnitude of the step size, we have to acknowledge this change at each step of outer integration; therefore the approximate value of the integral (4.113) is calculated from:

$$
\begin{aligned}
S = \frac{hk_0}{4} & \left[f(a, c(a)) + 2\sum_{i=1}^{n-1} f(x_i, c(x_i)) + f(b, c(b)) \right] \\
& + \frac{h}{2}\sum_{i=1}^{n-1}\left\{ k_i \sum_{j=1}^{m_i-1}\left[f(a, y_j) + 2f(x_i, y_j) + f(b, y_j) \right] \right\} \\
& + \frac{hk_n}{4}\left[f(a, d(a)) + 2\sum_{i=1}^{n-1} f(x_i, d(x_i)) + f(b, d(b)) \right]
\end{aligned}
\tag{4.115}
$$

If writing a computer program for the evaluation of double integrals, it is not necessary to apply Eqs. (4.112) and (4.114) in such a program. As a matter of fact, any ordinary integration function may be applied to evaluate the inner integral at each value of the outer variable; then the same function is applied for the second time to calculate the outer integral. This algorithm can be similarly applied to the multiple integrals of any dimension. The MATLAB function *dblquad* evaluates the double integral of a function with fixed inner integral limits.

4.11 USING BUILT-IN MATLAB® FUNCTIONS

In MATLAB the function *diff(y)* returns forward finite differences of the vector y. This function can be used to estimate the derivative of a function. For this purpose, it is necessary to create a vector of function variables from a vector or equally spaces independent variable. Then, the derivative of the function can be estimated from *diff(y)/h*. For example, the derivative of $y = sin(x)$ can be estimated as follows:

```
>> h = .001; x = 0:h:pi;
>> dy = diff(sin(x))/h;
```

In *Symbolic Math Toolbox* the command *diff(y)* returns the formula of derivative of function y. Considering the preceding example, we have:

```
>> syms x
>> y = sin(x)
>> dy = diff(y)
```

To take the second derivative of y, use the following command:

```
>> d2y = diff(y,2)
```

There are three functions in MATLAB, *trapz.m*, *quad.m,* and *quad8.m*, that numerically evaluate the integral of a vector or a function using different Newton-Cotes formulas:

The function *trapz(x, y)* calculates the integral of y (vector of function values) with respect to x (vector of variables) using the trapezoidal rule.

The function *quad*("*file_name*", *a*, *b*) evaluates the integral of the function represented in the m-file *file_name.m*, over the interval [*a*, *b*] by Simpson's 1/3 rule.

The function *quad8*("*file_name*", *a*, *b*) evaluates the integral of the function introduced in the m-file *file_name.m* from *a* to *b* using the eight-interval (nine-point) Newton-Cotes formula.

In *Symbolic Math Toolbox* the command *int*(*y*) returns the formula of integral of function *y*. The command *int*(*y*, *a*, *b*) evaluates the definite integral of the function *y* from *a* to *b*.

4.12 SUMMARY

In some cases it is required to evaluate the derivative of a function from tabulated data. This situation occurs usually when data points are obtained experimentally. Moreover, there are cases in which although the function is given analytically, it is easier to evaluate its derivative numerically. An example of this situation is when the value of the derivative of a function is needed in a computer program. In either case numerical differentiation can be calculated based on the calculus of finite differences, developed in Chapter 3.

In the present chapter various formulas for first to fourth derivatives are obtained using forward, backward and central differences. In each case formulas with low-order and high-order errors are presented. These formulas are applicable to equally spaced data points. In general, differentiation by central differences is more accurate than by the other two methods. For the unequally spaced points, or when the derivative is to be estimated at a point between the base points, the spline differentiation can be employed.

It is not possible to evaluate all definite integrals by analytical methods. In addition, there are cases that although the integral can be computed analytically, it is more convenient to evaluate it numerically. Like the previous part, numerical integration formulas are developed in this chapter based on the calculus of finite differences. For the equally spaced points, Newton-Cotes formulas are derived in this chapter. Specifically, formulas of trapezoidal, Simpson's 1/3, and Simpson's 3/8 rules are presented, which use 2, 3, and 4 base points, respectively. The trapezoidal rule is very easy to implement. Simpson's 1/3 and Simpson's 3/8 rules are accurate of the same order; thus Simpson's 1/3 rule is usually favored in practice because it has a simpler formula.

The Gauss quadrature method uses unequally spaced base points to achieve higher accuracy in the Newton-Cote formulas of the same number of base points. In the Gauss-Legendre quadrature the base points are determined from the roots of the Legendre polynomials. For the unequally spaced points, spline integration can be employed.

At the end of the chapter, built-in MATLAB functions which perform differentiation and integration are described.

PROBLEMS

4.1 Derive the equation which expresses the third derivative of *y* in terms of backward finite differences, with

(a) Error of order h

(b) Error of order h^2

4.2 Repeat Problem 4.1 using forward finite differences.

4.3 Derive the equations for the first, second, and third derivatives of y in terms of backward finite differences with an error of order h^3.

4.4 Repeat Problem 4.3, using forward finite differences.

4.5 Derive the equation which expresses the third derivative of y in terms of central finite differences, with:

(a) Error of order h^2

(b) Error of order h^4

4.6 Derive the equations for the first, second, and third derivatives of y in terms of central finite differences with an error of order h^6.

4.7 Velocity profiles of solids in a bed of sand particles fluidized with air at the superficial velocity of 1 m/s are given in Tables P4.7a and b. Calculate the axial gradient of velocities (i.e., $\partial V_z/\partial z$ and $\partial V_r/\partial z$). Plot the z-averaged gradients versus radial position and compare their order of magnitude.

4.8 Consider the decomposition of N_2O_5:

$$2N_2O_5 \rightarrow NO_2 + O_2$$

Table P4.8 shows the concentration of N_2O_5 as a function of time at 45°C. Calculate the rate of reaction with an error of order $O(h^2)$.

4.9 Use the four-point Gauss-Legendre quadrature formula for calculating the following integral:

$$I = \int_0^\varphi \frac{d\theta}{\sqrt{1 - k^2 \sin^2\theta}}$$

where $k^2 = 0.8$ and $\varphi = \pi/2$. Compare the results with the trapezoidal rule and Simpson's 3/8 rule using the same number of points.

4.10 In studying the mixing characteristics of chemical reactors a sharp pulse of a nonreacting tracer is injected into the reactor at time $t = 0$. The concentration of material in the effluent

Table P4.7a Radial velocity profile (mm/s).

		Radial position, mm							
		4.7663	14.2988	23.8313	33.3638	42.8962	52.4288	61.9612	71.4938
Axial position, mm	25	−13.09	−37.66	−52.41	−54.44	−58.21	−41.35	−23.97	−7.21
	75	−15.81	−15.99	−27.81	−25.37	−22.3	−11.1	−2.26	1.63
	125	1.77	1.17	3.45	5.5	1.63	−1.79	−0.26	1.09
	175	1.43	−0.57	4.86	2.44	0.2	−0.65	0.35	2.21
	225	−5.07	−7.26	−18.43	−18.17	−17.3	−10	−2.65	0.29
	275	13.11	16.51	19.32	21	20.29	15.64	0.98	−9.81
	325	11.7	34.5	58.3	71.44	73.49	64.88	50.91	19.14
	375	8.18	25.29	31.18	37.07	30.05	2.61	−17.06	−15.88
	425	3.35	−0.39	−18	−42.22	−57.42	−82.36	−69.34	−17.35
	475	−27.05	−22.25	−49.45	−79.45	−110.08	−116.62	−128.25	−76.49

Table P4.7b Axial velocity profile (mm/s).

		Radial position, mm							
		4.7663	**14.2988**	**23.8313**	**33.3638**	**42.8962**	**52.4288**	**61.9612**	**71.4938**
Axial	25	93.33	74.12	69.35	43.68	18.8	−6.9	−21.56	−22.65
position,	75	244.73	217.07	177.09	103.79	16.87	−39.74	−74.91	−59.48
mm	125	304.34	260.58	201.15	118.82	22.76	−52.23	−82.86	−51.9
	175	308.81	281.67	209.18	133.9	53.88	−51.92	−98.47	−41.94
	225	379.66	328.52	279.3	165.61	53.25	−65.97	−133.92	−46.69
	275	416.08	366.96	314.09	203.08	44.97	−76.93	−160.04	−91.33
	325	184.46	157.25	111.99	63.23	1.03	−63.66	−71.23	−31.4
	375	55.74	−12.28	−18.74	−47.26	−42.1	−9.95	125.57	271.16
	425	−67.81	−118.77	−108.46	−89.68	9.24	61.78	175.43	309.21
	475	−136.25	−32.33	−65.5	−111.72	38.74	115.6	84.88	191.37

Table P4.8 Concentration of N_2O_5 against time.

Time (min)	$[N_2O_5]$ (mol/L)
0	0.01756
20	0.00933
40	0.00531
60	0.00295
80	0.00167
100	0.00094
120	0.00014

from the reactor is measured as a function of time $c(t)$. The residence time distribution (RTD) function for the reactor is defined as:

$$E(t) = \frac{c(t)}{\int_0^\infty c(t)dt}$$

and the cumulative distribution function is defined as:

$$F(t) = \int_0^t E(t)dt$$

The mean residence time of the reactor is calculated from:

$$t_m = \frac{V}{q} = \int_0^\infty tE(t)dt$$

where V is the volume of the reactor and q is the flow rate. The variance of the RTD function is defined by:

$$\sigma^2 = \int_0^\infty (t - t_m)E(t)dt$$

The exit concentration data shown in Table P4.10 were obtained from a tracer experiment studying the mixing characteristics of a continuous flow reactor. Calculate the RTD function, cumulative distribution function, mean residence time, and the variance of the RTD function of this reactor.

4.11 The following catalytic reaction is carried out in an isothermal circulating fluidized bed reactor:

$$A_{(g)} \rightarrow B_{(g)}$$

For a surface-reaction limited mechanism, in which both A and B are absorbed on the surface of the catalyst, the rate law is:

$$-r_A = \frac{k_1 C_A}{1 + k_2 C_A + k_3 C_B}$$

where r_A is the rate of the reaction in $\text{kmol/m}^3 \cdot \text{s}$, C_A and C_B are concentrations of A and B, respectively, in kmol/m^3, and k_1, k_2, and k_3 are constants.

Assume that the solids move in plug flow at the same velocity as the gas (U). Evaluate the height of the reactor at which the conversion of A is 60%. Additional data are as follows:

$$C_{A0} = 0.2 \ \text{kmol/m}^3 \quad C_{B0} = 0 \quad U = 7.5 \ \text{m/s}$$
$$k_1 = 8 \ \text{s}^{-1} \quad k_2 = 3 \ \text{m}^3/\text{kmol} \quad k_3 = 0.01 \ \text{m}^3/\text{kmol}$$

4.12 A gaseous feedstock containing 40% A, 40% B, and 20% inert will be processed in a reactor where the following chemical reaction takes place:

$$A + 2B \rightarrow C$$

The reaction rate is:

$$-r_A = kC_A C_B^2$$

where
$k = 0.01 \ \text{s}^{-1} \ (\text{L}^2 \ \text{gmol}^{-2}\text{s}^{-1})$ at 500°C
$C_A =$ concentration of A, gmol/L
$C_B =$ concentration of B, gmol/L

Choose a basis of 100 gmol of feed and assume that all gases behave as ideal gases. Calculate the following:

Table P4.10 Time-concentration data.

Time (s)	c(t) (mg/L)	Time (s)	c(t) (mg/L)
0	0	5	5
1	2	6	2
2	4	7	1
3	7	8	0
4	6		

(a) The time needed to produce a product containing 11.8% B in a batch reactor operating at 500°C and at a constant pressure of 10 atm.

(b) The time needed to produce a product containing 11.8% B in a batch reactor operating at 500°C and constant volume. The temperature of the reactor is 500°C and the initial pressure is 10 atm.

4.13 A mixture of 28% of SO_2 in the air with a mass flow rate of 250 tons/day enters a tubular reactor and is converted into SO_3 through the following reaction:

$$SO_2 + \frac{1}{2}O_2 \rightarrow SO_3$$

The pressure of the reactor is 1485 kPa and its temperature is 227°C. The rate of consumption of SO_2 in such conditions is given as ($A = SO_2$):

$$-r_A = \frac{2(1-X)(0.54 - 0.5X)}{(1-0.14X)^2} \frac{mol}{L.s}$$

where X is the conversion of SO_2. Calculate the volume of the reactor for 50% conversion of SO_2 from:

$$V = F_{A_0} \int_0^X \frac{dX}{-r_A}$$

where F_{A0} is the molar flow rate of A at the inlet. Use the five-point Gauss-Legendre quadrature formula.

4.14 The velocity distribution in a turbulent pipe flow is:

$$\frac{V}{V_{max}} = \left(1 - \frac{r}{R}\right)^{1/7}$$

where r is the radial position from the center of the pipe, R is the radius of the pipe, V is the fluid velocity, and V_{max} is the maximum velocity of the fluid. Use Simpson's 1/3 rule to calculate the volumetric flow rate of fluid from:

$$Q = \int_0^R 2\pi r V dr$$

Additional data are $R = 5$ cm, $V_{max} = 3$ m/s.

4.15 The probability density of the Student's t is given as (see Chapter 8):

$$p(t) = \frac{1}{\sqrt{\nu\pi}} \frac{\Gamma[(\nu+1)/2]}{\Gamma(\nu/2)} \left(1 + \frac{t^2}{\nu}\right)^{-(\nu+1)/2}$$

The cumulative distribution for this density function is calculated from:

$$P(t) = \int_{-\infty}^t p(t)dt$$

Plot the cumulative distribution of Student's t function for $\nu = 5$.

4.16 Consider the following formula:

$$y = \int_0^x \frac{\sin t}{t} \, dt$$

Plot the function $y(x)$.

4.17 Derive the numerical approximation of double integrals using Simpson's 1/3 rule in both dimensions.

REFERENCES

[1] Larachi, F.; Chaouki, J.; Kennedy, G.; Dudukovic, M. P. Radioactive Particle Tracking in Multiphase Reactors: Principles and Applications. In *Non-invasive Monitoring of Multiphase Flows;* Chaouki, J., Larachi, F., Dudukovic, M. P., eds.; Elsevier Science B. V.: Amsterdam, 1997.

[2] Hanselman, D. C.; Littlefield, B. *Mastering MATLAB 7;* Prentice Hall: Upper Saddle River, NJ, 2004.

[3] Chapra, S. C.; Canale, R. P. *Numerical Methods for Engineers*, 7th ed.; McGraw-Hill: New York, 2014.

[4] Carnahan, B.; Luther, H. A.; Wilkes, J. O. *Applied Numerical Methods;* John Wiley & Sons: New York, 1969.

[5] Bird, R. B.; Stewart, W. E.; Lightfoot, E. N. *Transport Phenomena*, 2nd ed.; John Wiley & Sons: Hoboken, NJ, 2006.

ORDINARY DIFFERENTIAL EQUATIONS: INITIAL VALUE PROBLEMS

5

CHAPTER OUTLINE

MOTIVATION

Ordinary differential equations arise from the study of the dynamics of physical and chemical systems that have one independent variable. The latter may be either the space variable x or the time variable t, depending on the assumptions of the model. The sets of simultaneous first-order ordinary differential

Applied Numerical Methods for Chemical Engineers. DOI: https://doi.org/10.1016/B978-0-12-822961-3.00005-4

equations are the most commonly encountered types of problems in the analysis of multicomponent and/or multistage operations. Closed-form solutions for such sets of equations are not usually obtainable. However, numerical methods have been thoroughly developed for the solution of sets of simultaneous differential equations. In this chapter we discuss the most useful techniques for the solution of such problems. Higher-order differential equations can be reduced to the first order by a series of substitutions. Initial value problems of ordinary differential equations are developed in this chapter.

5.1 INTRODUCTION

Numerous cases can be set forward in which a physicochemical process can be presented as ordinary differential equations. For example, when a chemical reaction of the type:

$$A + B \underset{k_2}{\overset{k_1}{\rightleftharpoons}} C + D \overset{k_3}{\longrightarrow} E \tag{5.1}$$

takes place in a reactor, the material balance takes the form:

$$\text{Input} + \text{Generation} = \text{Output} + \text{Accumulation} \tag{5.2}$$

For a batch reactor, the input and output terms are zero; therefore the material balance simplifies to:

$$\text{Accumulation} = \text{Generation} \tag{5.3}$$

Assuming that reaction (5.1) takes place in the liquid phase with negligible change in volume, Eq. (5.3) written for each component of the reaction will be:

$$\frac{dC_A}{dt} = -k_1 C_A C_B + k_2 C_C C_D$$

$$\frac{dC_B}{dt} = -k_1 C_A C_B + k_2 C_C C_D$$

$$\frac{dC_C}{dt} = k_1 C_A C_B - k_2 C_C C_D - k_3 C_C^n C_D^m \tag{5.4}$$

$$\frac{dC_D}{dt} = k_1 C_A C_B - k_2 C_C C_D - k_3 C_C^n C_D^m$$

$$\frac{dC_E}{dt} = k_3 C_C^n C_D^m$$

where C_A, C_B, C_C, C_D, and C_E represent the concentrations of the five chemical components of this reaction. This is a set of simultaneous first-order nonlinear ordinary differential equations, which describes the dynamic behavior of the chemical reaction. With the methods to be developed in this chapter, these equations, with a set of initial conditions, can be integrated to obtain the time profiles of all the concentrations.

Consider the growth of a microorganism, say a yeast, in a continuous fermentor of the type shown in Fig. 5.1. The volume of the liquid in the fermentor is V. The flow rate of nutrients into the fermentor is F_{in} and the flow rate of products out of the fermentor is F_{out}. The unsteady state material balance for the cells X is:

$$F_{in} X_{in} + r_X V = F_{out} X_{out} + \frac{d(VX)}{dt} \tag{5.5}$$

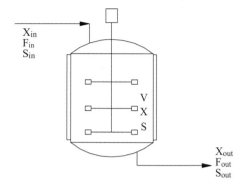

FIG. 5.1

Continuous fermentor.

The material balance for the substrate S is given by:

$$F_{in}S_{in} + r_s V = F_{out}S_{out} + \frac{d(VS)}{dt} \tag{5.6}$$

The overall volumetric balance is:

$$F_{in} = F_{out} + \frac{dV}{dt} \tag{5.7}$$

If we assume that the fermentor is perfectly mixed, that is, the concentrations at every point in the fermentor are the same, making it a continuous stirred tank reactor (CSTR), then,

$$X = X_{out}$$
$$S = S_{out} \tag{5.8}$$

and the equations simplify to:

$$\frac{d(VX)}{dt} = (F_{in}X_{in} - F_{out}X) + r_X V \tag{5.9}$$

$$\frac{d(VS)}{dt} = (F_{in}S_{in} - F_{out}S) + r_S V \tag{5.10}$$

$$\frac{dV}{dt} = F_{in} - F_{out} \tag{5.11}$$

Further assumptions are made that the flow rates in and out of the fermentor are identical and that the rates of cell formation and substrate utilization are given by:

$$r_X = \frac{\mu_{max}SX}{K + S} \tag{5.12}$$

and

$$r_S = -\frac{1}{Y_s}\frac{\mu_{max}SX}{K + S} \tag{5.13}$$

The set of equations becomes:

$$\frac{dX}{dt} = \left(\frac{F_{out}}{V}\right)(X_{in} - X) + \frac{\mu_{max}SX}{K + S} \tag{5.14}$$

$$\frac{dS}{dt} = \left(\frac{F_{out}}{V}\right)(S_{in} - S) - \frac{1}{Y_S}\frac{\mu_{max}SX}{K + S} \tag{5.15}$$

This is a set of simultaneous ordinary differential equations, which describe the dynamics of a continuous culture fermentation.

The dynamic behavior of a distillation column may be examined by making material balances around each stage of the column. Fig. 5.2 shows a typical stage n with a liquid flow into the stage L_{n+1} and out of stage L_n, and vapor flows into the stage V_{n-1} and out of the stage V_n. The liquid holdup on the stage is designated as H_n. There is no generation of material in this process, so the unsteady state material balance (Eq. 5.2) becomes:

$$\frac{dH_n}{dt} = V_{n-1} + L_{n+1} - V_n - L_n \tag{5.16}$$

The liquid and vapor in this operation are multicomponent mixtures of k components. The mole fractions of each component in the liquid and vapor phases are designated by x_i and y_i, respectively. Therefore the material balance for the ith component is:

$$\frac{d(H_n x_{i,n})}{dt} = V_{n-1}y_{i,n-1} + L_{n+1}x_{i,n+1} - V_n y_{i,n} - L_n x_{i,n} \tag{5.17}$$

The concentrations of liquid and vapor leaving stage n are related through the equilibrium relationship:

$$y_{i,n} = f(x_{i,n}) \tag{5.18}$$

If the assumptions of constant molar overflow and negligible change in vapor flow are made, then $V_{n-1} = V_n$. The change in liquid flow is:

$$\tau\frac{dL_n}{dt} = L_{n-1} - L_n \tag{5.19}$$

where τ is the hydraulic time constant.

The foregoing equations applied to each stage in a multistage separation process result in a large set of simultaneous ordinary differential equations. In all the foregoing examples the systems were

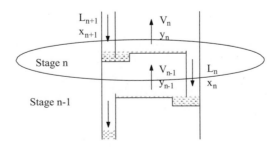

FIG. 5.2

Material balance around stage n of a distillation column.

chosen so that the models resulted in sets of *simultaneous first-order ordinary differential equations*. These are the most commonly encountered types of problems in the analysis of multicomponent and/or multistage operations. *Closed-form* solutions for such sets of equations are not usually obtainable. However, *numerical methods* have been highly developed for the solution of sets of simultaneous differential equations. In this chapter we discuss the most useful techniques for the solution of such problems. We first show that higher-order differential equations can be reduced to first order by a series of substitutions.

In this chapter we develop numerical solutions for initial value problems of ordinary differential equations. The initial value problem is defined with a set of ordinary differential equations in their canonical form:

$$\frac{d\mathbf{y}}{dx} = \mathbf{f}(x, \mathbf{y}) \tag{5.20}$$

with the vector of initial conditions given by:

$$\mathbf{y}(x_0) = \mathbf{y_0} \tag{5.21}$$

Note that a problem whose dependent variables and/or their derivatives are all known at the final value of the independent variable (rather than the initial value) is identical to the initial value problem because only the direction of integration must be reversed. Therefore, the term initial value problem refers to both cases.

5.2 CLASSIFICATIONS OF ORDINARY DIFFERENTIAL EQUATIONS

Ordinary differential equations are classified according to their *order*, their *linearity*, and their *boundary conditions*. The order of a differential equation is the order of the highest derivative present in that equation. The following are examples of first-, second-, and third-order differential equations:

$$\text{First order}: \frac{dy}{dx} + y = kx \tag{5.22}$$

$$\text{Second order}: \frac{d^2y}{dx^2} + y\frac{dy}{dx} = kx \tag{5.23}$$

$$\text{Third order}: \frac{d^3y}{dx^3} + a\frac{d^2y}{dx^2} + b\left(\frac{dy}{dx}\right)^2 = kx \tag{5.24}$$

Ordinary differential equations may be categorized as *linear* and *nonlinear* equations. A differential equation is nonlinear if it contains products of the dependent variable, its derivatives, or both. For example, Eqs. (5.23) and (5.24) are nonlinear because they contain the terms $y(dy/dx)$ and $(dy/dx)^2$, respectively, while Eq. (5.22) is linear. The general form of a linear differential equation of order n may be written as:

$$b_0(x)\frac{d^ny}{dx^n} + b_1(x)\frac{d^{n-1}y}{dx^{n-1}} + \cdots + b_{n-1}(x)\frac{dy}{dx} + b_n(x)y = R(x) \tag{5.25}$$

To obtain a unique solution of an *n*th-order differential equation or of a set of *n* simultaneous first-order differential equations, it is necessary to specify *n* values of the dependent variables (or their derivatives) at specific values of the independent variable.

Ordinary differential equations may be classified as *initial value* problems or *boundary-value* problems. In initial value problems the values of the dependent variables and/or their derivatives are *all* known at the initial value of the independent variable. In boundary value problems the dependent variables and/or their derivatives are known at more than one point of the independent variable. If some of the dependent variables (or their derivatives) are specified at the initial value of the independent variable, and the remaining variables (or their derivatives) are specified at the final value of the independent variable, then this is a *two-point boundary value* problem.

5.3 TRANSFORMATION TO CANONICAL FORM

Numerical integration of ordinary differential equations is most conveniently performed when the system consists of a set of *n* simultaneous first-order ordinary differential equations of the form:

$$\frac{dy_1}{dx} = f_1(y_1, y_2, \ldots, y_n, x)$$
$$\frac{dy_2}{dx} = f_2(y_1, y_2, \ldots, y_n, x)$$
$$\vdots$$
$$\frac{dy_n}{dx} = f_n(y_1, y_2, \ldots, y_n, x)$$

(5.26)

This is called the *canonical* form of the equations. When the initial conditions are given at a common point x_0:

$$y_1(x_0) = y_{1,0}$$
$$y_2(x_0) = y_{2,0}$$
$$\vdots$$
$$y_n(x_0) = y_{n,0}$$

(5.27)

Then, the system of Eqs. (5.26) have solutions of the form:

$$y_1 = F_1(x)$$
$$y_2 = F_2(x)$$
$$\vdots$$
$$y_n = F_n(x)$$

(5.28)

This problem can be condensed into matrix notation, where the system equations are represented by:

$$\frac{d\mathbf{y}}{dx} = \mathbf{f}(x, \mathbf{y})$$

(5.20)

The vector of initial conditions is:

$$\mathbf{y}(x_0) = \mathbf{y}_0$$

(5.21)

and the vector of solutions is:

$$y = F(x) \tag{5.29}$$

Differential equations of higher order, or systems containing equations of mixed order, can be transformed to the canonical form by a series of substitutions. For example, consider the nth-order differential equation:

$$\frac{d^n z}{dx^n} = G\left(z, \frac{dz}{dx}, \frac{d^2 z}{dx^2}, \dots, \frac{d^{n-1} z}{dx^{n-1}}, x\right) \tag{5.30}$$

The following transformations:

$$
\begin{aligned}
z &= y_1 \\
\frac{dz}{dx} &= \frac{dy_1}{dx} = y_2 \\
\frac{d^2 z}{dx^2} &= \frac{dy_2}{dx} = y_3 \\
&\vdots \\
\frac{d^{n-1} z}{dx^{n-1}} &= \frac{dy_{n-1}}{dx} = y_n \\
\frac{d^n z}{dx^n} &= \frac{dy_n}{dx}
\end{aligned}
\tag{5.31}
$$

when substituted into the nth-order Eq. (5.30), give the equivalent set of n first-order equations of the canonical form:

$$
\begin{aligned}
\frac{dy_1}{dx} &= y_2 \\
\frac{dy_2}{dx} &= y_3 \\
&\vdots \\
\frac{dy_n}{dx} &= G(y_1, y_2, y_3, \dots, y_n, x)
\end{aligned}
\tag{5.32}
$$

If the right-hand side of the differential equations is not a function of the independent variable, that is,

$$\frac{dy}{dx} = f(y) \tag{5.33}$$

then the set is *autonomous*. A *nonautonomous* set may be transformed into an autonomous set by an appropriate substitution (see Example 5.1(b) and (d)). If the functions $f(y)$ are linear in terms of y, then the equations can be written in matrix form:

$$y' = Ay \tag{5.34}$$

Solutions for linear sets of ordinary differential equations are developed in Section 5.4. The methods for the solution of nonlinear sets are discussed in Section 5.5.

A more restricted form of a differential equation is:

$$\frac{dy}{dx} = f(x) \tag{5.35}$$

where $f(x)$ are functions of the independent variable only. Solution methods for these equations were developed in Chapter 4.

The next example demonstrates the technique for converting higher-order linear and nonlinear differential equations to canonical form.

EXAMPLE 5.1 TRANSFORM THE FOLLOWING ORDINARY DIFFERENTIAL EQUATIONS INTO THEIR CANONICAL FORM

Apply the transformation defined by Eqs. (5.31) and (5.32) to the following ordinary differential equations:

(a) $\dfrac{d^4z}{dt^4} + 5\dfrac{d^3z}{dt^3} - 2\dfrac{d^2z}{dt^2} - 6\dfrac{dz}{dt} + 3z = 0$

(b) $\dfrac{d^4z}{dt^4} + 5\dfrac{d^3z}{dt^3} - 2\dfrac{d^2z}{dt^2} - 6\dfrac{dz}{dt} + 3z = e^{-t}$

(c) $\dfrac{d^3z}{dx^3} + z^2\dfrac{d^2z}{dx^2} - \left(\dfrac{dz}{dx}\right)^3 - 2z = 0$

(d) $\dfrac{d^3z}{dt^3} + t^3\dfrac{d^2z}{dt^2} - t^2\dfrac{dz}{dt} + 5z = 0$

Solution

(a)

$$\frac{d^4z}{dt^4} + 5\frac{d^3z}{dt^3} - 2\frac{d^2z}{dt^2} - 6\frac{dz}{dt} + 3z = 0$$

Apply the transformation according to Eqs. (5.31):

$$z = y_1$$
$$\frac{dz}{dt} = \frac{dy_1}{dt} = y_2$$
$$\frac{d^2z}{dt^2} = \frac{dy_2}{dt} = y_3$$
$$\frac{d^3z}{dt^3} = \frac{dy_3}{dt} = y_4$$
$$\frac{d^4z}{dt^4} = \frac{dy_4}{dt}$$

Make these substitutions into Eq. (a) to obtain the following four equations:

$$\frac{dy_1}{dt} = y_2$$
$$\frac{dy_2}{dt} = y_3$$
$$\frac{dy_3}{dt} = y_4$$
$$\frac{dy_4}{dt} = -3y_1 + 6y_2 + 2y_3 - 5y_4$$

(Continued)

EXAMPLE 5.1　(CONTINUED)

This is a set of linear ordinary differential equations that can be represented in matrix form:

$$y' = Ay \tag{5.34}$$

where matrix A is given by:

$$A = \begin{bmatrix} 0 & 1 & 0 & 0 \\ 0 & 0 & 1 & 0 \\ 0 & 0 & 0 & 1 \\ -3 & 6 & 2 & -5 \end{bmatrix}$$

The method of obtaining the solution of sets of linear ordinary differential equations is discussed in Section 5.4.

(b)
$$\frac{d^4z}{dt^4} + 5\frac{d^3z}{dt^3} - 2\frac{d^2z}{dt^2} - 6\frac{dz}{dt} + 3z = e^{-t}$$

The presence of the term e^t on the right-hand side of this equation makes it a nonhomogeneous equation. The left-hand side is identical to that of Eq. (a) so that the transformations of part (a) are applicable. An additional transformation is needed to replace the e^{-t} term. This transformation is:

$$y_5 = e^{-t}$$
$$\frac{dy_5}{dx} = e^{-t} = -y_5$$

Make the substitutions into Eq. (b) to obtain the following set of five linear ordinary differential equations:

$$\frac{dy_1}{dt} = y_2$$
$$\frac{dy_2}{dt} = y_3$$
$$\frac{dy_3}{dt} = y_4$$
$$\frac{dy_4}{dt} = -3y_1 + 6y_2 + 2y_3 - 5y_4 + y_5$$
$$\frac{dy_5}{dt} = -y_5$$

which also condenses into the matrix form of Eq. (5.34), with the matrix A given as:

$$A = \begin{bmatrix} 0 & 1 & 0 & 0 & 0 \\ 0 & 0 & 1 & 0 & 0 \\ 0 & 0 & 0 & 1 & 0 \\ -3 & 6 & 2 & -5 & 1 \\ 0 & 0 & 0 & 0 & -1 \end{bmatrix}$$

(Continued)

EXAMPLE 5.1 (CONTINUED)

(c)

$$\frac{d^3z}{dx^3} + z^2 \frac{d^2z}{dx^2} - \left(\frac{dz}{dx}\right)^3 - 2z = 0$$

Apply the following transformations:

$$z = y_1$$
$$\frac{dz}{dx} = \frac{dy_1}{dx} = y_2$$
$$\frac{d^2z}{dx^2} = \frac{dy_2}{dx} = y_3$$
$$\frac{d^3z}{dx^3} = \frac{dy_3}{dx}$$

Make the substitutions into Eq. (c) to obtain the set:

$$\frac{dy_1}{dx} = y_2$$
$$\frac{dy_2}{dx} = y_3$$
$$\frac{dy_3}{dx} = 2y_1 + y_2^3 - y_1^2 y_3$$

This is a set of *nonlinear* differential equations that cannot be expressed in matrix form. The methods of solution of nonlinear differential equations are developed in Section 5.5.

(d)

$$\frac{d^3z}{dt^3} + t^3 \frac{d^2z}{dt^2} - t^2 \frac{dz}{dt} + 5z = 0$$

Apply the following transformations:

$$z = y_1$$
$$\frac{dz}{dt} = \frac{dy_1}{dt} = y_2$$
$$\frac{d^2z}{dt^2} = \frac{dy_2}{dt} = y_3$$
$$\frac{d^3z}{dt^3} = \frac{dy_3}{dt}$$
$$y_4 = t$$
$$\frac{dy_4}{dt} = 1$$

(Continued)

EXAMPLE 5.1 (CONTINUED)

Make the substitutions into Eq. (d) to obtain the set:

$$\frac{dy_1}{dt} = y_2$$

$$\frac{dy_2}{dt} = y_3$$

$$\frac{dy_3}{dt} = -5y_1 + y_4^2 y_2 - y_4^3 y_3$$

$$\frac{dy_4}{dt} = 1$$

This is a set of autonomous nonlinear differential equations. Note that the foregoing set of substitutions converted the nonautonomous Eq. (d) to a set of autonomous equations.

Professionally written robust codes for solving ordinary differential equations numerically (e.g., MATLAB® *ode45*) are restricted to a scalar (i.e., single) or systems of first-order ordinary differential equations. That fact should be motivation enough for going through all this trouble to convert a higher order (or nonautonomous) ordinary differential equation into a system of first-order ordinary differential equations. With a good understanding of the discussion so far, the reader should be able to transform any higher-order ordinary differential equation into a matrix of first-order ordinary differential equations, write the resulting matrix into a MATLAB function, and then include that function in the list of input arguments to a call to *ode45*. It turns out that there is no need to write such a code. There is already a MATLAB function, called *odeToVectorField*, to do just that.

5.4 LINEAR ORDINARY DIFFERENTIAL EQUATIONS

The analysis of many *physicochemical* systems yields mathematical models that are sets of *linear* ordinary differential equations with constant coefficients and can be reduced to the form:

$$y' = Ay \tag{5.34}$$

with given initial conditions:

$$y(0) = y_0 \tag{5.36}$$

Such conditions abound in chemical engineering. The unsteady state material and energy balances of multiunit processes, without chemical reaction, often yield such differential equations.

Sets of linear ordinary differential equations with constant coefficients have closed-form solutions that can be readily obtained from the eigenvalues and eigenvectors of matrix A. To develop this solution, let us first consider a single linear differential equation of the type:

$$\frac{dy}{dt} = ay \tag{5.37}$$

with the given initial condition:

$$y(0) = y_0 \tag{5.38}$$

Eq. (5.37) is essentially the scalar form of the matrix set of Eq. (5.34). The solution of the scalar equation can be obtained by separating the variables and integrating both sides of the equation:

$$\int_{y_0}^{y} \frac{dy}{y} = \int_{0}^{t} a\,dt$$
$$\ln \frac{y}{y_0} = at \tag{5.39}$$
$$y = e^{at} y_0$$

In an analogous fashion the matrix set can be integrated to obtain the solution:

$$\mathbf{y} = e^{\mathbf{A}t} \mathbf{y_0} \tag{5.40}$$

In this case $\mathbf{y}$ and $\mathbf{y_0}$ are *vectors* of the dependent variables and the initial conditions, respectively. The term $e^{\mathbf{A}t}$ is the matrix exponential function, which was defined by Eq. (2.83):

$$e^{\mathbf{A}t} = \mathbf{I} + \mathbf{A}t + \frac{\mathbf{A}^2 t^2}{2!} + \frac{\mathbf{A}^3 t^3}{3!} + \cdots \tag{5.41}$$

It can be demonstrated that Eq. (5.40) is a solution of Eq. (5.34) by differentiating it:

$$\frac{d\mathbf{y}}{dt} = \frac{d}{dt}\left(e^{\mathbf{A}t}\right)\mathbf{y_0}$$
$$= \frac{d}{dt}\left(\mathbf{I} + \mathbf{A}t + \frac{\mathbf{A}^2 t^2}{2!} + \frac{\mathbf{A}^3 t^3}{3!} + \cdots\right)\mathbf{y_0}$$
$$= \left(\mathbf{A} + \mathbf{A}^2 t + \frac{\mathbf{A}^3 t^2}{2!} + \cdots\right)\mathbf{y_0}$$
$$= \mathbf{A}\left(\mathbf{I} + \mathbf{A}t + \frac{\mathbf{A}^2 t^2}{2!} + \cdots\right)\mathbf{y_0}$$
$$= \mathbf{A}\left(e^{\mathbf{A}t}\right)\mathbf{y_0}$$
$$= \mathbf{A}\mathbf{y}$$

The solution of the set of linear ordinary differential equations is very cumbersome to evaluate in the form of Eq. (5.40) because it requires the evaluation of the infinite series of the exponential term $e^{\mathbf{A}t}$. However, this solution can be modified, by further algebraic manipulation, to express it in terms of the eigenvalues and eigenvectors of the matrix $\mathbf{A}$.

In Chapter 2 we showed that a nonsingular matrix $\mathbf{A}$ of order n has n eigenvectors and n non-zero eigenvalues, whose definitions are given by:

$$\begin{aligned}
\mathbf{A}\mathbf{x}_1 &= \lambda_1 \mathbf{x}_1 \\
\mathbf{A}\mathbf{x}_2 &= \lambda_2 \mathbf{x}_2 \\
&\vdots \\
\mathbf{A}\mathbf{x}_n &= \lambda_n \mathbf{x}_n
\end{aligned} \tag{5.42}$$

All these eigenvectors and eigenvalues can be represented in a compact form as follows:

$$\mathbf{A}\mathbf{X} = \mathbf{X}\mathbf{\Lambda} \tag{5.43}$$

where the columns of matrix X are the individual eigenvectors:

$$X = [x_1, x_2, x_3, \ldots, x_n] \tag{5.44}$$

and Λ is a diagonal matrix with the eigenvalues of A on its diagonal:

$$\Lambda = \begin{bmatrix} \lambda_1 & 0 & 0 & \cdots & 0 \\ 0 & \lambda_2 & 0 & \cdots & 0 \\ 0 & 0 & \lambda_3 & \cdots & 0 \\ \cdots & \cdots & \cdots & \cdots & \cdots \\ 0 & 0 & 0 & \cdots & \lambda_n \end{bmatrix} \tag{5.45}$$

If we postmultiply each side of Eq. (5.43) by X^{-1}, we obtain:

$$AXX^{-1} = A = X\Lambda X^{-1} \tag{5.46}$$

Squaring Eq. (5.46),

$$A^2 = [X\Lambda X^{-1}][X\Lambda X^{-1}] = X\Lambda^2 X^{-1} \tag{5.47}$$

Similarly, raising Eq. (5.46) to any power n we obtain:

$$A^n = X\Lambda^n X^{-1} \tag{5.48}$$

Starting with Eq. (5.41) and replacing the matrices A, A^2, ..., A^n with their equivalents from Eqs. (5.46)−(5.48), we obtain:

$$e^{At} = I + X\Lambda X^{-1}t + X\Lambda^2 X^{-1}\frac{t^2}{2!} + \cdots \tag{5.49}$$

The identity matrix I can be premultiplied by X and postmultiplied by X^{-1} without changing it. Therefore Eq. (5.49) rearranges to:

$$e^{At} = X\left(I + \Lambda t + \frac{\Lambda^2 t^2}{2!} + \cdots\right)X^{-1} \tag{5.50}$$

which simplifies to:

$$e^{At} = Xe^{\Lambda t}X^{-1} \tag{5.51}$$

where the exponential matrix $e^{\Lambda t}$ is defined as:

$$e^{\Lambda t} = \begin{bmatrix} e^{\lambda_1 t} & 0 & 0 & \cdots & 0 \\ 0 & e^{\lambda_2 t} & 0 & \cdots & 0 \\ 0 & 0 & e^{\lambda_3 t} & \cdots & 0 \\ \cdots & \cdots & \cdots & \cdots & \cdots \\ 0 & 0 & 0 & \cdots & e^{\lambda_n t} \end{bmatrix} \tag{5.52}$$

The solution of the linear differential equations can now be expressed in terms of eigenvalues and eigenvectors by combining Eqs. (5.40) and (5.51):

$$y = [Xe^{\Lambda t}X^{-1}]y_0 \tag{5.53}$$

The eigenvalues and eigenvectors of matrix A can be calculated using the techniques developed in Chapter 1 or simply by applying the built-in function *eig* in MATLAB. This is demonstrated in Example 5.2.

EXAMPLE 5.2 A SEQUENTIAL CHEMICAL REACTION SYSTEM

Develop a general MATLAB function to solve the set of linear differential equations. Apply this function to determine the concentration profiles of all components of the following chemical reaction system:

$$A \underset{k_2}{\overset{k_1}{\rightleftarrows}} B \underset{k_4}{\overset{k_3}{\rightleftarrows}} C$$

Assume that all steps are first-order reactions and write the set of linear ordinary differential equations that describe the kinetics of these reactions. Solve the problem numerically for the following values of the kinetic rate constants:

$$k_1 = 1 \text{ min}^{-1} \quad k_2 = 0 \text{ min}^{-1} \quad k_3 = 2 \text{ min}^{-1} \quad k_4 = 3 \text{ min}^{-1}$$

The value of $k_2 = 0$ reveals that the first reaction is irreversible in this special case. The initial concentrations of the three components are:

$$C_{A_0} = 1 \quad C_{B_0} = 0 \quad C_{C_0} = 0$$

Plot the graph of concentrations versus time.

Method of solution

Assuming that all steps are first-order reactions, the set of differential equations that give the rate of formation of each compound is the following:

$$\frac{dC_A}{dt} = -k_1 C_A + k_2 C_B$$
$$\frac{dC_B}{dt} = k_1 C_A - k_2 C_B - k_3 C_B + k_4 C_C$$
$$\frac{dC_C}{dt} = k_3 C_B - k_4 C_C$$

In matrix form this set reduces to:

$$\dot{c} = Kc$$

where

$$\dot{c} = \begin{bmatrix} \dfrac{dC_A}{dt} \\ \dfrac{dC_B}{dt} \\ \dfrac{dC_C}{dt} \end{bmatrix} \quad c = \begin{bmatrix} C_A \\ C_B \\ C_C \end{bmatrix}$$

(Continued)

EXAMPLE 5.2 (CONTINUED)

and

$$K = \begin{bmatrix} -k_1 & k_2 & 0 \\ k_1 & -k_2 - k_3 & k_4 \\ 0 & k_3 & -k_4 \end{bmatrix}$$

The solution of a set of linear ordinary differential equations can be obtained either by applying Eq. (5.40):

$$c = e^{Kt}c_0$$

or by Eq. (5.53):

$$c = \left[Xe^{\Lambda t}X^{-1} \right]c_0$$

where the matrix X consists of the eigenvectors of K and c_0 is the vector of initial concentrations:

$$c_0 = \begin{bmatrix} C_{A_0} \\ C_{D_0} \\ C_{C_0} \end{bmatrix}$$

Program description

The programs and functions developed in this example can be found in: https://www.elsevier.com/books-and-journals/book-companion/9780128229613

The MATLAB function *LinearODE.m* solves a set of linear ordinary differential equations. The first part of the function checks the number of inputs and their sizes or values. The next section of the function performs the solution of the set of ordinary differential equations that can be done by either the matrix exponential method (Eq. 5.40) or the eigenvector method (Eq. 5.53). The method of the solution may be introduced to the function through the fifth input argument. The default method of solution is the matrix exponential method.

The main program *Example5_2.m* solves the particular problem posed in this example by applying *LinearODE.m*. This program gets the required input data, including the rate constants and initial concentrations of the components, from the keyboard. Then, it builds the matrix of coefficients and the vector of times at which the concentrations are to be calculated. In the last section the program asks the user to select the method of solution and calls the function *LinearODE* to solve the set of equations for obtaining the concentrations and plots the results. The reader may try another method of solution and repeat solving the set of linear differential equations in this part.

(Continued)

EXAMPLE 5.2 (CONTINUED)

Input and results

```
>> Example5_2
A->B, k1 = 1
B->A, k2 = 0
B->C, k3 = 2
C->B, k4 = 3
```
Initial concentration of A = 1
Initial concentration of B = 0
Initial concentration of C = 0
Maximum time = 5
Time interval = 0.1
1) Matrix exponential method
2) Eigenvector method
0) Exit
Choose the method of solution: 2
Choose the method of solution: 0

Discussion of results

The results of the solution to this problem are shown in Fig. E5.2. It is seen from this figure that, as expected for this special case, after a long enough time, all of component A is consumed and the components B and C satisfy the equilibrium condition $C_B/C_C = k_4/k_3$. These results also confirm the conservation of mass principle:

$$C_{A_0} + C_{B_0} + C_{C_0} = C_A + C_B + C_C$$

Because both methods of solution are exact, results obtained by these methods would be identical. However, when dealing with a large number of equations and/or a long time vector, the matrix exponential method is appreciably faster in the MATLAB environment than the eigenvector method. This is because the exponential of a matrix is performed by the built-in MATLAB function *expm*, while the eigenvector method involves several element-by-element operations when building the matrix $e^{\Lambda t}$. The reader is encouraged to verify the difference between the methods by repeating the solution and choosing a smaller time interval, say 0.001, and applying this to both solution methods.

It should be mentioned that different options in MATLAB for the solution of linear ordinary differential equations: (1) exact analytical solutions that result in expressions for the evolution of the dependent variable(s) as a function of the independent variable, obtained using the *Symbolic Math Toolbox*; (2) exact numerical results obtained via analytical matrix methods, as in the preceding example; and (3) approximate numerical results obtained using numerical methods such as Runge-Kutta, to be discussed in a subsequent section.

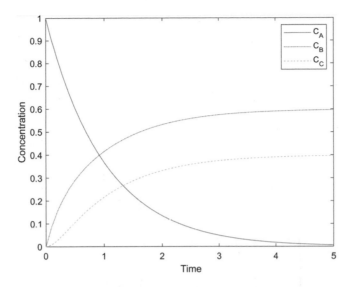

FIG. E5.2

Concentration profiles.

5.5 NONLINEAR ORDINARY DIFFERENTIAL EQUATIONS

As mentioned earlier, an initial value problem is a set of ordinary differential equations in their canonical form:

$$\frac{dy}{dx} = f(x, y) \tag{5.20}$$

with the initial conditions given by:

$$y(x_0) = y_0 \tag{5.21}$$

If the function f is nonlinear with respect to y, the set of ordinary differential equations is nonlinear. To be able to illustrate these methods graphically, we treat y as a single variable rather than as a vector of variables. The formulas developed for the solution of a single differential equation are readily expandable to those for a set of differential equations that must be solved *simultaneously*. This concept is demonstrated in Section 5.5.4.

We begin the development of these methods by first rearranging Eq. (5.20) and integrating both sides between the limits of $x_i \leq x \leq x_{i+1}$ and $y_i \leq y \leq y_{i+1}$:

$$\int_{y_i}^{y_{i+1}} dy = \int_{x_i}^{x_{i+1}} f(x, y) dx \tag{5.54}$$

The left side integrates readily to obtain:

$$y_{i+1} - y_i = \int_{x_i}^{x_{i+1}} f(x, y)dx \qquad (5.55)$$

One method for integrating Eq. (5.55) is to take the left-hand side of this equation and use finite differences for its approximation. This technique works directly with the tangential trajectories of the dependent variable y rather than with the areas under the function $f(x, y)$. This is the technique applied in Sections 5.5.1 and 5.5.2.

In Chapter 4 we developed the integration formulas by first replacing the function $f(x)$ with an interpolating polynomial and then evaluating the integral $f(x)dx$ between the appropriate limits. A similar technique could be followed here to integrate the right-hand side of Eq. (5.55). This approach is followed in Section 5.5.3.

5.5.1 THE EULER AND MODIFIED EULER METHODS

One of the earliest techniques developed for the solution of ordinary differential equations is the *Euler method*. This is simply obtained by recognizing that the left side of Eq. (5.55) is the first forward finite difference of y at position i:

$$y_{i+1} - y_i = \Delta y_i \qquad (5.56)$$

which, when rearranged, gives a "forward marching" formula for evaluating y:

$$y_{i+1} = y_i + \Delta y_i \qquad (5.57)$$

The forward difference term Δy_i is obtained from Eq. (3.53) applied to y at position i:

$$\Delta y_i = hDy_i + \frac{h^2 D^2 y_i}{2} + \frac{h^3 D^3 y_i}{6} + \cdots \qquad (5.58)$$

In the Euler method the foregoing series is truncated after the first term to obtain:

$$\Delta y_i = hDy_i + O(h^2) \qquad (5.59)$$

The combination of Eqs. (5.57) and (5.59) gives the *explicit Euler formula* for integrating differential equations:

$$y_{i+1} = y_i + hDy_i + O(h^2) \qquad (5.60)$$

The derivative Dy_i is replaced by its equivalent y'_i or $f(x_i, y_i)$ to give the more commonly used form of the explicit Euler method[1]:

$$y_{i+1} = y_i + hf(x_i, y_i) + O(h^2) \qquad (5.61)$$

This equation simply states that the next value of y is obtained from the current value by moving a step of width h in the tangential direction of y, and that direction is calculated based on the current, known value of the dependent variable. This is demonstrated graphically in Fig. 5.3a. This Euler

---

[1] From here on the term y'_i and $f(x_i, y_i)$ will be used interchangeably. The reader should remember that these are equal to each other through the differential Eq. (5.20).

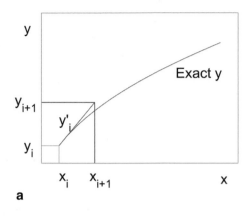
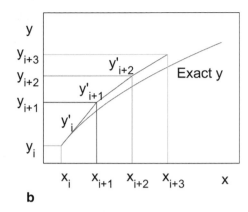

FIG. 5.3

The explicit Euler method of integration. (a) Single step. (b) Several steps.

formula is rather inaccurate because it has a truncation error of only $O(h^2)$. If h is large the trajectory of y can quickly deviate from its true value, as shown in Fig. 5.3b.

The accuracy of the Euler method can be improved by using a combination of forward and backward differences. Note that the first forward difference of y at i is equal to the first backward difference of y at $(i + 1)$:

$$\Delta y_i = y_{i+1} - y_i = \nabla y_{i+1} \tag{5.62}$$

Therefore the forward marching formula in terms of backward differences is:

$$y_{i+1} = y_i + \nabla y_{i+1} \tag{5.63}$$

The backward difference term ∇y_{i+1} is obtained from Eq. (3.32) applied to y at position $(i + 1)$:

$$\nabla y_{i+1} = hDy_{i+1} - \frac{h^2 D^2 y_{i+1}}{2} + \frac{h^3 D^3 y_{i+1}}{6} - \dots \tag{5.64}$$

Combining Eqs. (5.63) and (5.64):

$$y_{i+1} = y_i + hf(x_{i+1}, y_{i+1}) + O(h^2) \tag{5.65}$$

This is called the *implicit Euler formula* (or backward Euler) because it involves the calculation of function f at an unknown value of y_{i+1}. Eq. (5.65) can be viewed as taking a step forward from position i to $(i + 1)$ in a gradient direction that must be evaluated at $(i + 1)$.

Implicit equations cannot be solved individually but must be set up as sets of simultaneous algebraic equations. When these sets are linear, the problem can be solved by the application of the Gauss elimination methods developed in Chapter 2. If the set consists of nonlinear equations, the problem is much more difficult and must be solved using Newton's method for simultaneous nonlinear algebraic equations developed in Chapter 1.

In the case of the Euler methods the problem can be simplified by first applying the explicit method to *predict* a value y_{i+1}:

$$(y_{i+1})_{Pr} = y_i + hf(x_i, y_i) + O(h^2) \tag{5.66}$$

and then using this predicted value in the implicit method to get a *corrected* value:

$$(y_{i+1})_{Cor} = y_i + hf\left(x_{i+1}, (y_{i+1})_{Pr}\right) + O(h^2) \tag{5.67}$$

This combination of steps is known as the *Euler predictor-corrector* (or *modified Euler*) method, whose application is demonstrated graphically in Fig. 5.4. Correction by Eq. (5.67) may be applied more than once until the corrected value converges; that is, the difference between the two consecutive corrected values becomes less than the convergence criterion. However, not much more accuracy is achieved after the second application of the corrector.

The explicit, as well as the implicit, forms of the Euler methods have an error of order (h^2). However, when used in combination with each other, as predictor-corrector, their accuracy is enhanced, yielding an error of order (h^3). This conclusion can be reached by adding Eqs. (5.57) and (5.63):

$$y_{i+1} = y_i + \frac{1}{2}(\Delta y_i + \nabla y_{i+1}) \tag{5.68}$$

and using Eqs. (5.58) and (5.64) to obtain:

$$y_{i+1} = y_i + \frac{h}{2}[f(x_i, y_i) + f(x_{i+1}, y_{i+1})] + O(h^3) \tag{5.69}$$

The terms of the order (h^2) cancel out because they have opposite signs, thus giving a formula of higher accuracy. Eq. (5.69) is essentially the same as the trapezoidal rule (Eq. 5.73), with the only difference being in the way the function is evaluated at (x_{i+1}, y_{i+1}).

It has been shown that the Euler implicit formula is more stable than the explicit one [1]. The stability of these methods will be discussed in Section 5.8.

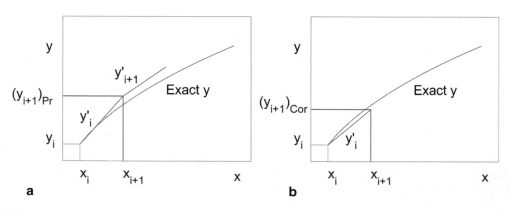

FIG. 5.4

The Euler predictor-corrector method. (a) Value of y_{i+1} is predicted and y'_{i+1} is calculated. (b) Value of y_{i+1} is corrected.

It can be seen by writing Eq. (5.69) in the form:

$$y_{i+1} = y_i + \frac{1}{2}hf(x_i, y_i) + \frac{1}{2}hf(x_{i+1}, y_{i+1}) + O(h^3) \tag{5.70}$$

that this Euler method uses the weighted trajectories of the function y evaluated at two positions that are located one full step of width h apart and weighted equally. In this form Eq. (5.70) is also known as the Crank-Nicolson method.

Eq. (5.70) can be written in a more general form as:

$$y_{i+1} = y_i + w_1 k_1 + w_2 k_2 \tag{5.71}$$

where, in this case

$$k_1 = hf(x_i, y_i) \tag{5.72}$$

$$k_2 = hf(x_i + c_2 h, y_i + a_{21} k_1) \tag{5.73}$$

The choice of the weighting factors w_1 and w_2 and the positions i and $(i + 1)$ at which to evaluate the trajectories is dictated by the accuracy required of the integration formula, that is, by the number of terms retained in the infinite series expansion.

This concept forms the basis for a whole series of integration formulas, with increasingly higher accuracies, for ordinary differential equations. These are discussed in the following section.

EXAMPLE 5.3 USING EULER AND MODIFIED EULER METHODS FOR CALCULATING THE VARIATION OF CONCENTRATION IN A BATCH REACTOR

A chemical reaction takes place in a batch reactor. The rate of reaction is given by:

$$-r_A = -\frac{dC_A}{dt} = \frac{k_1 C_A}{1 + k_2 C_A}$$

The constant values are $k_1 = 1.5$ and $k_2 = 0.1$ and the initial concentration of A in the reactor is 1 (all values in consistent units). Use Euler and modified Euler methods to determine the variation of concentration of A against time in the reactor. Consider the time step $h = 0.25$.

Method of solution

In this problem the independent variable is time (t) and the dependent variable is the concentration of A (C_A). Therefore Eq. (5.67) can be written as $(t \equiv x$ and $C_A \equiv y)$:

$$\frac{dC_A}{dt} = f(t, C_A) = -\frac{k_1 C_A}{1 + k_2 C_A}$$

With the following initial condition:

$$C_A(0) = 1$$

We solve this initial value problem by both Euler and modified Euler methods. The solution will be shown in detail for the first two time steps by hand and then until $t = 2$ in Excel.

(Continued)

EXAMPLE 5.3 (CONTINUED)

Euler method: Eq. (5.61) is used in this case for solving the problem. The flowchart of the Euler method is shown in Fig. E5.3a.

Initial value is:

$$i = 0, \ t_0 = 0, \ C_{A,0} = 1$$

In the next time step ($i = 1$, $t_1 = t_0 + h = 0.25$) the concentration is calculated as follows:

$$C_{A,1} = C_{A,0} + h\left(-\frac{k_1 C_{A,0}}{1 + k_2 C_{A,0}}\right) = 1 + 0.25\left(-\frac{1.5 \times 1}{1 + 0.1 \times 1}\right) = 0.6591$$

For the next time step ($i = 2$, $t_2 = t_1 + h = 0.5$) we have:

$$C_{A,2} = C_{A,1} + h\left(-\frac{k_1 C_{A,1}}{1 + k_2 C_{A,1}}\right) = 0.6591 + 0.25\left(-\frac{1.5 \times 0.6591}{1 + 0.1 \times 0.6591}\right) = 0.4272$$

Calculations of the next steps can be performed in a similar manner. Fig. E5.3b illustrates the calculations in Excel.

Modified Euler method: Eqs. (5.66) and (5.67) are used for solving the differential equation. The flowchart of the modified Euler method is shown in Fig. E5.3c.

Initial value is:

$$i = 0, \ t_0 = 0, \ C_{A,0} = 1$$

In the next time step ($i = 1$, $t_1 = t_0 + h = 0.25$) the concentration is calculated predicted by Eq. (5.66):

$$(C_{A,1})_{Pr} = C_{A,0} + h\left(-\frac{k_1 C_{A,0}}{1 + k_2 C_{A,0}}\right) = 1 + 0.25\left(-\frac{1.5 \times 1}{1 + 0.1 \times 1}\right) = 0.6591$$

Then, we correct this value using Eq. (5.67):

$$C_{A,1} = (C_{A,1})_{Cor} = C_{A,0} + h\left(-\frac{k_1 (C_{A,1})_{Pr}}{1 + k_2 (C_{A,1})_{Pr}}\right) = 1 + 0.25\left(-\frac{1.5 \times 0.6591}{1 + 0.1 \times 0.6591}\right) = 0.7681$$

For the next time step ($i = 2$, $t_2 = t_1 + h = 0.5$), these two values are:

$$(C_{A,2})_{Pr} = C_{A,1} + h\left(-\frac{k_1 C_{A,1}}{1 + k_2 C_{A,1}}\right) = 0.7681 + 0.25\left(-\frac{1.5 \times 0.7681}{1 + 0.1 \times 0.7681}\right) = 0.5006$$

$$C_{A,2} = (C_{A,2})_{Cor} = C_{A,1} + h\left(-\frac{k_1 C_{A,1}}{1 + k_2 C_{A,1}}\right) = 0.7681 + 0.25\left(-\frac{1.5 \times 0.5006}{1 + 0.1 \times 0.5006}\right) = 0.5893$$

Calculations of the next steps can be performed in a similar manner. Fig. E5.3d illustrates the calculations in Excel.

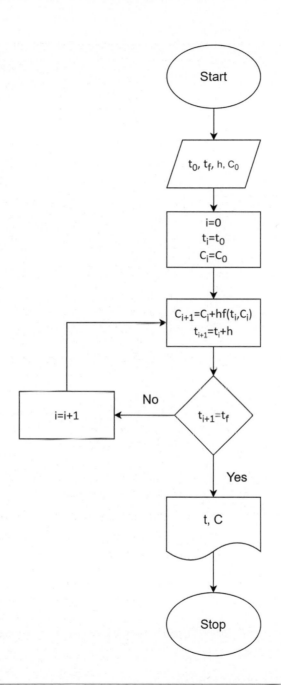

FIG. E5.3a

Flowchart of the explicit Euler method for solving ordinary differential equations.

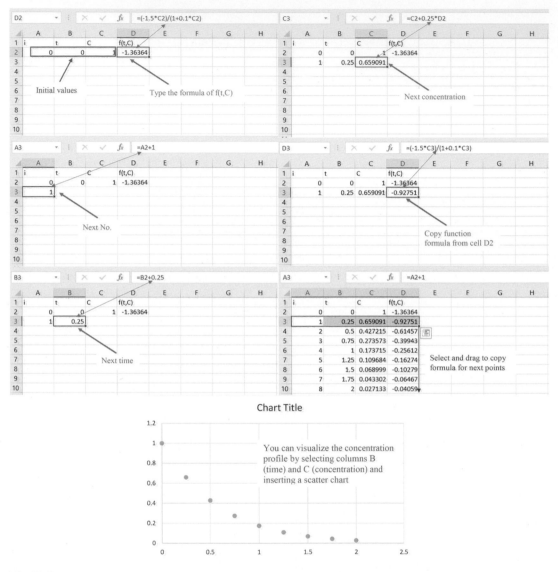

FIG. E5.3b

Calculation of evolution of concentration in a batch reactor in Excel by Euler method.

5.5.2 THE RUNGE-KUTTA METHODS

The most widely used methods of integration for ordinary differential equations are the series of methods called Runge-Kutta second, third, and fourth orders, plus a number of other techniques that are variations on the Runge-Kutta theme. These methods are based on the concept of weighted

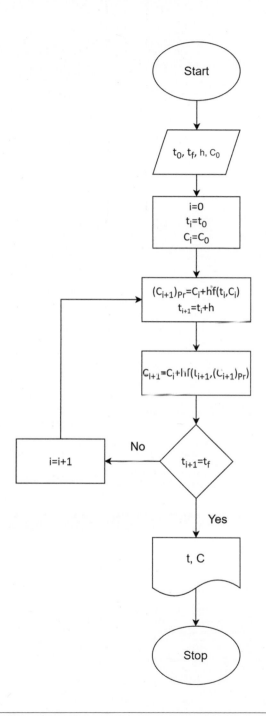

FIG. E5.3c

Flowchart of the modified Euler method for solving ordinary differential equations. Note that the corrector formula may be applied more than once (not shown in this figure).

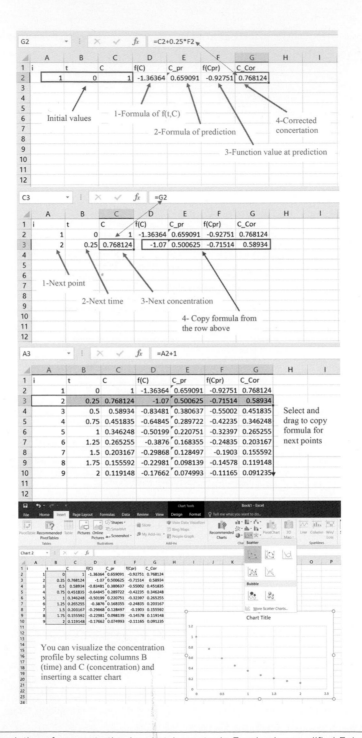

FIG. E5.3d

Calculation of the evolution of concentration in a batch reactor in Excel using modified Euler method.

trajectories formulated at the end of Section 5.5.1. In a more general fashion the forward marching integration formula for the differential Eq. (5.20) is given by the recurrence equation:

$$y_{i+1} = y_i + w_1 k_1 + w_2 k_2 + w_3 k_3 + \cdots + w_m k_m \tag{5.74}$$

Each of the trajectories k_i are evaluated by:

$$
\begin{aligned}
k_1 &= hf(x_i, y_i) \\
k_2 &= hf(x_i + c_2 h, y_i + a_{21} k_1) \\
k_3 &= hf(x_i + c_3 h, y_i + a_{31} k_1 + a_{32} k_2) \\
&\vdots \\
k_m &= hf(x_i + c_m h, y_i + a_{m1} k_1 + a_{m2} k_2 + \cdots + a_{mm-1} k_{m-1})
\end{aligned}
\tag{5.75}
$$

These equations can be written in a compact form as:

$$y_{i+1} = y_i + \sum_{i=1}^{m} w_i k_i \tag{5.76}$$

$$k_j = hf\left(x_i + c_j h, y_i + \sum_{l=1}^{j-1} a_{jl} k_l\right) \tag{5.77}$$

where $c_1 = 0$ and $a_{1j} = 0$. The value of m, which determines the complexity and accuracy of the method, is set when $(m + 1)$ terms are retained in the infinite series expansion of y_{i+1}:

$$y_{i+1} = y_i + hy'_i + \frac{h^2 y''_i}{2!} + \frac{h^3 y'''_i}{3!} + \cdots \tag{5.78}$$

or

$$y_{i+1} = y_i + hDy_i + \frac{h^2 D^2 y_i}{2!} + \frac{h^3 D^3 y_i}{3!} + \cdots \tag{5.79}$$

The procedure for deriving the Runge-Kutta methods can be divided into five steps that are demonstrated in the text that follows in the derivation of the *second-order Runge-Kutta* formulas.

Step 1: Choose the value of m, which fixes the accuracy of the formula to be obtained. For second-order Runge-Kutta, $m = 2$. Truncate the series (5.79) after the $(m + 1)$th term:

$$y_{i+1} = y_i + hDy_i + \frac{h^2 D^2 y_i}{2!} + O(h^3) \tag{5.80}$$

Step 2: Replace each derivative of y in (5.80) by its equivalent in f, remembering that f is a function of both x and $y(x)$:

$$Dy_i = f_i \tag{5.81}$$

$$
\begin{aligned}
D^2 y_i = \frac{df}{dx} &= \left(\frac{\partial f}{\partial x}\frac{dx}{dx} + \frac{\partial f}{\partial y}\frac{dy}{dx}\right) \\
&= (f_x + ff_y)_i
\end{aligned}
\tag{5.82}
$$

Combine Eqs. (5.80)–(5.82) and regroup the terms:

$$y_{i+1} = y_i + hf_i + \frac{h^2}{2}f_{x_i} + \frac{h^2}{2}f_i f_{y_i} + O(h^3) \tag{5.83}$$

Step 3: Write Eq. (5.76) with m terms in the summation:

$$y_{i+1} = y_i + w_1 k_1 + w_2 k_2 \tag{5.84}$$

where

$$k_1 = hf(x_i, y_i) \tag{5.85}$$

$$k_2 = hf(x_i + c_2 h, y_i + a_{21} k_1) \tag{5.86}$$

Step 4: Expand the f function in the Taylor series and just keep the first-order terms:

$$f(x_i + c_2 h, y_i + a_{21} k_1) = f_i + c_2 hf_{x_i} + a_{21} hf_{y_i} f_i + O(h^2) \tag{5.87}$$

Combine Eqs. (5.84)–(5.87) and regroup the terms:

$$y_{i+1} = y_i + (w_1 + w_2)hf_i + (w_2 c_2)h^2 f_{x_i} + (w_2 a_{21})h^2 f_i f_{y_i} + O(h^3) \tag{5.88}$$

Step 5: For Eqs. (5.83) and (5.88) to be identical, the coefficients of the corresponding terms must be equal to one another. This results in a set of simultaneous nonlinear algebraic equations in the unknown constants w_j, c_j, and a_{jl}. For this second-order Runge-Kutta method, there are three equations and four unknowns:

$$\begin{aligned} w_1 + w_2 &= 1 \\ w_2 c_2 &= \frac{1}{2} \\ w_2 a_{21} &= \frac{1}{2} \end{aligned} \tag{5.89}$$

It turns out that there are always more unknowns than equations. The degree of freedom allows us to choose some of the parameters. For second-order Runge-Kutta, there is one degree of freedom. For third- and fourth-order Runge-Kutta, there are two degrees of freedom. For fifth-order Runge-Kutta, there are at least five degrees of freedom. This freedom of choice of parameters gives rise to a very large number of different forms of the Runge-Kutta formulas. It is usually desirable to choose first the values of the c_j constants, thus fixing the positions along with the independent variable, where the functions:

$$f\left(x_i + c_j h, y_i \sum_{l=1}^{j-1} a_{jl} k_l\right)$$

are to be evaluated. An important consideration in choosing the free parameters is to minimize the *round off error* of the calculation. Discussion of the effect of the round off error will be given in Section 5.8.

For the second-order Runge-Kutta method, which we are currently deriving, let us choose $c_2 = 1$. The rest of the parameters are evaluated from Eqs. (5.89):

$$\begin{aligned} w_1 = w_2 &= \frac{1}{2} \\ a_{21} &= 1 \end{aligned} \tag{5.90}$$

With this set of parameters, the second-order Runge-Kutta formula is:

$$
\begin{aligned}
y_{i+1} &= y_i + \frac{1}{2}(k_1 + k_2) + O(h^3) \\
k_1 &= hf(x_i, y_i) \\
k_2 &= hf(x_i + h, y_i + k_1)
\end{aligned}
\tag{5.91}
$$

This method is essentially identical to the Crank-Nicolson method (see Eq. 5.70).

A different version of the second-order Runge-Kutta is obtained by choosing to evaluate the function at the midpoints (i.e., $c_2 = 1/2$). This yields the formula:

$$
\begin{aligned}
y_{i+1} &= y_i + k_2 + O(h^3) \\
k_1 &= hf(x_i, y_i) \\
k_2 &= hf\left(x_i + \frac{1}{2}h, y_i + \frac{1}{2}k_1\right)
\end{aligned}
\tag{5.92}
$$

Higher-order Runge-Kutta formulas are derived in an analogous manner. Several of these are listed in Table 5.1. The fourth-order Runge-Kutta, which has an error of $O(h^5)$, is probably the most widely used numerical integration method for ordinary differential equations. The reason for the popularity of using this formula is the good balance between function evaluations and accuracy of the formula, but no iterative convergences as with the implicit Euler.

EXAMPLE 5.4 USING THE FOURTH-ORDER RUNGE-KUTTA METHOD FOR CALCULATING THE VARIATION OF CONCENTRATION IN A BATCH REACTOR

Repeat Example 5.3 and solve the mass balance differential equation by the fourth-order Runge-Kutta method to determine the variation of concentration in a batch reactor.

Method of solution

Equation of the fourth-order Runge-Kutta given in Table 5.1. is used for solving the problem. The solution will be shown in detail for the first two time steps by hand and then until $t = 2$ in Excel. The flowchart of the Euler method is shown in Fig. E5.4a.

Initial value is:

$$
i = 0, \ t_0 = 0, \ C_{A,0} = 1
$$

In the next time step ($i = 1$, $t_1 = t_0 + h = 0.25$) the concentration is calculated as follows:

$$
k_1 = hf(t_0, C_0) = 0.25\left(-\frac{1.5 \times 1}{1 + 0.1 \times 1}\right) = -0.3409
$$

$$
k_2 = hf\left(t_0 + \frac{h}{2}, C_0 + \frac{k_1}{2}\right) = 0.25\left(-\frac{1.5 \times (1 - 0.3409/2)}{1 + 0.1 \times (1 - 0.3409/2)}\right) = -0.2873
$$

$$
k_3 = hf\left(t_0 + \frac{h}{2}, C_0 + \frac{k_2}{2}\right) = 0.25\left(-\frac{1.5 \times (1 - 0.2873/2)}{1 + 0.1 \times (1 - 0.2873/2)}\right) = -0.2958
$$

$$
k_4 = hf(t_0 + h, C_0 + k_3) = 0.25\left(-\frac{1.5 \times (1 - 0.2958)}{1 + 0.1 \times (1 - 0.2958)}\right) = -0.2467
$$

(Continued)

Table 5.1 Summary of the Runge-Kutta integration formulas.

Second-order

$$y_{i+1} = y_i + \frac{1}{2}(k_1 + k_2) + O(h^3)$$
$$k_1 = hf(x_i, y_i)$$
$$k_2 = hf(x_i + h, y_i + k_1)$$

Third-order

$$y_{i+1} = y_i + \frac{1}{6}(k_1 + 4k_2 + k_3) + O(h^4)$$
$$k_1 = hf(x_i, y_i)$$
$$k_2 = hf\left(x_i + \frac{h}{2}, y_i + \frac{k_1}{2}\right)$$
$$k_3 = hf(x_i + h, y_i + 2k_2 - k_1)$$

Fourth-order

$$y_{i+1} = y_i + \frac{1}{6}(k_1 + 2k_2 + 2k_3 + k_4) + O(h^5)$$
$$k_1 = hf(x_i, y_i)$$
$$k_2 = hf\left(x_i + \frac{h}{2}, y_i + \frac{k_1}{2}\right)$$
$$k_3 = hf\left(x_i + \frac{h}{2}, y_i + \frac{k_2}{2}\right)$$
$$k_4 = hf(x_i + h, y_i + k_3)$$

Fifth-order

$$y_{i+1} = y_i + \frac{1}{90}(7k_1 + 32k_3 + 12k_4 + 32k_5 + 7k_6) + O(h^6)$$
$$k_1 = hf(x_i, y_i)$$
$$k_2 = hf\left(x_i + \frac{h}{2}, y_i + \frac{k_1}{2}\right)$$
$$k_3 = hf\left(x_i + \frac{h}{4}, y_i + \frac{3k_1}{16} + \frac{k_2}{16}\right)$$
$$k_4 = hf\left(x_i + \frac{h}{2}, y_i + \frac{k_3}{2}\right)$$
$$k_5 = hf\left(x_i + \frac{3h}{4}, y_i - \frac{3k_2}{16} + \frac{6k_3}{16} + \frac{9k_4}{16}\right)$$
$$k_6 = hf\left(x_i + h, y_i + \frac{k_1}{7} + \frac{4k_2}{7} + \frac{6k_3}{7} - \frac{12k_4}{7} + \frac{8k_5}{7}\right)$$

Runge-Kutta-Fehlberg

$$y_{i+1} = y_i + \left(\frac{25}{216}k_1 + \frac{1408}{2565}k_3 + \frac{2197}{4104}k_4 - \frac{1}{5}k_5\right) + O(h^5)$$
$$k_1 = hf(x_i, y_i)$$
$$k_2 = hf\left(x_i + \frac{h}{4}, y_i + \frac{k_1}{4}\right)$$
$$k_3 = hf\left(x_i + \frac{3}{8}h, y_i + \frac{3}{32}k_1 + \frac{9}{32}k_2\right)$$
$$k_4 = hf\left(x_i + \frac{12}{13}h, y_i + \frac{1932}{2197}k_1 - \frac{7200}{2197}k_2 + \frac{7296}{2197}k_3\right)$$
$$k_5 = hf\left(x_i + h, y_i + \frac{439}{216}k_1 - 8k_2 + \frac{3860}{513}k_3 - \frac{845}{4104}k_4\right)$$
$$k_6 = hf\left(x_i + \frac{h}{2}, y_i - \frac{8}{27}k_1 + 2k_2 - \frac{3544}{2565}k_3 + \frac{1859}{4104}k_4 - \frac{11}{40}k_5\right)$$
$$T_E \simeq \frac{1}{360}k_1 - \frac{128}{4275}k_3 - \frac{2197}{75240}k_4 + \frac{1}{50}k_5 + \frac{2}{55}k_6$$

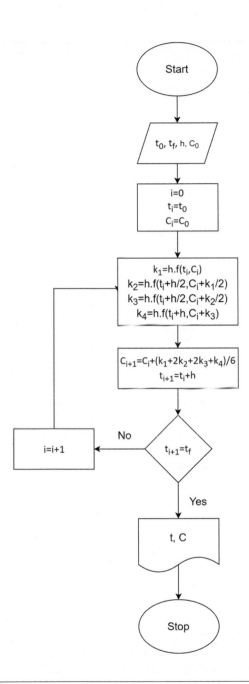

FIG. E5.4a

Flowchart of the fourth-order Runge-Kutta method for solving ordinary differential equations.

EXAMPLE 5.4 (CONTINUED)

$$C_1 = C_0 + \frac{1}{6}(k_1 + 2k_2 + 2k_3 + k_4)$$
$$= 1 + \frac{1}{6}(-0.3409 + 2(-0.2873) + 2(-0.2958) - 0.2467) = 0.7077$$

For the next time step ($i = 2$, $t_2 = t_1 + h = 0.5$), the concentration is evaluated as follows:

$$k_1 = hf(t_1, C_1) = 0.25\left(-\frac{1.5 \times 0.7077}{1 + 0.1 \times 0.7077}\right) = -0.2479$$

$$k_2 = hf\left(t_1 + \frac{h}{2}, C_1 + \frac{k_1}{2}\right) = 0.25\left(-\frac{1.5 \times (0.7077 - 0.2479/2)}{1 + 0.1 \times (0.7077 - 0.2479/2)}\right) = -0.2068$$

$$k_3 = hf\left(t_1 + \frac{h}{2}, C_1 + \frac{k_2}{2}\right) = 0.25\left(-\frac{1.5 \times (0.7077 - 0.2068/2)}{1 + 0.1 \times (0.7077 - 0.2068/2)}\right) = -0.2137$$

$$k_4 = hf(t_1 + h, C_1 + k_3) = 0.25\left(-\frac{1.5 \times (0.7077 - 0.2137)}{1 + 0.1 \times (0.7077 - 0.2137)}\right) = -0.1765$$

$$C_2 = C_1 + \frac{1}{6}(k_1 + 2k_2 + 2k_3 + k_4)$$

$$= 0.7077 + \frac{1}{6}(-0.2479 + 2(-0.2068) + 2(-0.2137) - 0.1765) = 0.4968$$

Calculations of the next steps can be performed in a similar manner. Fig. E5.4b illustrates the calculations in Excel.

5.5.3 THE ADAMS AND ADAMS-MOULTON METHODS

The Runge-Kutta family of integration techniques, developed earlier, are called *single-step* methods. The value of y_{i+1} is obtained from y_i and the trajectories of y within the single step from (x_i, y_i) to (x_{i+1}, y_{i+1}). This procedure marches forward, taking a single step of width h, over the entire interval of integration. These methods are very suitable for solving initial value problems because they are *self-starting* from a given initial point of integration.

Other categories of integration techniques, called *multistep* methods, have been developed. These compute the value of y_{i+1} using several previously unknown, or calculated, values of y (e.g., y_i, y_{i-1}, y_{i-2}) as the base points. For this reason, the multiple-step methods are *non–self-starting*. For the solution of initial value problems, where only y_0 is known, the multiple-step methods must be "primed" by first using a self-starting procedure to obtain the requisite number of base points. There are several multiple-step methods, and two of these, the Adams and Adams-Moulton methods, will be covered in this section.

Once again, let us start by evaluating y_{i+1} by integrating the derivative function over the interval $[x_i, x_{i+1}]$:

$$y_{i+1} - y_i = \int_{x_i}^{x_{i+1}} f(x, y)dx \tag{5.55}$$

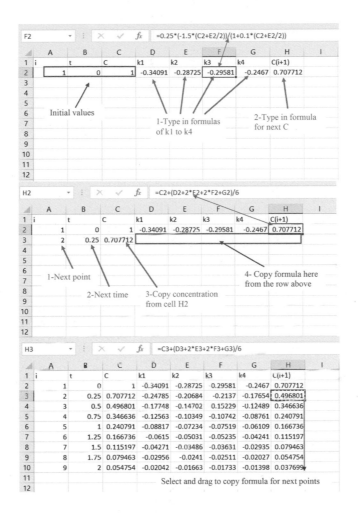

FIG. E5.4b

Calculation of evolution of concentration in a batch reactor in Excel by fourth-order Runge-Kutta method.

To evaluate the right-hand side of Eq. (5.55), $f(x, y)$ may be approximated by an nth-degree polynomial. In the Adams method a quadratic polynomial is passed through the three past points, that is, (x_{i-2}, y_{i-2}), (x_{i-1}, y_{i-1}), and (x_i, y_i), and is used to extrapolate the value of $f(x_{i+1}, y_{i+1})$. If we choose a uniform step size, a second-degree backward Gregory-Newton interpolating polynomial may be applied to this problem and Eq. (5.55) becomes:

$$y_{i+1} = y_i + \int_{x_i}^{x_{i+1}} \left[f_i - \frac{(x - x_i)}{h} \nabla f_i + \frac{(x - x_i)(x - x_{i+1})}{2!h^2} \nabla^2 f_i \right] dx + \int_{x_i}^{x_{i+1}} R_n(x) dx \qquad (5.93)$$

where $f_i = f(x_i, y_i)$ and it may be considered a function of x only. Noting that $x_{i+1} - x_i = h$, Eq. (5.93) reduces to:

$$y_{i+1} = y_i + h\left(f_i + \frac{1}{2}\nabla f_i + \frac{5}{12}\nabla^2 f_i\right) + O(h^4) \tag{5.94}$$

This equation would be easier to use by expanding the backward differences in terms of the function values (Eqs. 3.24 and 3.25). Replacing the backward differences followed by further rearrangements results in the following formula known as the Adams method for the solution of the ordinary differential equations:

$$y_{i+1} = y_i + \frac{h}{12}[23f(x_i, y_i) - 16f(x_{i-1}, y_{i-1}) + 5f(x_{i-2}, y_{i-2})] + O(h^4) \tag{5.95}$$

Eq. (5.95) shows that prior to evaluating y_{i+1}, the values of the function at three points before that have to be known. Because in an initial value problem, only the value of the function at the start of the solution interval is known, another two succeeding values should be calculated by a single-step method, such as Runge-Kutta. The solution of the ordinary differential equation from the fourth point may then be continued with Eq. (5.95).

To derive the Adams-Moulton technique, we repeat the same procedure by applying a third-degree interpolating polynomial (using four past points) instead of a second-degree polynomial to approximate $f(x, y)$ in Eq. (5.55). This procedure results in the prediction of y_{i+1}:

$$(y_{i+1})_{Pr} = y_i + \frac{h}{24}[55f(x_i, y_i) - 59f(x_{i-1}, y_{i-1}) + 37f(x_{i-2}, y_{i-2}) - 9f(x_{i-3}, y_{i-3})] + O(h^5) \tag{5.96}$$

In the Adams-Moulton method we do not stop here and correct y_{i+1} before moving to the next step. The value of y_{i+1} calculated from Eq. (5.96) is a good approximation of the dependent variable at position $(i + 1)$; therefore almost the correct value of $f(x_{i+1}, y_{i+1})$ may be evaluated from $f(x_{i+1}, (y_{i+1})_{Pr})$ at this stage. We now interpolate the function $f(x, y)$, using a cubic Gregory-Newton backward interpolating polynomial, over the range from x_{i-2} to x_{i+1} and calculate the corrected value of y_{i+1} by the integral of Eq. (5.55):

$$(y_{i+1})_{Cor} = y_i + \frac{h}{24}\left[9f\left(x_{i+1}, (y_{i+1})_{Pr}\right) + 19f(x_i, y_i) - 5f(x_{i-1}, y_{i-1}) + f(x_{i-2}, y_{i-2})\right] + O(h^5) \tag{5.97}$$

Eqs. (5.96) and (5.97) should be used as the predictor and corrector, respectively. Correction by Eq. (5.97) may be applied more than once until the corrected value converges; that is, the difference between the two consecutive corrected values becomes less than the convergence criterion. However, two applications of the corrector are probably optimum in terms of computer time and the accuracy gained. Once again, the solution of the ordinary differential equation by this technique may start from the fifth point; therefore some other technique should be applied at the beginning of the solution to evaluate y_1 to y_3.

5.5.4 SIMULTANEOUS DIFFERENTIAL EQUATIONS

It was mentioned at the beginning of Section 5.5 that the methods of solution of a single differential equation are readily adaptable to solving sets of simultaneous differential equations. To illustrate this, we use the set of n simultaneous ordinary differential equations:

$$\frac{dy_1}{dx} = f_1(x, y_1, y_2, \ldots, y_n)$$

$$\frac{dy_2}{dx} = f_2(x, y_1, y_2, \ldots, y_n) \tag{5.98}$$

$$\vdots$$

$$\frac{dy_n}{dx} = f_n(x, y_1, y_2, \ldots, y_n)$$

and expand, for example, the fourth-order Runge-Kutta formulas to:

$$y_{i+1,j} = y_{ij} + \frac{1}{6}\left(k_{1j} + 2k_{2j} + 2k_{3j} + k_{4j}\right) \quad j = 1, 2, \ldots, n$$

$$k_{1j} = hf_j(x_i, y_{i1}, y_{i2}, \ldots, y_{in})$$

$$k_{2j} = hf_j\left(x_i + \frac{h}{2}, y_{i1} + \frac{k_{11}}{2}, y_{i2} + \frac{k_{12}}{2}, \ldots, y_{in} + \frac{k_{1n}}{2}\right)$$

$$k_{3j} = hf_j\left(x_i + \frac{h}{2}, y_{i1} + \frac{k_{21}}{2}, y_{i2} + \frac{k_{22}}{2}, \ldots, y_{in} + \frac{k_{2n}}{2}\right)$$

$$k_{4j} = hf_j(x_i + h, y_{i1} + k_{31}, y_{i2} + k_{32}, \ldots, y_{in} + k_{3n})$$

This method is easily programmable using nested loops. In MATLAB the values of k and y_i can be put in vectors and easily perform Eq. (5.99) in matrix form.

EXAMPLE 5.5 SOLUTION OF THE NONISOTHERMAL PLUG-FLOW REACTOR

Write general MATLAB functions for integrating simultaneous nonlinear differential equations using the Euler, Euler predictor-corrector (modified Euler), Runge-Kutta, Adams, and Adams-Moulton methods. Apply these functions for the solution of differential equations that simulate a nonisothermal plug flow reactor, as described in the text that follows [2].

Vapor-phase cracking of acetone, described by the following endothermic reaction,

$$CH_3COCH_3 \rightarrow CH_2CO + CH_4$$

takes place in a jacketed tubular reactor. Pure acetone enters the reactor at a temperature of $T_0 = 1035$ K and pressure of $P_0 = 162$ kPa and the temperature of external gas in the heat exchanger is constant at $T_a = 1150$ K. Other data are as follows:

$$\text{Volumetric flowrate:} v_0 = 0.002 \text{ m}^3/\text{s}$$

$$\text{Volume of the reactor:} V_R = 1 \text{ m}^3$$

$$\text{Overall heat transfer coefficient:} U = 110 \text{ W/m}^2.\text{K}$$

$$\text{Heat transfer area: } a = 150 \text{ m}^2/\text{m}^3 \text{ reactor}$$

$$\text{Reaction constant: } k = 3.58 \text{ exp}\left[34222\left(\frac{1}{1035} - \frac{1}{T}\right)\right] \text{s}^{-1}$$

(Continued)

EXAMPLE 5.5 (CONTINUED)

Heat of reaction: $\Delta H_R = 80770 + 6.8(T - 298) - 5.75 \times 10^{-3}(T^2 - 298^2)$

$$- 1.27 \times 10^{-6}(T^3 - 298^3) \text{J/mol}$$

Heat capacity of acetone: $C_{p_A} = 26.63 + 0.1830T - 45.86 \times 10^{-6}T^2 \text{ J/mol.K}$

Heat capacity of ketene: $C_{p_B} = 20.04 + 0.0945T - 30.95 \times 10^{-6}T^2 \text{ J/mol.K}$

Heat capacity of methane: $C_{p_C} = 13.39 + 0.0770T - 18.71 \times 10^{-6}T^2 \text{ J/mol.K}$

Determine the temperature profile of the gas along the length of the reactor. Assume constant pressure throughout the reactor.

Method of solution

To calculate the temperature profile in the reactor, we have to solve the material balance and energy balance equations simultaneously:

$$\text{Mole balance: } \frac{dX}{dV} = \frac{-r_A}{F_{A_0}}$$

$$\text{Energy balance: } \frac{dT}{dV} = \frac{UA(T_a - T) + r_A \Delta H_R}{F_{A_0}(C_{p_A} + X\Delta C_p)}$$

where X is the conversion of acetone, V is the volume of the reactor, $F_A = C_A v_0$ is the molar flow rate of acetone at the inlet, T is the temperature of the reactor, $\Delta C_p = C_{pB} + C_{pC} - C_{pA}$, and C_{A0} is the concentration of acetone vapor at the inlet. The reaction rate is given as:

$$-r_A = kC_{A_0}\frac{1 - X}{1 + X} \cdot \frac{T_0}{T}$$

To introduce the pair of differential equations as a MATLAB function, the following definitions are assumed:

$$y_1 = X(V) \text{ and } y_2 = T(V)$$

Therefore

$$\frac{dy_1}{dV} = f_1(V, y_1, y_2)$$
$$\frac{dy_2}{dV} = f_2(V, y_1, y_2)$$

(Continued)

EXAMPLE 5.5 (CONTINUED)

Program description

The programs and functions developed in this example can be found in: https://www.elsevier.com/books-and-journals/book-companion/9780128229613

Five general MATLAB functions are written for the solution of a set of simultaneous nonlinear ordinary differential equations: *Euler.m*, *MEuler.m*, *RK.m*, *Adams.m*, and *AdamsMoulton.m*. All these functions consist of two main sections. The first part is an initialization in which specific input arguments are checked and some vectors to be used in the second part are initiated. The next section of the function is the solution of the set of nonlinear ordinary differential equations according to the specified method that is done simultaneously in vector form. Brief descriptions of the method of solution of these five functions are given in the text that follows.

Euler.m: The Euler method: This function solves the set of differential equations based on Eq. (5.61).

MEuler.m: The Euler predictor-corrector (modified Euler) method: This function solves the set of differential equations based on Eqs. (5.66) and (5.67).

RK.m: The Runge-Kutta methods: This function is capable of solving the set of differential equations by a second-, third-, fourth-, or fifth-order Runge-Kutta method. The formulas that appeared in Table 5.1 are used for calculating a Runge-Kutta solution of the differential equations.

Adams.m: The Adams method: This function solves the set of differential equations using Eq. (5.95). The required starting points are evaluated by the third-order Runge-Kutta (using the function *RK.m*) that has the same order of truncation error as the Adams method.

AdamsMoulton.m: The Adams-Moulton method: This function solves the set of differential equations using Eqs. (5.96) and (5.97). The required starting points are evaluated by fourth-order Runge-Kutta (using the function *RK.m*) that has the same order of truncation error as the Adams-Moulton method.

The first input argument to all these method functions is the name of the MATLAB function containing the set of differential equations. Note that the first input argument to this function has to be the independent variable, even if it is not used explicitly in the equations. It is important that this function returns the values of the derivatives (f_i) as a column vector. The other inputs to the method functions are the initial and final values of the independent variable, interval width, and the initial value of the dependent variable. In *RK.m* the order of the method may also be specified. It is possible to pass, through the foregoing functions, additional arguments to the m-file describing the set of differential equations.

Input and results

>> **Example5_5**
1) Euler
2) Modified Euler
3) Runge-Kutta
4) Adams
5) Adams-Moulton

(Continued)

EXAMPLE 5.5 (CONTINUED)

6) Comparison of methods

0) End

Choose the method of solution: 0

Do you want to repeat the solution with different input data (0/1)? 0

Discussion of results

The mole and energy balance equations are solved by three a different methods of different order of error: Euler $[O(h^2)]$, second-order Runge-Kutta $[O(h^3)]$, and Adams $[O(h^4)]$. Graphical results are given in Figs. E5.5a and b. In the beginning the temperature of the reactor decreases because the reaction is endothermic. However, it starts to increase steadily at about 10% of the length of the reactor due to the heat transfer from the hot gas circulation around the reactor.

It can be seen from Figs. E5.5a and b that there are visible differences between the three methods in the temperature profile where the temperature reaches the minimum. This region is where the change in the derivative of temperature (energy balance formula) is greater than the other parts of the curve and as a result, different techniques for approximation of this derivative give different values for it. The reader is encouraged to repeat this example with different methods of solution and step sizes.

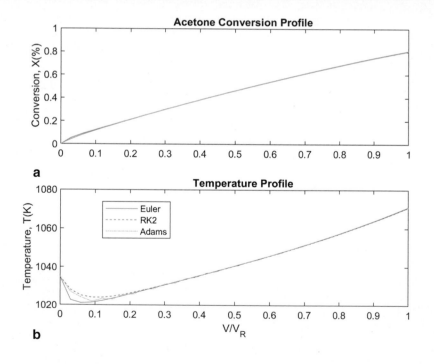

FIG. E5.5

Conversion and temperature profiles: (a) Conversion; (b) Temperature.

5.6 USING BUILT-IN MATLAB® FUNCTIONS

There are several functions in MATLAB for the solution of a set of ordinary differential equations. These solvers, along with their method of solution, are listed in Table 5.2. The solver that one would want to try first on a problem is *ode45*. The statement $[x, y] = ode45('y\_prime', [x_0, x_f], y_0)$ solves the set of ordinary differential equations described in the MATLAB function *y_prime.m*, from x_0 to x_f, with the initial values given in the vector y_0, and returns the values of independent and dependent variables in the vectors x and y, respectively. The vector of the dependent variable, x, is not equally spaced because the function controls the step size. If the solution is required at specified points of x, the interval $[x_0, x_f]$ should be replaced by a vector containing the values of the independent variable at these points. For example, $[x, y] = ode45('y\_prime', [x_0, h, x_f], y_0)$ returns the solution of the set of ordinary differential equations from x_0 to x_f at intervals of the width h. The vector x in this case would be monotonic (with the exception of, perhaps, its last interval). The basic syntax for applying the other MATLAB ordinary differential equation solvers is the same as that described earlier for *ode45*.

The function *y_prime.m* should return the value of the derivative(s) as a column vector. The first input to this function has to be the independent variable, x, even if it is not explicitly used in the definition of the derivative. The second input argument to *y_prime* is the vector of the dependent variable, y. It is possible to pass additional parameters to the derivative function. It should be noted, however, that in this case the third input to *y_prime.m* has to be an empty variable, flag, and the additional parameters are introduced in the fourth argument.

5.7 DIFFERENCE EQUATIONS AND THEIR SOLUTIONS

The application of forward, backward, or central finite differences in the solution of differential equations transforms these equations to *difference equations* of the form:

$$f(y_k, y_{k+1}, \ldots, y_{k+n}) = 0 \tag{5.100}$$

Table 5.2 Ordinary differential equation solvers in MATLAB.

Solver	Application	Method of solution
ode23	Nonstiff equations	Runge-Kutta lower-order (second-order: three stages)
ode45	Nonstiff equations	Runge-Kutta higher-order (fourth-order: five stages)
ode113	Nonstiff equations	Adams-Bashforth-Moulton of varying order ($1-12$)
ode23s	Stiff equations	Modified Rosenbrock with interpolation
ode23t	Moderately stiff equations	Trapezoidal rule with interpolation
ode23tb	Stiff equations	Implicit Runge-Kutta (first-stage trapezoidal rule: second-stage backward differentiation of order 2)
ode15s	Stiff equations	Implicit, multistep of varying order ($1-5$)
ode15i	Fully implicit equations	Varying order

In addition, difference equations are obtained from the application of material balances on multi-stage operations, such as distillation and extraction.

Depending on their origin, difference equations may be linear or nonlinear, homogeneous or nonhomogeneous, and with constant or variable coefficients. For the purposes of this book, it will be necessary to discuss only the methods of solution of *homogeneous linear difference equations with constant coefficients*.

The *order* of a difference equation is the difference between the highest and lowest subscript of the dependent variable in the equation; that is, it is the number of finite steps spanned by the equation. The order of Eq. (5.100) is given by:

$$Order = (k+n) - k = n \tag{5.101}$$

The process of obtaining y_k is called *solving the difference equation*. The methods of obtaining such solutions are analogous to those used in finding analytical solutions of differential equations. As a matter of fact, the theory of difference equations is parallel to the corresponding theory of differential equations. Difference equations resemble ordinary differential equations. For example, the following is a second-order homogeneous linear ordinary *differential* equation:

$$y'' + 3y' - 4y = 0 \tag{5.102}$$

while this is a second-order homogeneous linear *difference* equation:

$$y_{k+2} + 3y_{k+1} - 4y_k = 0 \tag{5.103}$$

The solution of the differential Eq. (5.102) can be obtained from the methods of differential calculus applied as follows:

1. Replace the derivatives in (5.102) with the differential operators:

$$D^2 y_k + 3D y_k - 4y_k = 0$$

2. Factor out the y:

$$\left(D^2 + 3D - 4\right)y = 0$$

3. Find the roots of the *characteristic equation*:

$$D^2 + 3D - 4 = 0$$

These roots are called the *eigenvalues* of the differential equation. In this case they are:

$$\lambda_1 = 1 \text{ and } \lambda_2 = 4$$

4. Construct the solution of the homogeneous differential equation as follows:

$$y = C_1 e^{\lambda_1 x} + C_2 e^{\lambda_2 x}$$

$$= C_1 e^{(1)x} + C_2 e^{(-4)x} \tag{5.104}$$

where C_1 and C_2 are constants that must be evaluated from the boundary conditions of the differential equation.

Similarly, the solution of the difference Eq. (5.103) can be obtained by using the shift operator E:

1. Replace each term of Eq. (5.103) with its equivalent using the shift operator:

$$E^2 y_k + 3E y_k - 4y_k = 0$$

2. Factor out the y_k:

$$(E^2 + 3E - 4)y_k = 0$$

3. Find the roots of the characteristic equation:

$$E^2 + 3E - 4 = 0$$

These roots are $\lambda_1 = 1$ and $\lambda_2 = -4$.

4. Construct the solution of the homogeneous difference equation as follows:

$$\begin{aligned} y_k &= C_1 \lambda_1^k + C_2 \lambda_2^k \\ &= C_1(1)^k + C_2(-4)^k \end{aligned} \tag{5.105}$$

where C_1 and C_2 are constants that must be evaluated from the boundary conditions of the difference equation.

In the foregoing case both eigenvalues were real and distinct. When the eigenvalues are real and repeated, the solution for a second-order equation with both roots identical is formed as follows:

$$y_k = (C_1 + C_2 k)\lambda^k \tag{5.106}$$

For an nth-order equation, which has m repeated roots (λ_m) and one distinct root (λ_n), the general formulation of the solution is obtained by superposition:

$$y_k = (C_1 + C_2 k + C_3 k^2 + \cdots + C_m k^{m-1})\lambda_m^k + C_n \lambda_n^k \tag{5.107}$$

In the case where the characteristic equation contains two complex roots,

$$\lambda_1 = \alpha + \beta i \text{ and } \lambda_2 = \alpha - \beta i \tag{5.108}$$

the solution is:

$$y_k = C_1(\alpha + \beta i)^k + C_2(\alpha - \beta i)^k \tag{5.109}$$

This solution may be also expressed in terms of trigonometric quantities by using the trigonometric (polar) form of complex numbers:

$$\alpha \pm \beta i = r(\cos\theta \pm i \sin\theta) \tag{5.110}$$

This is obtained by showing the complex number as a vector in the complex plane represented in Fig. 5.5. The *modulus r* of the complex number is obtained from the Pythagorean theorem:

$$r = \sqrt{\alpha^2 + \beta^2} \tag{5.111}$$

The values of α and β are expressed in terms of the *phase angle* θ:

$$\alpha = r \cos\theta \tag{5.112}$$

$$\beta = r \sin\theta \tag{5.113}$$

and the phase angle is given by:

$$\theta = \tan^{-1}\frac{\beta}{\alpha} \tag{5.114}$$

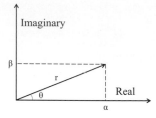

FIG. 5.5

Representation of a real number in a plane.

Substituting Eq. (5.110) in Eq. (5.109) and using de Moivre's theorem:

$$(\cos\theta \pm i \sin \theta)^k = \cos k\theta \pm i \sin k\theta \tag{5.115}$$

we obtain the solution of the difference equation as:

$$y_k = r^k \left[C_1' \cos k\theta + C_2' \sin k\theta \right] \tag{5.116}$$

where

$$C_1' = C_1 + C_2 \ \text{ and } \ C_2' = (C_1 - C_2)i.$$

It can be concluded from the foregoing discussion that the solution of homogeneous linear difference equations with constant coefficients is of the form:

$$y_k = f(k, \lambda) \tag{5.117}$$

where k is the forward-marching counter and λ is the vector of eigenvalues of the characteristic equation. The stability and convergence of these solutions depend on the values of the eigenvalues. The following stability cases apply to the solutions of difference equations:

1. The solution is stable, converging without oscillations, when:
   a. All the eigenvalues are real distinct and have absolute values less than, or equal to, unity:

$$\lambda = \text{real distinct}$$

$$|\lambda| \leq 1$$

   b. The eigenvalues are real, but repeated, and have absolute values less than unity:

$$\lambda = \text{real repeated}$$

$$|\lambda| < 1$$

2. The solution is stable, converging with damped oscillations, when:
   a. Complex distinct eigenvalues are present and the moduli of the eigenvalues are less than, or equal to, unity:

$$\lambda = \text{complex distinct}$$

$$|r| \leq 1$$

b. Complex repeated eigenvalues are present and the moduli of the eigenvalues are less than unity:

$$\lambda = \text{complex repeated}$$
$$|r| < 1$$

3. The solution is unstable and nonoscillatory, when:
   a. All the eigenvalues are real distinct and one or more of these have absolute values greater than unity.

$$\lambda = \text{real distinct}$$
$$|\lambda| > 1.0$$

   b. The eigenvalues are real but repeated, and one or more of these have absolute values equal to, or greater than, unity:

$$\lambda = \text{real repeated}$$
$$|\lambda| \geq 1$$

4. The solution is unstable and oscillatory, when:
   a. Complex distinct eigenvalues are present and the moduli of one or more of these are greater than unity:

$$\lambda = \text{complex distinct}$$
$$|r| > 1$$

   b. Complex repeated eigenvalues are present and the moduli of one or more of these are equal to, or greater than, unity:

$$\lambda = \text{complex repeated}$$
$$|r| \geq 1$$

The numerical solutions of ordinary and partial differential equations are based on the finite difference formulation of these differential equations. Therefore, the stability and convergence considerations of finite difference solutions have important implications on the numerical solutions of differential equations. This topic will be discussed in more detail in Section 5.8 and Chapter 7.

5.8 PROPAGATION, STABILITY, AND CONVERGENCE

Topics of paramount importance in the numerical integration of differential equations are the *error propagation*, *stability*, and *convergence* of these solutions. Two types of stability considerations enter in the solution of ordinary differential equations: *inherent stability* (or instability) and *numerical stability* (or instability). Inherent stability is determined by the mathematical formulation of the problem and depends on the eigenvalues of the Jacobian matrix of the differential equations. On the other hand, numerical stability is a function of the error propagation in the numerical integration method. The behavior of error propagation depends on the values of the characteristic roots of the difference equations that yield the numerical solution. In this section we concern ourselves with numerical stability considerations as they apply to the numerical integration of ordinary differential equations.

There are three types of errors that are present in the application of numerical methods. These are the *truncation error*, the *round off error*, and the *propagation error*. Truncation error is a function of the number of terms that are retained in the approximation of the solution from the infinite series expansion and may be reduced by retaining a larger number of terms in the series or by reducing the step size of integration h. The plethora of available numerical methods of integration of ordinary differential equations provides a choice of increasingly higher accuracy (lower truncation error) at an escalating cost in the number of arithmetic operations to be performed and with the concomitant accumulation of round off errors.

Computers carry numbers using a finite number of significant figures. A round off error is introduced in the calculation when the computer rounds up or down (or just chops) the number to n significant figures. Round off errors may be reduced significantly by the use of double precision. However, even a very small round off error may affect the accuracy of the solution, especially in numerical integration methods that march forward (or backward) for hundreds or thousands of steps, each step is performed using rounded numbers.

The truncation and round off errors in numerical integration accumulate and propagate, creating the propagation error, which, in some cases, may grow in exponential or oscillatory patterns, thus causing the calculated solution to deviate drastically from the correct solution.

Fig. 5.6 illustrates the propagation of error in the Euler integration method. Starting with a known initial condition y_0, the method calculates the value y_1, which contains the truncation error for this step and a small round off error introduced by the computer. The error has been magnified to illustrate it more clearly. The next step starts with y_1 as the initial point and calculates y_2. But because y_1 already contains truncation and round off errors, the value obtained for y_2 contains these

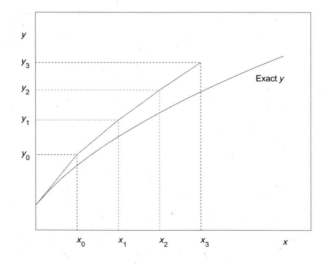

FIG. 5.6

Error propagation of the Euler method.

errors propagated, in addition to the new truncation and round off errors from the second step. The same process occurs in subsequent steps.

Error propagation in numerical integration methods is a complex operation that depends on several factors. Round off error, which contributes to propagation error, is entirely determined by the accuracy of the computer being used. The truncation error is fixed by choice of the method being applied, by the step size of integration, and by the values of the derivatives of the functions being integrated. For these reasons, it is necessary to examine the error propagation and stability of each method individually and in connection with the differential equations to be integrated. Some techniques work well with one class of differential equations but fail with others.

In the sections that follow we examine systematically the error propagation and stability of several numerical integration methods and suggest ways of reducing these errors by the appropriate choice of step size and integration algorithm.

5.8.1 STABILITY AND ERROR PROPAGATION OF EULER METHODS

Let us consider the initial value differential equation in the linear form:

$$\frac{dy}{dx} = \lambda y \tag{5.118}$$

where the initial condition is given as:

$$y(x_0) = y_0 \tag{5.119}$$

We assume that λ is real and y_0 is finite. The analytical solution of this differential equation is:

$$y(x) = y_0 e^{\lambda x} \tag{5.120}$$

This solution is *inherently stable* for $\lambda < 0$. Under these conditions:

$$\lim_{x \to \infty} y(x) = 0 \tag{5.121}$$

Next, we examine the stability of the numerical solution for the explicit Euler method. Momentarily we ignore the truncation and round off errors. Applying Eq. (5.60), we obtain the recurrence equation:

$$y_{n+1} = y_n + h\lambda y_n \tag{5.122}$$

which rearranges to the following first-order homogeneous difference equation:

$$y_{n+1} - (1 + h\lambda)y_n = 0 \tag{5.123}$$

Using the methods described in Section 5.7, we obtain the characteristic equation:

$$E - (1 + h\lambda) = 0 \tag{5.124}$$

whose root is:

$$\mu_1 = (1 + h\lambda) \tag{5.125}$$

From this, we obtain the solution of the difference Eq. (5.123) as:

$$y_n = C(1 + h\lambda)^n \tag{5.126}$$

The constant C is calculated from the initial condition, at $x = x_0$:

$$n = 0, \quad y_n = y_0 = C \tag{5.127}$$

Therefore, the final form of the solution is:

$$y_n = y_0 (1 + h\lambda)^n \tag{5.128}$$

The differential equation is an initial value problem; therefore, n can increase without bound. Because the solution y is a function of $(1 + h\lambda)^n$, its behavior is determined by the value of $(1 + h\lambda)$. A numerical solution is said to be *absolutely stable* if:

$$\lim_{n \to \infty} y_n = 0 \tag{5.129}$$

The solution of the differential Eq. (5.118) using the explicit Euler method is absolutely stable if:

$$|1 + h\lambda| \leq 1 \tag{5.130}$$

Because $(1 + h\lambda)$ is the root of the characteristic Eq. (5.124), an alternative definition of absolute stability is:

$$|\mu_i| \leq 1 \quad i = 1, 2, \ldots, k \tag{5.131}$$

where more than one root exists in the multistep numerical methods.

Returning to the problem at hand, the inequality (5.130) is rearranged to:

$$-2 \leq h\lambda \leq 0 \tag{5.132}$$

This inequality sets the limits of the integration step size for a stable solution as follows: Because h is positive, then $\lambda < 0$ and,

$$h \leq \frac{2}{|\lambda|} \tag{5.133}$$

Inequality (5.133) is a finite *general stability boundary*, and for this reason, the explicit Euler method is called *conditionally stable*. Any method with an infinite general stability boundary can be called *unconditionally stable*.

At the outset of our discussion, we assumed that λ was real to simplify the derivation. This assumption is not necessary: λ can be a complex number. In the earlier discussion of the stability of difference equations (Section 5.7), we mentioned that a solution is stable converging with damped oscillations when complex roots are present and the moduli of the roots are less than or equal to unity:

$$|r| \leq 1 \tag{5.134}$$

The two inequalities (5.132) and (5.134) describe the circle with a radius of unity on the complex plane shown in Fig. 5.7. Because the explicit Euler method can be categorized as a first-order Runge-Kutta method, the corresponding curve in this figure is marked by RK1. The set of values of $h\lambda$ inside the circle yields stable numerical solutions of Eq. (5.118) using the Euler integration method.

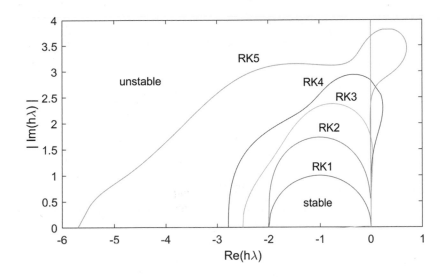

FIG. 5.7

Stability region in the complex plane for Runge-Kutta methods of order 1 (explicit Euler), 2, 3, 4, and 5.

We now return to the consideration of the truncation and round off errors of the Euler method and develop a difference equation that describes the propagation of the error in the numerical solution. We work with the nonlinear form of the initial value problem:

$$\frac{dy}{dx} = f(x, y) \tag{5.20}$$

where the initial condition is given by:

$$y(x_0) = y_0 \tag{5.21}$$

We define the accumulated error of the numerical solution at step $n + 1$ as:

$$\varepsilon_{n+1} = y_{n+1} - y(x_{n+1}) \tag{5.135}$$

where $y(x_{n+1})$ is the *exact* value of y, and y_{n+1} is the *calculated* value of y at x_{n+1}. We then write the exact solution $y(x_{n+1})$ as a Taylor series expansion, showing as many terms as needed for the Euler method:

$$y(x_{n+1}) = y(x_n) + hf(x_n, y(x_n)) + T_{E,n+1} \tag{5.136}$$

where $T_{E,n+1}$ is the local truncation error for step $n + 1$. We also write the calculated value y_{n+1} obtained from the implicit Euler formula:

$$y_{n+1} = y_n + hf(x_n, y_n) + R_{E,n+1} \tag{5.137}$$

where $R_{E,n+1}$ is the round off error introduced by the computer in step $(n + 1)$.

Combining Eqs. (5.135)−(5.137) we have:

$$\varepsilon_{n+1} = y_n - y(x_n) + h[f(x_n, y_n) - f(x_n, y(x_n))] - T_{E,n+1} + R_{E,n+1} \tag{5.138}$$

which simplifies to:

$$\varepsilon_{n+1} = \varepsilon_n + h[f(x_n, y_n) - f(x_n, y(x_n))] - T_{E,n+1} + R_{E,n+1} \tag{5.139}$$

The mean-value theorem:

$$f(x_n, y_n) - f(x_n, y(x_n)) = \frac{\partial f}{\partial y}\Big|_{\alpha, x_n}[y_n - y(x_n)] \qquad y_n < \alpha < y(x_n) \tag{5.140}$$

can be used to further modify the error Eq. (5.139) to:

$$\varepsilon_{n+1} - \left[1 + h\frac{\partial f}{\partial y}\Big|_{\alpha, x_n}\right]\varepsilon_n = -T_{E,n+1} + R_{E,n+1} \tag{5.141}$$

This is a *first-order nonhomogeneous difference equation with varying coefficients* that can be solved only by iteration. However, by making the following simplifying assumptions:

$$
\begin{aligned}
T_{E,n+1} &= T_E = \text{constant} \\
R_{E,n+1} &= R_E = \text{constant} \\
\frac{\partial f}{\partial y}\Big|_{\alpha, x_n} &= \lambda = \text{constant}
\end{aligned}
\tag{5.142}
$$

Eq. (5.141) simplifies to:

$$\varepsilon_{n+1} - (1 + h\lambda)\varepsilon_n = -T_E + R_E \tag{5.143}$$

whose solution is given by the sum of the homogeneous and particular solutions [3]:

$$\varepsilon_n = C_1(1 + h\lambda)^n + \frac{-T_E + R_E}{1 - (1 + h\lambda)} \tag{5.144}$$

Comparison of Eqs. (5.123) and (5.143) reveals that the characteristic equations for the solution y_n and the error ε_n are identical. The truncation and round off error terms in Eq. (5.143) introduce the particular solution. The constant C_1 is calculated by assuming that the initial condition of the differential equation has no error, that is, $\varepsilon_0 = 0$. The final form of the equation that describes the behavior of the propagation error is:

$$\varepsilon_n = \frac{-T_E + R_E}{h\lambda}[(1 + h\lambda)^n - 1] \tag{5.145}$$

A great deal of insight can be gained by examining Eq. (5.145) thoroughly. As expected, the value of $(1 + h\lambda)$ is the determining factor in the behavior of the propagation error. Consider first the case of a fixed finite step size h, with the number of integration steps increasing to a very large n. The limit on the error as $n \to \infty$ is:

$$\lim_{n \to \infty} |\varepsilon_n| = \frac{-T_E + R_E}{h\lambda} \quad \text{for} \quad |1 + h\lambda| < 1 \tag{5.146}$$

$$\lim_{n \to \infty} |\varepsilon_n| = \infty \quad \text{for} \quad |1 + h\lambda| > 1 \tag{5.147}$$

In the first situation (Eq. 5.146), $\lambda < 0$, $0 < h < 2/|\lambda|$, the error is bounded and the numerical solution is stable. The numerical solution differs from the exact solution by only the finite quantity $(-T_E + R_E)/h\lambda$, which is a function of the truncation error, the round off error, the step size, and the eigenvalue of the differential equation.

In the second situation (Eq. 5.147) $\lambda > 0$, $h > 0$, the error is unbounded and the numerical solution is unstable. For $\lambda > 0$, however, the exact solution is *inherently unstable*. For this reason, we introduce the concept of *relative error* defined as:

$$\text{relative error} = \frac{\varepsilon_n}{y_n} \tag{5.148}$$

Using Eqs. (5.128) and (5.145) we obtain the relative error as:

$$\frac{\varepsilon_n}{y_n} = \frac{-T_E + R_E}{y_0 h \lambda}\left[1 - \frac{1}{(1+h\lambda)^n}\right] \tag{5.149}$$

The relative error is bounded for $\lambda > 0$ and unbounded for $\lambda < 0$. So we conclude that for inherently stable differential equations, the absolute propagation error is the pertinent criterion for numerical stability, while for inherently unstable differential equations, the relative propagation error must be investigated.

Let us now consider a fixed interval of integration, $0 \leq x \leq \alpha$, so that:

$$h = \frac{\alpha}{n} \tag{5.150}$$

and we increase the number of integration steps to a very large n. This, of course, causes $h \to 0$. A numerical method is said to be *convergent* if:

$$\lim_{h \to 0} |\varepsilon_n| = 0 \tag{5.151}$$

In the absence of round off error, the Euler method, and most other integration methods, are convergent because:

$$\lim_{h \to 0} T_E = 0 \tag{5.152}$$

and

$$\lim_{h \to 0} |\varepsilon_n| = 0 \tag{5.151}$$

However, round off error is *never* absent in numerical calculations. As $h < 0$, the round off error is the crucial factor in the propagation of error:

$$\lim_{h \to 0} |\varepsilon_n| = R_E \lim_{h \to 0} \frac{(1+h\lambda)^n - 1}{h\lambda} \tag{5.153}$$

Application of L'Hôpital's rule shows that the round off error propagates unbounded as the number of integration steps becomes very large:

$$\lim_{h \to 0} \varepsilon_n = R_E[\infty] \tag{5.154}$$

This is a dilemma in the numerical methods: A smaller step size of integration reduces the truncation error but requires a large number of steps, thereby increasing the round off error.

A similar analysis of the *implicit Euler method* (backward Euler) results in the following two equations for the solution:

$$y_{n+1} = \frac{y_0}{(1-h\lambda)^n} \tag{5.155}$$

Table 5.3 Real stability boundaries.

Method	Boundary
Explicit Euler	$-2 \leq h\lambda \leq 0$
Implicit Euler	$\begin{cases} 0 < h < \infty & \text{for} \quad \lambda < 0 \\ -2 \leq h\lambda \leq 0 & \text{for} \quad \lambda > 0 \end{cases}$
Modified Euler (predictor-corrector)	$-1.077 \leq h\lambda \leq 0$
Second-order Runge-Kutta	$-2 \leq h\lambda \leq 0$
Third-order Runge-Kutta	$-2.5 \leq h\lambda \leq 0$
Fourth-order Runge-Kutta	$-2.875 \leq h\lambda \leq 0$
Fifth-order Runge-Kutta	$-5.7 \leq h\lambda \leq 0$
Adams	$-0.546 \leq h\lambda \leq 0$
Adams-Moulton	$-1.285 \leq h\lambda \leq 0$

and the propagation error:

$$\varepsilon_{n+1} = \frac{-T_E + R_E}{h\lambda}(1 - h\lambda)\left[\frac{1}{(1-h\lambda)^n} - 1\right] \tag{5.156}$$

For $\lambda < 0$ and $0 < \lambda < \infty$, the solution is stable:

$$\lim_{n \to \infty} y_n = 0 \tag{5.157}$$

and the error is bounded:

$$\lim_{n \to \infty} \varepsilon_n = -\frac{-T_E + R_E}{h\lambda}(1 - h\lambda) \tag{5.158}$$

No limitation is placed on the step size; therefore, the implicit Euler method is *unconditionally stable* for $\lambda < 0$. On the other hand, when $\lambda > 0$, the following inequality must be true for a stable solution:

$$|1 - h\lambda| \leq 1 \tag{5.159}$$

This imposes the limit on the step size:

$$-2 \leq h\lambda \leq 0 \tag{5.160}$$

It can be concluded that the implicit Euler method has a wider range of stability than the explicit Euler method (Table 5.3).

5.8.2 STABILITY AND ERROR PROPAGATION OF RUNGE-KUTTA METHODS

Using methods parallel to those of the previous section, the recurrence equations and the corresponding roots for the Runge-Kutta methods can be derived [4]. For the differential Eq. (5.118), these are:

Second-order Runge-Kutta:

$$y_{n+1} = \left(1 + h\lambda + \frac{1}{2}h^2\lambda^2\right)y_n \tag{5.161}$$

$$\mu_1 = 1 + h\lambda + \frac{1}{2}h^2\lambda^2 \tag{5.162}$$

Third-order Runge-Kutta:

$$y_{n+1} = \left(1 + h\lambda + \frac{1}{2}h^2\lambda^2 + \frac{1}{6}h^3\lambda^3\right)y_n \tag{5.163}$$

$$\mu_1 = 1 + h\lambda + \frac{1}{2}h^2\lambda^2 + \frac{1}{6}h^3\lambda^3 \tag{5.164}$$

Fourth-order Runge-Kutta:

$$y_{n+1} = \left(1 + h\lambda + \frac{1}{2}h^2\lambda^2 + \frac{1}{6}h^3\lambda^3 + \frac{1}{24}h^4\lambda^4\right)y_n \tag{5.165}$$

$$\mu_1 = 1 + h\lambda + \frac{1}{2}h^2\lambda^2 + \frac{1}{6}h^3\lambda^3 + \frac{1}{24}h^4\lambda^4 \tag{5.166}$$

Fifth-order Runge-Kutta:

$$y_{n+1} = \left(1 + h\lambda + \frac{1}{2}h^2\lambda^2 + \frac{1}{6}h^3\lambda^3 + \frac{1}{24}h^4\lambda^4 + \frac{1}{120}h^5\lambda^5 + \frac{0.5625}{720}h^6\lambda^6\right)y_n \tag{5.167}$$

$$\mu_1 = 1 + h\lambda + \frac{1}{2}h^2\lambda^2 + \frac{1}{6}h^3\lambda^3 + \frac{1}{24}h^4\lambda^4 + \frac{1}{120}h^5\lambda^5 + \frac{0.5625}{720}h^6\lambda^6 \tag{5.168}$$

The last term in the right-hand side of Eqs. (5.167) and (5.168) is specific to the fifth-order Runge-Kutta that appears in Table 5.1 and varies for different fifth-order formulas. The condition for absolute stability:

$$|\mu_i| \le 1 \quad i = 1, 2, \ldots, k \tag{5.131}$$

applies to all the foregoing methods. The absolute real stability boundaries for these methods are listed in Table 5.3, and the regions of stability in the complex plane are shown in Fig. 5.7. In general, as the order increases, so do the stability limits.

5.8.3 STABILITY AND ERROR PROPAGATION OF MULTISTEP METHODS

Using methods parallel to those of the previous section, the recurrence equations and the corresponding roots for the modified Euler, Adams, and Adams-Moulton methods can be derived [4]. For the differential Eq. (5.118), these are:

Modified Euler (the combination of predictor and corrector):

$$y_{n+1} = \left(1 + h\lambda + \frac{1}{2}h^2\lambda^2\right)y_n \tag{5.169}$$

$$\mu_1 = 1 + h\lambda + \frac{1}{2}h^2\lambda^2 \tag{5.170}$$

Adams:

$$y_{n+1} = \left(1 + \frac{23}{12}h\lambda\right)y_n - \frac{4h\lambda}{3}y_{n-1} + \frac{5h\lambda}{12}y_{n-2} \tag{5.171}$$

$$\mu^3 - \left(1 + \frac{23}{12}h\lambda\right)\mu^2 + \left(\frac{4}{3}h\lambda\right)\mu - \frac{5}{12}h\lambda = 0 \tag{5.172}$$

Adams-Moulton (the combination of predictor and corrector):

$$y_{n+1} = \left(1 + \frac{7h\lambda}{6} + \frac{55h^2\lambda^2}{64}\right)y_n - \left(\frac{5h\lambda}{24} + \frac{59h^2\lambda^2}{64}\right)y_{n-1}$$
$$+ \left(\frac{h\lambda}{24} + \frac{37h^2\lambda^2}{64}\right)y_{n-2} - \frac{9h^2\lambda^2}{64}y_{n-3} \tag{5.173}$$

$$\mu^4 - \left(1 + \frac{7h\lambda}{6} + \frac{55h^2\lambda^2}{64}\right)\mu^3 + \left(\frac{5h\lambda}{24} + \frac{59h^2\lambda^2}{64}\right)\mu^2$$
$$- \left(\frac{h\lambda}{24} + \frac{37h^2\lambda^2}{64}\right)\mu + \frac{9h^2\lambda^2}{64} = 0 \tag{5.174}$$

The condition for absolute stability:

$$|\mu_i| \leq 1 \quad i = 1, 2, \ldots, k \tag{5.131}$$

applies to all the foregoing methods. The absolute real stability boundaries for these methods are also listed in Table 5.3, and the regions of stability in the complex plane are shown in Fig. 5.8.

5.9 STEP SIZE CONTROL

The discussion of stability analysis in the previous sections made the simplifying assumption that the value of λ remains constant throughout the integration. This is true for linear equations such as Eq. (5.118); however, for the nonlinear Eq. (5.20), the value of λ may vary considerably over the interval of integration. The step size of integration must be chosen using the maximum possible value of λ, thus resulting in the minimum step size. This, of course, will guarantee stability at the expense of computation time. For problems in which computation time becomes excessive, it is possible to develop strategies for automatically adjusting the step size at each step of the integration.

A simple test for checking the step size is to do the calculations at each interval twice: Once with the initial choice of step size and then repeat the calculations over the same interval with a smaller step size, usually half that of the first one. If, at the end of the interval, the difference between the predicted value of y by both approaches is less than the specified convergence criterion, the step size may be increased. Otherwise, a larger than acceptable difference between the

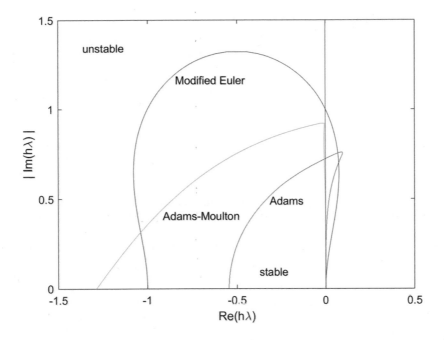

FIG. 5.8

Stability region in the complex plane for the modified Euler (Euler predictor-corrector), Adams, and Adams-Moulton methods

two calculated y values suggests that the step size is large and should be shortened to achieve an acceptable truncation error.

Another method of controlling the step size is to obtain an estimation of the truncation error at each interval. A good example of such an approach is the Runge-Kutta-Fehlberg method (see Table 5.1) that provides the estimation of the local truncation error. This error estimate can be easily introduced into the computer program and let the program automatically change the step size at each point until the desired accuracy is achieved.

As mentioned before, the optimum number of applications of the corrector is two. Therefore, in the case of using a predictor-corrector method, if the convergence is achieved before the second corrected value, the step size may be increased. On the other hand, if the convergence is not achieved after the second application of the corrector, the step size should be reduced.

5.10 STIFF DIFFERENTIAL EQUATIONS

In Section 5.8 we showed that the stability of the numerical solution of differential equations depends on the value of $h\lambda$, and that λ together with the stability boundary of the method determine the step size of integration. In the case of the linear differential equation:

$$\frac{dy}{dx} = \lambda y \tag{5.118}$$

λ is the eigenvalue of that equation, and it remains constant throughout the integration. The nonlinear differential equation:

$$\frac{dy}{dx} = f(x, y) \tag{5.20}$$

can be linearized at each step using the mean-value theorem (5.140), so that λ can be obtained from the partial derivative of the function with respect to y:

$$\lambda = \frac{\partial f}{\partial y}\Big|_{\alpha, x_n} \tag{5.175}$$

The value of λ is no longer a constant but varies in magnitude at each step of the integration.

This analysis can be extended to a set of simultaneous nonlinear differential equations:

$$\begin{aligned}
\frac{dy_1}{dx} &= f_1(x, y_1, y_2, \ldots, y_n) \\
\frac{dy_2}{dx} &= f_2(x, y_1, y_2, \ldots, y_n) \\
&\vdots \\
\frac{dy_n}{dx} &= f_n(x, y_1, y_2, \ldots, y_n)
\end{aligned} \tag{5.98}$$

Linearization of the set produces the Jacobian matrix:

$$\boldsymbol{J} = \begin{bmatrix} \dfrac{\partial f_1}{\partial y_1} & \cdots & \dfrac{\partial f_1}{\partial y_n} \\ & \vdots & \\ \dfrac{\partial f_n}{\partial y_1} & \cdots & \dfrac{\partial f_n}{\partial y_n} \end{bmatrix} \tag{5.176}$$

The eigenvalues $\{\lambda_i \mid i = 1, 2, \ldots, n\}$ of the Jacobian matrix are the determining factors in the stability analysis of the numerical solution. The step size of integration is determined by the stability boundary of the method and the maximum eigenvalue.

When the eigenvalues of the Jacobian matrix of the differential equations are all of the same order of magnitude, no unusual problems arise in the integration of the set. However, when the maximum eigenvalue is several orders of magnitude larger than the minimum eigenvalue, the equations are said to be *stiff*. The *stiffness ratio* (*SR*) of such a set is defined as:

$$SR = \frac{\max\limits_{1 \leq i \leq n} \left|\text{Real}(\lambda_i)\right|}{\min\limits_{1 \leq i \leq n} \left|\text{Real}(\lambda_i)\right|} \tag{5.177}$$

The step size of integration is determined by the largest eigenvalue and the final time of integration is usually fixed by the smallest eigenvalue; therefore, integration of differential equations using explicit methods may be time intensive. Finlayson [1] recommends using implicit methods for integrating stiff differential equations to reduce computation time. The Gear method is also recommended for solving the set of stiff ordinary differential equations [5].

The MATLAB functions *ode15s*, *ode23s*, *ode23t*, and *ode23tb* are solvers suitable for the solution of stiff ordinary differential equations (see Table 5.2).

5.11 SUMMARY

An initial value problem is a set of first-order differential equations for which all boundary conditions are given at a specific independent variable. A higher-order equation differential equation can also be transformed into its canonical form, that is, a set of first-order differential equations. In this chapter we first provided the solution of a set of the linear ordinary differential equation. The solution can be obtained by either the matrix exponential method or the eigenvector method.

Various methods for the solution of nonlinear ordinary differential equations are also presented in this chapter. These methods are developed based on Taylor expansions and calculus of finite differences. The simplest solutions are different variations of the Euler method in which the terms after the first-order derivative are truncated in the expansion. Explicit, implicit, and predictor-corrector formulas are derived.

In the explicit formula the dependent variable in the next point is calculated from the previous point. In the implicit scheme the equation cannot be solved individually but must be set up as a set of simultaneous algebraic equations.

The explicit formula is easier to implement but is conditionally stable (i.e., the solution becomes unstable if the step is large). On the other hand, the implicit formula is unconditionally stable but is more complicated to implement.

The Euler predictor-corrector method, in which the explicit and implicit formula is applied in turn, benefits from the advantages of these two methods and skips the disadvantages. In fact, this method is as easy to implement as the explicit method but does not have the stability problem as in the implicit formula.

Runge-Kutta methods are developed next. The fourth-order Runge-Kutta formula is probably the most widely used numerical integration method for ordinary differential equations.

Unlike Euler and Runge-Kutta methods, which use only the previous point in the calculation, multistep methods use several previously calculated values. Adams and Adams-Moulton are the multistep methods covered in this chapter. All aforementioned methods, although presented for a single ordinary differential equation, can be easily applied to sets of equations.

Built-in MATLAB functions that solve the initial value problem of the ordinary differential equation are described. Also, a section is devoted to the solution of difference equations. At the end of the error propagation, stability, and convergence of the presented methods are discussed and comments are given on step size control strategies.

PROBLEMS

5.1 Derive the second-order Runge-Kutta method of Eq. (5.92) using central differences.

5.2 The solution of the following second-order linear ordinary differential equation should be determined using numerical techniques:

$$\frac{d^2x}{dt^2} - 3\frac{dx}{dt} - 10x = 0$$

The initial conditions for this equation are, at $t = 0$:

$$x\Big|_0 = 3 \quad \text{and} \quad \frac{dx}{dt}\Big|_0 = 15$$

(a) Transform the foregoing differential equation into a set of first-order linear differential equations with appropriate initial conditions.

(b) Find the solution using eigenvalues and eigenvectors, and evaluate the variables in the range $0 \leq t \leq 1.0$.

(c) Use the fourth-order Runge-Kutta method to verify the results of part (b).

5.3 A radioactive material (A) decomposes according to the series reaction:

$$A \xrightarrow{k_1} B \xrightarrow{k_2} C$$

where k_1 and k_2 are the rate constants and B and C are the intermediate and final products, respectively. The rate equations are:

$$\frac{dC_A}{dt} = -k_1 C_A$$

$$\frac{dC_B}{dt} = k_1 C_A - k_2 C_B$$

$$\frac{dC_C}{dt} = k_2 C_B$$

where C_A, C_B, and C_C are the concentrations of materials A, B, and C, respectively. The values of the rate constants are:

$$k_1 = 3 \text{ s}^{-1} \quad k_2 = 1 \text{ s}^{-1}$$

Initial conditions are:

$$C_A(0) = 1 \text{ mol/m}^3 \quad C_B(0) = 0 \quad C_C(0) = 0$$

(a) Use the eigenvalue-eigenvector method to determine the concentrations C_A, C_B, and C_C as a function of time t.

(b) At time $t = 1$ s and $t = 10$ s, what are the concentrations of A, B, and C?

(c) Sketch the concentration profiles for A, B, and C.

5.4 Integrate the following differential equations:

$$\frac{dC_A}{dt} = -4C_A + C_B C_A(0) = 100$$

$$\frac{dC_B}{dt} = 4C_A - 4C_B C_B(0) = 0.0$$

for the period $0 \leq t \leq 5$, using:

(a) The Euler predictor-corrector method,

(b) The fourth-order Runge-Kutta method.

Which method would give a solution closer to the analytical solution? Why do these methods give different results?

5.5 Fig. P5.5 shows a chemical reaction between three species, whose concentrations are designated by Y_1, Y_2, Y_3, taking place in a batch reactor.

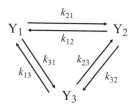

FIG. P5.5

The system of chemical reactions.

The equations describing the dynamics of this chemical reaction scheme, based on monomolecular kinetics, is a set of simultaneous linear ordinary differential equations.

(a) Derive these equations.

(b) Using the following set of kinetic rate constants and initial conditions, determine the solution of these equations.

$$k_{21} = 1 \quad k_{12} = 1.5 \quad k_{31} = 0.8 \quad k_{13} = 0.5 \quad k_{32} = 0.7 \quad k_{23} = 1.1$$
$$Y_1(0) = 1 \quad Y_2(0) = 0 \quad Y_3(0) = 0$$

(c) Sketch the time profiles of Y_1, Y_2, and Y_3.

5.6 In the study of fermentation kinetics, the logistic law:

$$\frac{dy_1}{dt} = k_1 y_1 \left(1 - \frac{y_1}{k_2}\right)$$

has been used frequently to describe the dynamics of cell growth. This equation is a modification of the logarithmic law:

$$\frac{dy_1}{dt} = k_1 y_1$$

The term $(1 - y_1/k_2)$ in the logistic law accounts for the cessation of growth due to a limiting nutrient.

The logistic law has been used successfully in modeling the growth of *Penicillium chrysogenum*, a penicillin-producing organism [6]. In addition, the rate of production of penicillin has been mathematically quantified by the equation:

$$\frac{dy_2}{dt} = k_3 y_1 - k_4 y_2$$

Penicillin (y_2) is produced at a rate proportional to the concentration of the cell (y_1) and is degraded by hydrolysis, which is proportional to the concentration of the penicillin itself.

(a) Discuss other possible interpretations of the logistic law.

(b) Show that k_2 is equivalent to the maximum cell concentration that can be reached under given conditions.

(c) Apply the fourth-order Runge-Kutta integration method to find the numerical solution of the cell and penicillin equations.

(d) Use the following constants and initial conditions:

$$k_1 = 0.03120 \quad k_2 = 47.70 \quad k_3 = 3.374 \quad k_4 = 0.01268$$

At $t = 0$, $y_1(0) = 5.0$, and $y_2(0) = 0.0$; the range of t is $0 \leq t \leq 212$ hours.

5.7 The conversion of glucose to gluconic acid is simple oxidation of the aldehyde group of the sugar to a carboxyl group. This transformation can be achieved by a microorganism in a fermentation process. The enzyme glucose oxidase, present in the microorganism, converts glucose to gluconolactone. In turn, the gluconolactone hydrolyzes to form the gluconic acid. The overall mechanism of the fermentation process that performs this transformation can be described as follows:

Cell growth:

$$\text{Glucose} + \text{Cells} \rightarrow \text{Cells}$$

Glucose oxidation:

$$\text{Glucose} + O_2 \xrightarrow{\text{Glucose oxidase}} \text{Gluconolactone} + H_2O_2$$

Gluconolactone hydrolysis:

$$\text{Gluconolactone} + H_2O_2 \rightarrow \text{Gluconic acid}$$

Peroxide decomposition:

$$H_2O_2 \xrightarrow{\text{Catalyst}} H_2O + \frac{1}{2}O_2$$

A mathematical model of the fermentation of the bacterium *Pseudomonas ovalis*, which produces gluconic acid, has been developed by Rai and Constantinides [7]. This model, which describes the dynamics of the logarithmic growth phases, can be summarized as follows:

Rate of cell growth:

$$\frac{dy_1}{dt} = b_1 y_1 \left(1 - \frac{y_1}{b_2}\right)$$

Rate of gluconolactone formation:

$$\frac{dy_2}{dt} = \frac{b_3 y_1 y_4}{b_4 + y_4} - 0.9082 b_5 y_2$$

Rate of gluconic acid formation:

$$\frac{dy_3}{dt} = b_5 y_2$$

Rate of glucose consumption:

$$\frac{dy_4}{dt} = -1.011 \left(\frac{b_3 y_1 y_4}{b_4 + y_4}\right)$$

where

y_1 = concentration of cell
y_2 = concentration of gluconolactone
y_3 = concentration of gluconic acid
y_4 = concentration of glucose
b_1 to b_5 = parameters of the system that are functions of temperature and pH

At the operating conditions of 30°C and pH 5.6, the values of the five parameters were determined from experimental data to be:

$$b_1 = 0.949 \quad b_4 = 37.51$$
$$b_2 = 3.439 \quad b_5 = 1.169$$
$$b_3 = 18.72$$

At these conditions, develop the time profiles of all variables, y_1 to y_4, for the period $0 \le t \le 9$ hours. The initial conditions at the start of this period are:

$$y_1(0) = 0.5 \text{ U.O.D.}/\text{mL} \quad y_3(0) = 0.0 \text{ mg}/\text{mL}$$
$$y_2(0) = 0.0 \text{ mg}/\text{mL} \quad y_4(0) = 50.0 \text{ mg}/\text{mL}$$

5.8 The best known mathematical representation of population dynamics between interacting species is the Lotka-Volterra model [8]. For the case of two competing species, these equations take the general form:

$$\frac{dN_1}{N_1 dt} = f_1(N_1, N_2) \quad \frac{dN_2}{N_2 dt} = f_2(N_1, N_2)$$

where N_1 is the population density of species 1 and N_2 is the population density of species 2. The functions f_1 and f_2 describe the specific growth rates of the two populations. Under certain assumptions, these functions can be expressed in terms of N_1, N_2, and a set of constants whose values depend on natural birth and death rates and on the interactions between the two species. Numerous examples of such interactions can be cited from ecological and microbiological studies. The predator-prey problem, which has been studied extensively, presents a very interesting ecological example of population dynamics. On the other hand, the interaction between bacteria and phages in a fermentor is a well-known nemesis to industrial microbiologists.

Let us now consider in detail the classical predator-prey problem, that is, the interaction between two wildlife species, namely, the prey, which is a herbivore, and the predator, a carnivore. These two animals coinhabit a region where the prey has an abundant supply of natural vegetation for food, and the predators depend on the prey for their entire supply of food. This is a simplification of the real ecological system where more than two species coexist and where predators usually feed on a variety of prey. The Lotka-Volterra equations have also been formulated for such complex systems; however, for the sake of this problem, our ecological system will contain only two interacting species. An excellent example of such an ecological system is Isle Royale National Park, a 210-square-mile archipelago in Lake Superior. The park comprises a single large island and many small islands that extend off the main island. According to a very interesting article in National Geographic [9], moose arrived on Isle Royale around 1900, probably swimming in from Canada. By 1930 their unchecked numbers approached 3000, ravaging vegetation. In 1949 across an ice bridge from Ontario came a predator: the wolf. Since 1958 the longest study of its kind [10−12] still seeks to define the complete cycle in the ebb and flow of predator and prey populations, with wolves fluctuating from 11 to 50 and moose from 500 to 2400.

To formulate the predator-prey problem, we make the following assumptions:

(a) In the absence of the predator the prey has a natural birth rate b and a natural death rate d. Because an abundant supply of natural vegetation for food is available and assuming that no catastrophic diseases plague the prey, the birth rate is higher than the death rate; therefore, the net specific growth rate α is positive; that is,

$$\frac{dN_1}{N_1 dt} = b - d = \alpha$$

(b) In the presence of the predator the prey is consumed at a rate proportional to the number of predators present βN_2, as shown in the following:

$$\frac{dN_1}{N_1 dt} = \alpha - \beta N_2$$

(c) In the absence of the prey the predator has a negative specific growth rate $(-\gamma)$ because the inevitable consequence of such a situation is the starvation of the predator:

$$\frac{dN_2}{N_2 dt} = -\gamma$$

(d) In the presence of the prey the predator has an ample supply of food that enables it to survive and produce at a rate proportional to the abundance of the prey δN_1. Under these circumstances, the specific growth rate of the predator is:

$$\frac{dN_2}{N_2 dt} = -\gamma + \delta N_1$$

The equations in parts (b) and (d) constitute the Lotka-Volterra model for the one-predator-one-prey problem. Rearranging these two equations to put them in the canonical form:

$$\frac{dN_1}{dt} = \alpha N_1 - \beta N_1 N_2 \tag{1}$$

$$\frac{dN_2}{dt} = -\gamma N_2 + \delta N_1 N_2 \tag{2}$$

This is a set of simultaneous first-order nonlinear ordinary differential equations. The solution of these equations first requires the determination of the constants α, β, γ, and δ, and the specification of boundary conditions. The latter could be either initial or final conditions. In population dynamics, it is more customary to specify initial population densities, as actual numerical values of the population densities may be known at some point in time, which can be called the initial starting time. However, it is conceivable that one may want to specify the final values of the population densities to be accomplished as targets in a well-managed ecological system. In this problem, we will specify the initial population densities of the prey and predator to be:

$$N_1(t_0) = N_1^0 \quad \text{and} \quad N_2(t_0) = N_2^0 \tag{3}$$

Eqs. (1) to (3) constitute the complete mathematical formulation of the predator-prey problem based on assumptions (a) to (d). Different assumptions would yield another set of differential equations (see Problem 5.9). In addition, the choice of constants and initial conditions influence the solution of the differential equations and generate a diverse set of qualitative behavior patterns for the two populations. Depending on the form of the differential equations and the values of the constants chosen, the solution patterns may vary from stable damped oscillations, where the species reach their respective stable symbiotic population densities, to highly unstable situations in which one of the species is driven to extinction while the other explodes to extreme population density.

The literature on the solution of the Lotka-Volterra problems is voluminous. Several references on this topic are given at the end of this chapter. A closed-form analytical solution of this system of nonlinear ordinary differential equations is not possible. The equations must be integrated numerically using any of the numerical integration methods covered in this chapter. However, before numerical integration is attempted, the stability of these equations must be examined thoroughly. In a recent treatise on this subject Vandermeer [13] examined the stability of the solutions of these equations around equilibrium points. These points are located by setting the derivatives in Eqs. (1) and (2) to zero:

$$\frac{dN_1}{dt} = \alpha N_1 - \beta N_1 N_2 = 0$$
$$\frac{dN_2}{dt} = -\gamma N_2 + \delta N_1 N_2 = 0$$

and rearranging these equations to obtain the values of N_1 and N_2 at the equilibrium point in terms of the constants:

$$N_1{}^* = \frac{\gamma}{\delta} \quad N_2{}^* = \frac{\alpha}{\beta}$$

where $*$ denotes the equilibrium values of the population densities. Vandermeer stated, "Sometimes only one point ($N_1{}^*$, $N_2{}^*$) will satisfy the equilibrium equations. At other times, multiple points will satisfy the equilibrium equations The neighborhood stability analysis is undertaken in the neighborhood of a single equilibrium point." The stability is determined by examining the eigenvalues of the Jacobian matrix evaluated at equilibrium:

$$\boldsymbol{J} = \begin{vmatrix} \left(\frac{\partial f_1}{\partial N_1}\right)^* & \left(\frac{\partial f_1}{\partial N_2}\right)^* \\ \left(\frac{\partial f_2}{\partial N_1}\right)^* & \left(\frac{\partial f_2}{\partial N_2}\right)^* \end{vmatrix}$$

where f_1 and f_2 are the right-hand sides of Eqs. (1) and (2), respectively.

The eigenvalues of the Jacobian matrix can be obtained by the solution of the following equation (as described in Chapter 2):

$$|\boldsymbol{J} - \lambda \boldsymbol{I}| = 0$$

For the problem of two differential equations, there are two eigenvalues that can possibly have both real and imaginary parts. These eigenvalues take the general form:

$$\lambda_1 = a_1 + b_1 i$$
$$\lambda_2 = a_2 + b_2 i$$

where $i = \sqrt{-1}$. The values of the real parts (a_1, a_2) and imaginary parts (b_1, b_2) determine the nature of the stability (or instability) in the neighborhood of the equilibrium points. These possibilities are summarized in Table P5.8a.

Many combinations of values of constants and initial conditions exist that would generate solutions to Eqs. (1) and (2). To obtain a realistic solution to these equations, we use the data of Allen [10] and Peterson [11] on the moose-wolf populations of Isle Royale National Park given in

Table P5.8a Stability conditions.

a_1, a_2	b_1, b_2	Stability analysis
Negative	Zero	Stable, nonoscillatory
Positive	Zero	Unstable, nonoscillatory
Opposite signs	Zero	Metastable, saddle point
Negative	Nonzero	Stable, oscillatory
Positive	Nonzero	Unstable, oscillatory
Zero	Nonzero	Neutrally stable, oscillatory

Table P5.8b Population of moose and wolves on Isle Royale.

Year	Moose	Wolves	Year	Moose	Wolves
1960	573	22	1980	705	50
1961	597	22	1981	544	30
1962	603	23	1982	972	14
1963	639	20	1983	900	23
1964	726	26	1984	1041	24
1965	762	28	1985	1062	22
1966	900	26	1986	1025	20
1967	1008	22	1987	1380	16
1968	1176	22	1988	1653	12
1969	1191	17	1989	1397	11
1970	1320	18	1990	1216	15
1971	1323	20	1991	1313	12
1972	1194	23	1992	1596	12
1973	1137	24	1993	1880	13
1974	1026	31	1994	1770	17
1975	915	41	1995	2422	16
1976	708	44	1996	1200	22
1977	573	34	1997	500	24
1978	905	40	1998	700	14
1979	738	43			

Table P5.8b. From these data, which cover the period 1959−98, we estimate the average values of the moose and wolf populations (over the entire 40-year period) and use these as equilibrium values:

$$N_1{}^* = \frac{\gamma}{\delta} = 1045 \quad N_2{}^* = \frac{\alpha}{\beta} = 23$$

In addition, we estimate the period of oscillation to be 25 years. This was based on the moose data; the wolf data show a shorter period. For this reason, we predict that the predator equation may not be a good representation of the data. Lotka has shown that the period of oscillation around the equilibrium point is approximated by:

$$\tau = \frac{2\pi}{\sqrt{\alpha\gamma}}$$

These three equations have four unknowns. By assuming the value of a to be 0.3 (this is an estimate of the net specific growth rate of the prey in the absence of the predator), the complete set of constants is:

$$\alpha = 0.3 \qquad \beta = 0.0130$$

$$\gamma = 0.2106 \quad \delta = 0.0002015$$

These initial conditions are taken from Allen [10] for 1959, the earliest date for which complete data are available. These are:

$$N_1(1959) = 522 \quad \text{and} \quad N_2(1959) = 20$$

Integrate the predator-prey equations for the period 1959–98 using the foregoing constants and initial conditions and compare the simulation with the actual data. Draw the phase plot of N_1 versus N_2 and discuss the stability of these equations with the aid of the phase plot.

5.9 It can be shown that whenever the Lotka-Volterra problem has the form of Eqs. (1) and (2) in Problem 5.8, the real parts of the eigenvalues of the Jacobian matrix are zero. This implies that the solution always has neutrally stable oscillatory behavior. This is explained by the fact that assumptions (a) to (d) of Problem 5.8 did not include the crowding effect each population may have on its own fertility or mortality. For example, Eq. (1) can be rewritten with the additional term εN^2:

$$\frac{dN_1}{dt} = \alpha N_1 - \beta N_1 N_2 - \varepsilon N_1^2$$

The new term introduces a negative density-dependence of the specific growth rate of the prey on its own population. This term can be viewed as either a contribution to the death rate or a reduction of the birth rate caused by overcrowding of the species.

In this problem modify the Lotka-Volterra equations by introducing the effect of overcrowding, account for at least one additional source of food for the predator (a second prey), or attempt to quantify other interferences you believe are important in the life cycle of these two species. Choose the constants and initial conditions of your equations carefully to obtain an ecologically feasible situation. Integrate the resulting equations and obtain the time profiles of the populations of all the species involved. In addition, draw phase plots of N_1 versus N_2, N_1 versus N_3, etc., and discuss the stability considerations with the aid of the phase plots.

5.10 A plug-flow reactor is to be designed to produce the product D from A according to the following reaction:

$$A \rightarrow D \qquad r_D = 60C_A \text{ molD/L.s}$$

In the operating condition of this reactor, the following undesired reaction also takes place:

$$A \rightarrow U \quad r_U = \frac{0.003 C_A}{1 + 10^5 C_A} \, \text{molU/L.s}$$

The undesired product U is a pollutant and it costs 10 $/mol U to dispose of it, while the desired product D has a value of 35 $/mol D. What size of the reactor should be chosen to obtain an effluent stream at its maximum value?

Pure reactant A with a volumetric flow rate of 15 L/s and molar flow rate of 0.1 mol/s enters the reactor. The value of A is 5 $/mol A.

5.11 A catalytic reaction takes place in a continuously stirred tank reactor (CSTR). The catalyst becomes deactivated during the reaction. The differential equation determining the changes of moles of component A during the reaction is given by:

$$\frac{dC_A}{dt} = -\frac{10 e^{-t/25} C_A^2}{1 + 2C_A}$$

The initial concentration of A in the reactor is 20.

Solve this differential equation by the second-order Runge-Kutta method up to $t = 2$. Use $h = 0.5$. Write a MATLAB program to solve this problem. This program should show graphically the evolution of C_A against t. The differential equation and the Runge-Kutta method should be introduced in separate MATLAB functions.

5.12 Write the unsteady-state mass balance for a continuously stirred tank reactor (CSTR) in which a second-degree reaction takes place. Considering the following values:

V = volume of the reactor ($10 \, \text{m}^3$)
$C_{A,in}$ = concentration of A in the feed (200 g/min)
Q = feed flow rate ($10 \, \text{m}^3/\text{min}$)
k = rate constant ($0.1 \, \text{m}^3/\text{g min}$)
C_0 = initial concentration of A in the reactor (0 g/min)

Solve the differential equation using the Euler method from $t = 0$ to $t = 1$ with $h = 0.25$. Write a MATLAB program to solve this equation with $h = 0.05$ up to $t = 1$.

5.13 The material balance in a catalyst slab, for which the reaction kinetic obeys the Langmuir-Hinshelwood model, is:

$$\frac{d^2 C}{dx^2} - \frac{\varphi C}{1 + \delta C + \gamma C^2} = 0$$

where x is the distance from the center of the slab and C is the concentration of the reactant. Considering the following boundary conditions:

$$x = 0, \quad dC/dx = 0; \quad C = 0.2$$

determine the concentration profile of the reactant in the catalyst.

Use the following constants:

$$\varphi = 0.5, \quad \delta = 0.2, \quad \gamma = 0.01$$

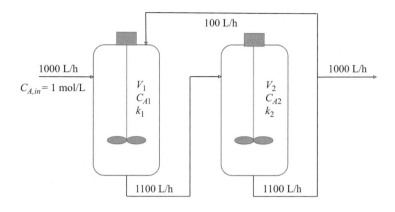

FIG. P5.14

Continuous stirred tank reactors in series.

Table P5.14 Reactor volumes, reaction rate constants, and initial concentrations.

Reactor I	Volume V_i (L)	Rate constant k_i (h^{-1})	Initial concentration $C_{Ai}(0)$ (mol/L)
1	800	0.2	0
2	1000	0.4	0

5.14 A chemical reaction takes place in a series of two continuously stirred tank reactors arranged as shown in Fig. P5.14. The chemical reaction is a first-order irreversible reaction of the type:

$$A \xrightarrow{k_i} B$$

The following assumptions can be made regarding this system:

i. The system is operating at an unsteady state.
ii. The reactions are in the liquid phase which is well mixed.
iii. There is no change in the volume or density of the liquid.
iv. The rate of generation of component A in each reactor is given by

$$r_i = -k_i V_i C_{Ai} \ \text{mol/h}$$

The temperature in each reactor is such that the value of k_i is different in each reactor. The values of V_i, k_i and initial concentrations are given in Table P5.14.

Based on this information, do the following:

(a) Set up the material balance equation for component A for each of the two reactors.
(b) Use the Runge-Kutta third order method to integrate this set of equations between the time of $t = 0$ and $t = 1$ hour.
(c) Sketch the profile of C_A for each of these reactors.

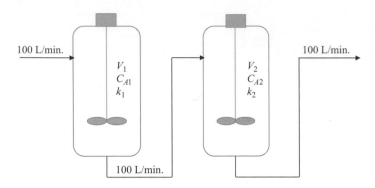

FIG. P5.15

Continuous stirred tank reactors in series.

Table P5.15 Reactor volumes and rate constants.			
Reactor i	Volume V_i (L)	Rate constant k_a (min^{-1})	Rate constant k_b (L/mol)
1	250	0.7	1.8
2	100	1	0.9

5.15 A chemical reaction takes place in a series of two continuous stirred tank reactors arranged as shown in Fig. P5.15. The rate of chemical reaction of component A is given by:

$$-r_A = \frac{k_a C_A}{1 + k_b C_A}$$

Initially, the system operates in a steady state, when the concentration of A in the feed is 1 mol/L. Then, the concentration is decreased to 0.5 mol/L. How does the concertation change against time in these reactors? Additional data are given in Table P5.15.

5.16 The chemical reaction $2A \rightarrow B$ takes place in the liquid phase in an ideal plug flow reactor of length 1 m. The reactant A enters the reactor with a concentration of 1.5 mol/m^3. The reactor operates isothermally and the superficial velocity of liquid in the reactor is 2 m/s. The rate of conversion of A is given by:

$$r_A = \frac{2C_A}{1 + 0.2 C_A C_B} \, mol/m^3 s$$

Write the steady-state material balance for both components and determine the concentration profiles of these components along the reactor from solving the related differential equations by the modified Euler method.

5.17 Obtain the solution of the difference Eq. (5.103) directly from the solution of the differential Eq. (5.102) by using the relationship $E = e^{hD}$.

REFERENCES

[1] Finlayson, B. A. *Nonlinear Analysis in Chemical Engineering;* McGraw-Hill: New York, 1980.

[2] Fogler, H. S. *Elements of Chemical Reaction Engineering*, 6th ed.; Pearson: New York, 2020.

[3] Elaydi, S. *An Introduction to Difference Equations;* Springer: New York, 2005.

[4] Lapidus, L.; Sienfeld, J. H. *Numerical Solution of Ordinary Differential Equations;* Academic Press: New York, 1971.

[5] Gear, G. W. *Numerical Initial Value Problems in Ordinary Differential Equations;* Prentice-Hall: Englewood Cliffs, NJ, 1971.

[6] Constantinides, A.; Spencer, J. L.; Gaden, E. L., Jr. Optimization of Batch Fermentation Processes, II. Optimum Temperature Profiles for Batch Penicillin Fermentations. *Biotechnology and Bioengineering* **1970,** *12*, 1081−1098.

[7] Rai, V. R.; Constantinides, A. Mathematical Modeling and Optimization of the Gluconic Acid Fermentation. *AIChE Symposium Series* **1973,** *69* (132), 114.

[8] Brauer, F.; Castillo-Chavez, C. *Mathematical Models in Population Biology and Epidemiology*, 2nd ed.; Springer: New York, 2011.

[9] Elliot, J. L. Isle Royale: A North Woods Park Primeval. *National Geographic* **April 1985,** *167*, 534.

[10] Allen, D. L. *Wolves of Minong;* University of Michigan Regional: Ann Arbor, 1993.

[11] Peterson, R. *The Wolves of Isle Royale, A Broken Balance;* University of Michigan Regional: Ann Arbor, 2007.

[12] Peterson, R. O. *Ecological Studies of Wolves on Isle Royale, Annual Reports;* Michigan Technological University: Houghton, Michigan, 1984 to 1998.

[13] Vandermeer, J. *Elementary Mathematical Ecology;* Medtech: Edgartown, MA, 2017.

ORDINARY DIFFERENTIAL EQUATIONS: BOUNDARY VALUE PROBLEMS

CHAPTER OUTLINE

MOTIVATION

Ordinary differential equations with boundary conditions specified at two or more points of the independent variable are classified as boundary value problems. There are many chemical engineering applications that result in ordinary differential equations of the boundary value type. The diversity of problems of the boundary value type has generated a variety of methods for their solution. The system equations in these problems could be linear or nonlinear, and the boundary conditions could be linear or nonlinear, separated or mixed, two-point or multipoint. In this chapter we have chosen to discuss algorithms that are applicable to the solution of nonlinear, as well as linear, boundary value problems. These are the Newton method, the finite-difference method, and the collocation methods.

6.1 INTRODUCTION

In this section we mention only a few examples of ordinary differential equations with split boundary values. In the study of diffusion with chemical reaction with chemical catalysis or enzyme catalysis we encounter a second-order differential equation with split boundary conditions. For

Applied Numerical Methods for Chemical Engineers. DOI: https://doi.org/10.1016/B978-0-12-822961-3.00006-6

example, consider a porous spherical catalyst pellet in which a first-order reaction takes place for converting molecules of A into products. Component A diffuses into the catalyst pellet and reacts on the porous surface of the catalyst simultaneously. Concertation of A, C_A, inside the pellet at the steady-state condition, can be obtained from solving the following second-order differential equation:

$$\frac{1}{r^2}\frac{d}{dr}\left(D_e r^2 \frac{dC_A}{dr}\right) - kC_A = 0 \tag{6.1}$$

where r is the radial position in the sphere, D_e is the effective diffusivity of fluid in the catalyst and k is the reaction rate constant. For the known concentration of A at the surface of the sphere, C_s, Eq. (6.1) should be solved subject to the following two boundary conditions:

$$r = 0, \qquad \frac{dC_A}{dr} = 0$$
$$r = R, \qquad C_A = C_s \tag{6.2}$$

The steady-state temperature distribution in a variable area fin such as the one shown in Fig. 6.1, can be obtained by solving the following differential equation:

$$\frac{d}{dx}\left(KA\frac{dT}{dx}\right) - hP(T - T_a) = 0 \tag{6.3}$$

Here, T is the temperature of the fin at distance x from the tip, K is the thermal conductivity, h is the convective heat transfer coefficient between fin and surrounding air, A is the cross-sectional area of the fin, P is the periphery of the fin, and T_a is the air temperature. In a variable area fin A and P can be functions of x. The temperature of the base of the fin is denoted as T_0. Also convection heat transfer takes place at the tip of the fin. Therefore Eq. (6.3) is subject to the following split boundary conditions:

$$x = L, \qquad T = T_0$$
$$x = 0, \qquad -K\frac{dT}{dx} = (T - T_a) \tag{6.4}$$

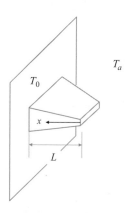

FIG. 6.1

Variable area fin.

When solving the boundary layer problem near a corner, it is necessary to solve numerically the following differential equation, subject to the specified boundary conditions [1]:

$$f''' + ff'' + \beta(1 - f'^2) = 0$$
$$\eta = 0, \quad f = f' = 0$$
$$\eta \to \infty, \quad f' = 1$$

(6.5)

The solution of this differential equation with the specified split boundary conditions gives $f(\eta)$, from which the stream function, then components of the velocity of the fluid in the boundary layer can be evaluated [1].

The diversity of problems of the boundary value type have generated a variety of methods for their solution. The system equations in these problems could be linear or nonlinear, and the boundary conditions could be linear or nonlinear, separated or mixed, two point or multipoint. Comprehensive discussions of the solutions of boundary value problems are given by Kubicek and Hlavacek [2] and by Aziz [3]. In this chapter we have chosen to discuss algorithms that are applicable to the solution of nonlinear, as well as linear, boundary value problems. These are the *Newton* method, the *finite-difference* method, and the *collocation* methods.

The canonical form of a two-point boundary value problem with linear boundary conditions is:

$$\frac{dy_i}{dx} = f_i(x, y_1, y_2, \ldots, y_n) \quad x_0 \leq x \leq x_f \quad j = 1, 2, \ldots, n$$

(6.6)

where the boundary conditions are split between the initial point x_0 and the final point x_f. The first r equations have initial conditions specified and the last $(n - r)$ equations have final conditions given:

$$y_j(x_0) = y_{j,0} \quad j = 1, 2, \ldots, r$$

(6.7)

$$y_j(x_f) = y_{j,f} \quad j = r + 1, \ldots, n$$

(6.8)

A second-order two-point boundary value problem may be expressed in the form:

$$\frac{d^2 y}{dx^2} = f\left(x, y, \frac{dy}{dx}\right) \quad x_0 \leq x \leq x_f$$

(6.9)

subject to the boundary conditions:

$$a_0 y(x_0) + b_0 y'(x_0) = c_0$$

(6.10)

$$a_f y(x_f) + b_f y'(x_f) = c_f$$

(6.11)

where the subscript 0 designates conditions at the left boundary (initial) and the subscript f identifies conditions at the right boundary (final). This problem can be transformed to the canonical form (6.6) by the appropriate substitutions described in Section 5.3.

6.2 THE SHOOTING METHOD

The shooting method converts the boundary value problem to an initial-value one to take advantage of the powerful algorithms available for the integration of initial-value problems (see Chapter 5). In this method the unspecified initial conditions of the system differential equations are guessed and the equations are integrated forward as a set of simultaneous initial-value differential equations. In the end the calculated final values are compared with the boundary conditions, and the guessed initial conditions are corrected if necessary. This procedure is repeated until the specified terminal values are achieved within a small convergence criterion. This general algorithm forms the basis for the family of shooting methods. These may vary in their choice of initial or final conditions and the integration of the equations in one direction or two directions. In this section we develop Newton's technique which is the most widely known of the shooting methods and can be applied successfully to the boundary value problem of any complexity as long as the resulting initial-value problem is stable and a set of good guesses for unspecified conditions can be made [2].

We first develop the Newton method for a set of two differential equations:

$$\frac{dy_1}{dx} = f_1(x, y_1, y_2)$$
$$\frac{dy_2}{dx} = f_2(x, y_1, y_2)$$

(6.12)

with split boundary conditions:

$$y_1(x_0) = y_{1,0}$$ (6.13)
$$y_2(x_f) = y_{2,f}$$ (6.14)

We guess the initial condition for y_2:

$$y_2(x_0) = \gamma$$ (6.15)

If the system equations are integrated forward, the two trajectories may look like those in Fig. 6.2. Because the value of $y_2(x_0)$ was only a guess, the trajectory of y_2 misses its target at x_f; that is, it does not satisfy the boundary condition of (6.14). For the given guess of γ, the calculated value of y_2 at x_f is designated as $y_2(x_f, \gamma)$. The desirable objective is to find the value of γ which forces $y_2(x_f, \gamma)$ to satisfy the specified boundary condition, that is,

$$y_2(x_f, \gamma) = y_{2,f}$$ (6.16)

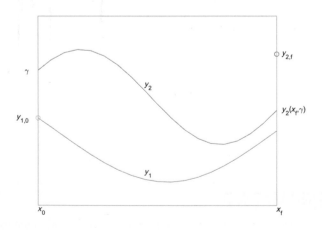

FIG. 6.2

Forward integration using a guessed initial condition γ. The symbol o designates the known boundary points.

Rearrange Eq. (6.16) to

$$\phi(\gamma) = y_2(x_f, \gamma) - y_{2,f} = 0 \tag{6.17}$$

The function $\phi(\gamma)$ can be expressed in a Taylor series around γ:

$$\phi(\gamma + \Delta\gamma) = \phi(\gamma) + \frac{\partial\phi}{\partial\gamma}\Delta\gamma + O\big[(\Delta\gamma)^2\big] \tag{6.18}$$

For the system to converge, that is, for the trajectory of y_2 to hit the specified boundary value of x_f,

$$\lim_{\Delta\gamma\to 0} \phi(\gamma + \Delta\gamma) = 0 \tag{6.19}$$

Therefore Eq. (6.18) becomes:

$$0 = \phi(\gamma) + \frac{\partial\phi}{\partial\gamma}\Delta\gamma + O\big[(\Delta\gamma)^2\big] \tag{6.20}$$

Truncation and rearrangement give:

$$\Delta\gamma = \frac{-\phi(\gamma)}{\Big[\frac{\partial\phi}{\partial\gamma}\Big]} \tag{6.21}$$

The reader should be able to recognize this equation as a form of the Newton-Raphson method of Chapter 1. Using the definition of $\phi(\gamma)$ (Eq. 6.17), taking its partial derivative, and combining with Eq. (6.21), we obtain:

$$\Delta\gamma = \frac{-\big[y_2(x_f, \gamma) - y_{2,f}\big]}{\Big[\frac{\partial y_2(x_f,\gamma)}{\partial\gamma}\Big]} = \frac{\delta y}{\Big[\frac{\partial y_2(x_f,\gamma)}{\partial\gamma}\Big]} \tag{6.22}$$

where δy is the difference between the specified final boundary value $y_{2,f}$ and the calculated final value $y_2(x_f, \gamma)$ obtained from using the guessed γ:

$$\delta y = -\big[y_2(x_f, \gamma) - y_{2,f}\big] \tag{6.23}$$

The value of $\Delta\gamma$ is the correction to be applied to the guessed γ to obtain a new guess:

$$(\gamma)_{new} = (\gamma)_{old} + \Delta\gamma \tag{6.24}$$

To avoid divergence, it may sometimes be necessary to take a fractional correction step by using relaxation that is,

$$(\gamma)_{new} = (\gamma)_{old} + \rho\Delta\gamma \quad 0 < \rho \le 1 \tag{6.25}$$

The solution of the set of differential equations continues with the new value of γ (calculated by Eq. 6.25) until $|\Delta\gamma| \le \varepsilon$.

The algorithm can now be generalized to apply to a set of n simultaneous system equations:

$$\frac{dy_i}{dx} = f_i(x, y_1, y_2, \ldots, y_n) \quad x_0 \le x \le x_f \quad j = 1, 2, \ldots, n \tag{6.6}$$

whose boundary conditions are split between the initial point and the final point. The first r equations have *initial* conditions specified, and the last $(n - r)$ equations have *final* conditions given:

$$y_j(x_0) = y_{j,0} \quad j = 1, 2, \ldots, r \tag{6.7}$$

$$y_j(x_f) = y_{j,f} \quad j = r + 1, \ldots, n \tag{6.8}$$

To apply Newton's procedure to integrate the system equations forward, the missing $(n - r)$ initial conditions are guessed as follows:

$$y_j(x_0) = \gamma_j \tag{6.26}$$

The system Eq. (6.6) with the given initial conditions (6.7) and the guessed initial conditions (6.26) are integrated simultaneously in the forward direction. At the right-hand boundary (x_f), the *Jacobian matrix* (equivalent to the derivative term in Eq. 6.22) is evaluated:

$$J(x_f, \gamma) = \begin{bmatrix} \left. \frac{\partial y_{r+1}}{\partial \gamma_{r+1}} \right|_{x_f} & \left. \frac{\partial y_{r+1}}{\partial \gamma_{r+2}} \right|_{x_f} & \cdots & \left. \frac{\partial y_{r+1}}{\partial \gamma_n} \right|_{x_f} \\ \vdots & \vdots & \ddots & \vdots \\ \left. \frac{\partial y_n}{\partial \gamma_{r+1}} \right|_{x_f} & \left. \frac{\partial y_n}{\partial \gamma_{r+2}} \right|_{x_f} & \cdots & \left. \frac{\partial y_n}{\partial \gamma_n} \right|_{x_f} \end{bmatrix} \tag{6.27}$$

The correction of the guessed initial values is implemented by the equation:

$$\Delta\gamma = [J(x_f, \gamma)]^{-1} \delta y \tag{6.28}$$

where the vector δy is the difference between the specified final boundary values and the calculated final values using the guessed initial conditions:

$$\delta y = - \begin{bmatrix} y_{r+1}(x_f, \gamma) - y_{r+1,f} \\ \vdots \\ y_n(x_f, \gamma) - y_{n,f} \end{bmatrix} \tag{6.29}$$

The new estimate of the guessed initial conditions is then evaluated from Eq. (6.24) in the vector form:

$$(\gamma)_{new} = (\gamma)_{old} + \rho\Delta\gamma \quad 0 < \rho \leq 1 \tag{6.25}$$

The shooting method algorithm using the Newton technique is shown in Fig. 6.3. Note that the number of differential equations with final boundary conditions is not, in any case, more than half of the total number of equations. In the case when final conditions are more than half the total number of differential equations, we may simply reverse the integrating direction and as a result, obtain a fewer number of final conditions.

EXAMPLE 6.1 FLOW OF A NON-NEWTONIAN FLUID

Write a general MATLAB® function for the solution of a boundary value problem by the shooting method using Newton's technique. Apply this function to find the velocity profile of a non-Newtonian fluid that is flowing through a circular tube as shown in Fig. E6.1a. Also, calculate the volumetric flow rate of the fluid. The viscosity of this fluid can be described by the Carreau model [4]:

$$\frac{\mu}{\mu_0} = \left[1 + (t_1 \dot{\gamma})^2 \right]^{(n-1)/2}$$

(Continued)

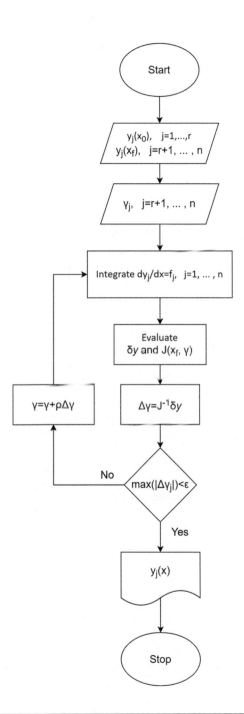

FIG. 6.3

Flowchart of the shooting method.

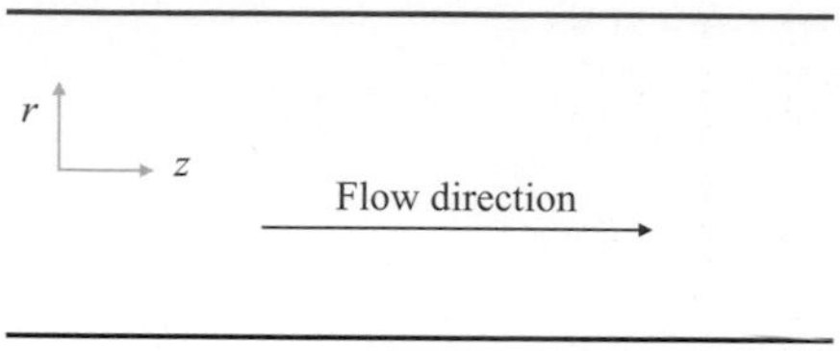

$$r \uparrow$$
$$\quad \llcorner \rightarrow z$$

Flow direction

FIG. E6.1a

Side view of a pipe.

EXAMPLE 6.1 (CONTINUED)

where μ is the viscosity of the fluid, μ_0 is the zero shear rate viscosity, $\dot{\gamma}$ is the shear rate, t_1 the characteristic time, and n is a dimensionless constant.

The momentum balance for this flow, assuming the tube is very long so that the end effect is negligible, results in:

$$\frac{d}{dr}(r\tau_{rz}) = -\frac{\Delta p}{L}r \tag{1}$$

where $\Delta P/L$ is the pressure drop gradient along the pipe and the shear stress is expressed as:

$$\tau_{rz} = -\mu\dot{\gamma} = -\mu\frac{dv_z(r)}{dr}$$

Therefore Eq. (1) is a second-order ordinary differential equation that should be solved with the following boundary conditions:

No slip at the wall: $r = R$, $v_z = 0$

Symmetry: $r = 0$, $\dfrac{dv_z}{dr} = 0$

The required data for the solution of this problem are:

$$\mu_0 = 102.0 \text{ Pa·s} \quad t_1 = 4.36 \times 10^{-2} \text{ s} \quad n = 0.375 \text{ } R = 0.1 \text{ m} \quad -\Delta p/L = 20 \text{ kPa/m}$$

Method of solution

First, we define the following two variables:

Dimensionless distance: $\eta = r/R$
Dimensionless velocity: $\phi = v_z/v^*$

where

$$v^* = \left(-\frac{-\Delta p}{L}\right)\frac{R^2}{\mu_0}$$

(Continued)

EXAMPLE 6.1 (CONTINUED)

Eq. (1) can be expanded and rearranged in its dimensionless form into the following second-order differential equation:

$$\frac{d^2\phi}{d\eta^2} = -\frac{\frac{1}{\eta}\frac{d\phi}{d\eta} + \left[1 + \lambda^2\left(\frac{d\phi}{d\eta}\right)^2\right]^{(1-n)/2}}{1 - \frac{(1-n)\lambda^2\left(\frac{d\phi}{d\eta}\right)}{\left[1 + \lambda^2\left(\frac{d\phi}{d\eta}\right)^2\right]}} \tag{2}$$

where

$$\lambda = \frac{tv^*}{R}$$

To obtain the canonical form of Eq. (2), we apply the following transformation:

$$y_1 = \frac{d\phi}{d\eta}$$
$$y_2 = \phi$$

The canonical form of Eq. (2) is then given as:

$$\frac{dy_1}{d\eta} = -\frac{\frac{1}{\eta}y_1 + \left[1 + \lambda^2 y_1^2\right]^{(1-n)/2}}{1 - \frac{(1-n)\lambda^2 y_1}{\left[1 + \lambda^2 y_1^2\right]}} \tag{3}$$

$$\frac{dy_2}{d\eta} = y_1 \tag{4}$$

The set of nonlinear ordinary differential Eqs. (3) and (4) should be solved with the following boundary conditions:

$$y_1(0) = y_{1,0} = 0 \tag{5}$$
$$y_2(1) = y_{2,f} = 0 \tag{6}$$

The initial value $y_1(0)$ is known, but the initial value $y_2(0)$ must be guessed. We designate this guess, in accordance with Eq. (6.26), as follows:

$$y_2(0) = \gamma = \frac{R^2}{4\mu_0 v*}\left(-\frac{\Delta p}{L}\right) = \frac{1}{4} \tag{7}$$

The right-hand side of Eq. (7) corresponds to the velocity of the fluid at the center of the pipe if it was a Newtonian fluid with the viscosity μ_0.

The complete set of equations for the solution of this two-point boundary value problem consists of:

1. The four system equations with their known boundary values [Eqs. (3), (4), (5), (6)]
2. The guessed initial condition for y_2 [Eq. (7)]

(Continued)

EXAMPLE 6.1 (CONTINUED)

3. Eq. (6.27) for construction of the Jacobian matrix
4. Eq. (6.29) for calculation of the δy vector
5. Eqs. (6.28) and (6.25) for correcting the guessed initial conditions

Once the velocity profile is determined, the flow rate of the fluid can be calculated from the following integral formula:

$$Q = \int_0^R 2\pi r v_z dr$$

Program description

The programs and functions developed in this example can be found in: https://www.elsevier.com/books-and-journals/book-companion/9780128229613

The MATLAB function *shooting.m* is developed to solve a set of first-order ordinary differential equations in a boundary value problem using the shooting method. The structure of this function is very similar to that of the function *Newton.m* developed in Example 2.7.

The function *shooting.m* begins with checking the input arguments. The inputs to the function are the name of the file containing the set of differential equations, lower and upper limits of integration interval, the integration step size, the vector of initial conditions, the vector of final conditions, the vector of guesses of initial conditions for those equations who have final conditions, the order of Runge-Kutta method, the relaxation factor, and the convergence criterion. From the foregoing list, introducing the integration step size, the order of the Runge-Kutta method, the relaxation factor, and the convergence criterion are optional and the function assumes default values for each of the foregoing variables, if necessary. The number of guessed initial conditions has to be equal to the number of final conditions; also, the number of equations should be equal to the total number of boundary conditions (initial and final). If these conditions are not met, the function gives a proper error message on the screen and stops execution.

The next section in the function is Newton's technique. This procedure begins with solving the set of differential equations by the Runge-Kutta method, using the known and guessed initial conditions, in the forward direction. It then sets the differentiation increment for the approximate initial conditions and consequently evaluates the elements of the Jacobian matrix, column wise, by differentiating using the forward finite-differences method. At the end of this section, the approximate initial conditions are corrected according to Eqs. (6.25) and (6.28). This procedure is repeated until the convergence is reached for all the final conditions.

It is important to note that the function *shooting.m* requires to receive the values of the set of ordinary differential equations at each point in a column vector with the values of the equations whose initial conditions are known to be at the top, followed by those whose final conditions are fixed. It is also important to pass the initial and final conditions to the function in the order which corresponds to the order of equations appearing in the file which introduces the ordinary differential equations.

(Continued)

EXAMPLE 6.1 (CONTINUED)

The main program *Example6_1.m* asks the reader to input the parameters required for the solution of the problem. The program then calls the function *shooting* to solve the set of equations and finally, it shows the value of the flow rate on the screen and plots the calculated velocity profile. The default values of the relaxation factor and the convergence criterion are used in this example.

The function *Ex6_1_func.m* evaluates the values of the set of Eqs. (3) and (4) at a given point. The first function evaluated is that of Eq. (3), the initial condition of which is known.

Input and results

>> **Example6_1**
Inside radius of the pipe (m) = 0.1
Pressure drop gradient (Pa/m) = 20e3
Zero shear rate viscosity of the fluid (Pa.s) = 102
Characteristic time of the fluid (s) = 4.36e-2
The exponent n from the power-law = 0.375
M-file containing the set of differential equations: 'Ex6_1_func'
Order of Runge-Kutta method = 4
Step size = 0.01
Volumetric flow rate = 7.43 L/s

(Continued)

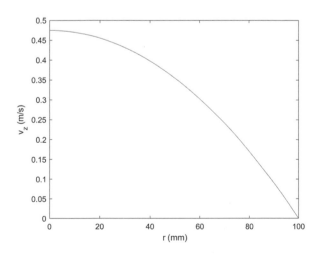

FIG. E6.1b

Velocity profile for non-Newtonian fluid.

EXAMPLE 6.1 (CONTINUED)

Discussion of results

The volumetric flow rate of the fluid in this condition is calculated to be 2.91 L/s and the velocity profile is shown in Fig. E6.1b. It should be noted that because there is the term $1/\eta$ in Eq. (2), the lower limit of numerical integration cannot be zero. A very small value close to zero should be used instead in such a situation. In the main program *Example6_1*, the lower limit of integration is set to $h/100$ which is negligible with respect to the dimension of the pipe.

EXAMPLE 6.2 USING THE SHOOTING METHOD FOR DETERMINING THE TEMPERATURE DISTRIBUTION IN A HOLLOW DISC

Determine the temperature distribution in the hollow disc shown in Fig. E6.2a. Inside and outside radii are 0.1 m and 0.2 m, respectively. Temperatures at internal and external surfaces are 450 K and 300 K, respectively. Solve this problem with the shooting method and use the explicit Euler with $h = 0.01$ m for forward integration.

The steady-state heat balance for this problem, assuming constant thermal conductivity of the disc material, results in:

$$\frac{d}{dr}\left(r\frac{dT}{dr}\right) = 0 \tag{1}$$

The boundary conditions for this differential equation are as follows:

$$r = 0.1, \; T = 450 \; K$$
$$r = 0.2, \; T = 300 \; K$$

Method of solution

First we expand Eq. (1):

$$r\frac{d^2T}{dr^2} + \frac{dT}{dr} = 0 \tag{2}$$

Next, we should obtain the canonical form of Eq. (2). For this purpose, we apply the following transformation:

$$y_1 = T$$
$$y_2 = \frac{dT}{dr}$$

The canonical form of Eq. (2) is then given as:

$$\begin{cases} \dfrac{dy_1}{dr} = y_2 \\ \dfrac{dy_2}{dr} = -\dfrac{y_2}{r} \end{cases} \tag{3}$$

(Continued)

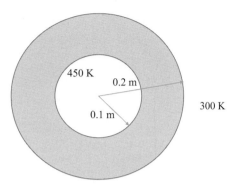

FIG. E6.2a

Hollow disc.

EXAMPLE 6.2 (CONTINUED)

The set of nonlinear ordinary differential Eq. (3) should be solved with the following boundary conditions:

$$\begin{cases} y_1(0.1) = y_{1,0} = 450 \\ y_1(0.2) = y_{1,f} = 300 \end{cases} \tag{4}$$

Fig. E6.2b demonstrates the calculation steps in Excel.

6.3 THE FINITE-DIFFERENCE METHOD

The *finite-difference* method replaces the derivatives in the differential equations with finite-difference approximations at each point in the interval of integration, thus converting the differential equations to a large set of simultaneous nonlinear algebraic equations. To demonstrate this method, we use, as before, the set of two differential equations:

$$\frac{dy_1}{dx} = f_1(x, y_1, y_2)$$
$$\frac{dy_2}{dx} = f_2(x, y_1, y_2) \tag{6.12}$$

with split boundary conditions:

$$y_1(x_0) = y_{1,0} \tag{6.13}$$

$$y_2(x_f) = y_{2,f} \tag{6.14}$$

Next we express the derivatives of y in terms of forward finite differences using Eq. (4.33):

$$\frac{dy_{1,i}}{dx} = \frac{1}{h}\left(y_{1,i+1} - y_{1,i}\right) + O(h) \tag{6.30a}$$

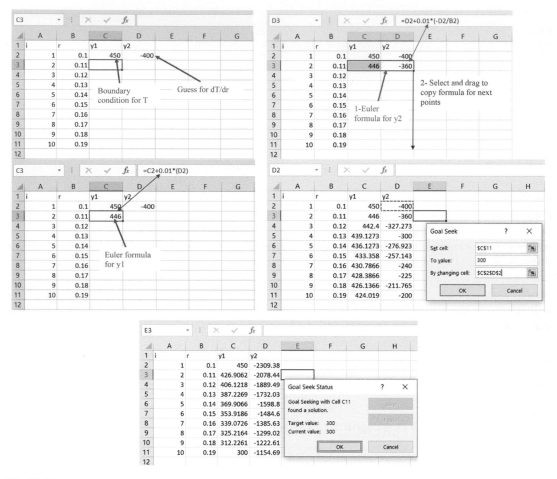

FIG. E6.2b

Calculating the temperature profile in a hollow disc by applying the shooting method in Excel.

$$\frac{dy_{2,i}}{dx} = \frac{1}{h}\left(y_{2,i+1} - y_{2,i}\right) + O(h) \tag{6.30b}$$

For higher accuracy, we could have used Eq. (4.38), which has an error of order (h^2), instead of Eq. (4.33). In either case the steps of obtaining the solution to the boundary value problem are identical.

Combining Eqs. (6.30) with (6.12) we obtain:

$$y_{1,i+1} - y_{1,i} = hf_1(x, y_{1,i}, y_{2,i}) \tag{6.31a}$$

$$y_{2,i+1} - y_{2,i} = hf_2(x, y_{1,i}, y_{2,i}) \tag{6.31b}$$

We divide the interval of integration into n segments of equal length and write Eqs. (6.31) for $i = 0, 1, 2, \ldots, n - 1$. These form a set of $2n$ simultaneous nonlinear algebraic equations in $(2n + 2)$ variables. The two boundary conditions provide values for two of these variables:

$$y_1(x_0) = y_{1,0} \tag{6.13}$$

$$y_2(x_f) = y_{2,f} \tag{6.14}$$

Therefore, the system of $2n$ equations in $2n$ unknown can be solved using Newton's method for simultaneous nonlinear algebraic equations, described in Chapter 1. It should be emphasized, however, that the problem of solving a large set of nonlinear algebraic equations is not a trivial task. It requires, first, a good initial guess of all the values of y_{ij}, and it involves the evaluation of the $(2n \times 2n)$ Jacobian matrix. Kubicek and Hlavacek [3] state that computational experience with the finite-difference technique has shown that for a practical engineering problem, this method is more difficult to apply than the shooting method. They recommend that the finite-difference method be used only for problems that are too unstable to integrate by the shooting methods. On the other hand, if the differential equations are *linear*, the resulting set of simultaneous algebraic equations will also be linear. In such a case the solution can be obtained by straightforward application of matrix inversion or the Gauss elimination procedure.

If we encounter a second-order differential equation we may reduce the number of unknowns by a factor of 2 if replacing the derivatives with appropriate finite difference formulas, and not transforming it to the canonical form. To explain this method, consider the following second-order differential equation:

$$\frac{d^2y}{dx^2} + P(x)\frac{dy}{dx} + Q(x)y = R(x) \tag{6.32}$$

subject to the boundary conditions:

$$a_0 y(x_0) + b_0 y'(x_0) = c_0 \tag{6.10}$$

$$a_f y(x_f) + b_f y'(x_f) = c_f \tag{6.11}$$

We express the second and first derivatives of y in terms of finite differences. Let us use the central difference method this time. These derivatives in terms of central finite differences are:

$$\frac{dy_i}{dx} = \frac{1}{2h}(y_{i+1} - y_{i-1}) + O(h^2) \tag{4.50}$$

$$\frac{d^2y_i}{dx^2} = \frac{1}{h^2}(y_{i+1} - 2y_i + y_{i-1}) + O(h^2) \tag{4.59}$$

Inserting these formulas into Eq. (6.32), we have:

$$\frac{1}{h^2}(y_{i+1} - 2y_i + y_{i-1}) + P(x_i)\frac{1}{2h}(y_{i+1} - y_{i-1}) + Q(x_i)y_i = R(x_i) \tag{6.33}$$

This equation can be rearranged as:

$$\left(\frac{1}{h^2} - \frac{P_i}{2h}\right)y_{i-1} + \left(-\frac{2}{h^2} + Q_i\right)y_i + \left(\frac{1}{h^2} + \frac{P_i}{2h}\right)y_{i+1} = R_i \tag{6.34}$$

Eq. (6.34) can be written for internal points; that is, for $i = 1, 2, \ldots, n - 1$. In other words Eq. (6.34) represents a set of $n - 1$ simultaneous linear equations.

The boundary conditions (6.10) and (6.11) also should be added to this set of equations. For this purpose, we should replace the first-order derivatives in these equations with the appropriate finite-difference formula. At the initial point ($i = 0$) we should use the forward differences and at the end-point ($i = n$) we should use the backward differences. Note that because for the internal points we have used the derivative formulas in terms of central differences with the error of order $O(h^2)$, the first-order derivatives at the initial and endpoints also should be of the same order. Therefore Eq. (4.38) should be used for the initial point:

$$\frac{dy_0}{dx} = \frac{1}{2h}(-y_2 + 4y_1 - 3y_0) + O(h^2) \tag{6.35}$$

and the boundary condition (6.10) will be rearranged as:

$$\left(a_0 - \frac{3b_0}{2h}\right)y_0 + \left(\frac{2b_0}{h}\right)y_1 - \left(\frac{b_0}{2h}\right)y_2 = c_0 \tag{6.36}$$

Similarly Eq. (4.21) should be used for the endpoint:

$$\frac{dy_n}{dx} = \frac{1}{2h}(3y_n - 4y_{n-1} + y_{n-2}) + O(h^2) \tag{6.37}$$

and the boundary condition (6.11) will be rearranged as:

$$\left(\frac{b_f}{2h}\right)y_{n-2} - \left(\frac{2b_f}{h}\right)y_{n-1} + \left(a_f + \frac{3b_f}{2h}\right)y_n = c_f \tag{6.38}$$

Now Eqs. (6.34), (6.36), and (6.38) form a set of n equations in n unknowns:

$$\begin{aligned}
&\left(a_0 - \frac{3b_0}{2h}\right)y_0 + \left(\frac{2}{h}\right)y_1 - \left(\frac{4}{h}\right)y_2 = c_0 \\
&\left(\frac{1}{h^2} - \frac{P_1}{2h}\right)y_0 + \left(-\frac{2}{h^2} + Q_1\right)y_1 + \left(\frac{1}{h^2} + \frac{P_1}{2h}\right)y_2 = R_1 \\
&\qquad\qquad\vdots \\
&\left(\frac{1}{h^2} - \frac{P_i}{2h}\right)y_{i-1} + \left(-\frac{2}{h^2} + Q_i\right)y_i + \left(\frac{1}{h^2} + \frac{P_i}{2h}\right)y_{i+1} = R_i \\
&\qquad\qquad\vdots \\
&\left(\frac{1}{h^2} - \frac{P_{n-1}}{2h}\right)y_{n-2} + \left(-\frac{2}{h^2} + Q_{n-1}\right)y_{n-1} + \left(\frac{1}{h^2} + \frac{P_{n-1}}{2h}\right)y_n = R_{n-1} \\
&\left(\frac{b_f}{2h}\right)y_{n-2} - \left(\frac{2b_f}{h}\right)y_{n-1} + \left(a_f + \frac{3b_f}{2h}\right)y_n = c_f
\end{aligned} \tag{6.39}$$

Owing to the form of the second-order differential Eq. (6.32), this is a set of linear algebraic equations that can be solved by one of the methods described in Chapter 2.

EXAMPLE 6.3 USING THE FINITE-DIFFERENCE METHOD FOR DETERMINING THE TEMPERATURE DISTRIBUTION IN A HOLLOW DISC

Repeat Example 6.2 and determine the temperature distribution in the hollow disc shown in Fig. E6.2a by the method of finite difference. Consider $h = 0.01$ m.

Method of solution

Expanding the differential equation of heat transfer and dividing it by r results in:

$$\frac{d^2T}{dr^2} + \frac{1}{r}\frac{dT}{dr} = 0 \tag{1}$$

Replacing the derivatives in this equation with the derivatives in terms of central finite differences of the error of order $O(h^2)$, we get:

$$\left(\frac{1}{h^2} - \frac{1}{2hr_i}\right)T_{i-1} - \left(\frac{2}{h^2}\right)T_i + \left(\frac{1}{h^2} + \frac{1}{2hr_i}\right)T_{i+1} = 0 \quad i = 1, 2, \ldots, 9 \tag{2}$$

Note that by comparing Eq. (1) with Eq. (6.32), we notice that $P = r$ and $Q = R = 0$, thus Eq. (2) can be also obtained from Eq. (6.34).

Because $h = 0.01$ and r varies from 0.1 to 0.2, there are 10 segments and 11 points in this problem. Therefore Eq. (2) represents 9 equations ($i = 1, 2, \ldots, 9$).

Moreover, the boundary conditions for this problem are:

$$T_0 = 450 \text{ K} \quad \text{and} \quad T_{10} = 300 \text{ K}$$

Therefore the set of algebraic equations for calculating the temperature profile in the hollow disc is:

$$\begin{bmatrix} 1 & 0 & 0 & \cdots & 0 & 0 & 0 \\ \left(\frac{1}{h^2} - \frac{1}{2hr_1}\right) & -\left(\frac{2}{h^2}\right) & \left(\frac{1}{h^2} + \frac{1}{2hr_1}\right) & \cdots & 0 & 0 & 0 \\ & \vdots & & \ddots & & \vdots & \\ 0 & 0 & 0 & \cdots & \left(\frac{1}{h^2} - \frac{1}{2hr_9}\right) & -\left(\frac{2}{h^2}\right) & \left(\frac{1}{h^2} + \frac{1}{2hr_9}\right) \\ 0 & 0 & 0 & \cdots & 0 & 0 & 1 \end{bmatrix} \begin{bmatrix} T_0 \\ T_1 \\ \vdots \\ T_9 \\ T_{10} \end{bmatrix} = \begin{bmatrix} 450 \\ 0 \\ \vdots \\ 0 \\ 300 \end{bmatrix}$$

It is also easy to show that:

$$r_i = 0.1 + ih \quad i = 0, 1, \ldots, 10$$

(Continued)

EXAMPLE 6.3 (CONTINUED)

Considering $h = 0.01$, we can evaluate the elements of the matrix of coefficients and solve this set of equations by one of the methods described in Chapter 2. Here, we only present the results after the solution in Table E6.3.

The temperature profile is shown graphically in Fig. E6.3.

Table E6.3 Temperature profile in the hollow disc.

r (m)	T (K)	r (m)	T (K)
0.10	450	0.16	348.30
0.11	429.38	0.17	335.18
0.12	410.55	0.18	322.81
0.13	393.23	0.19	311.10
0.14	377.20	0.20	300
0.15	362.27		

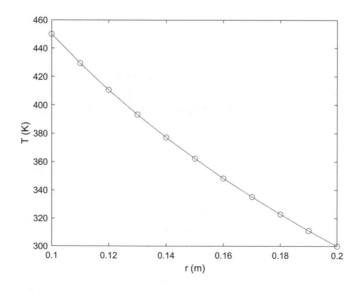

FIG. E6.3

Temperature profile in the hollow disc.

6.4 **COLLOCATION METHODS**

These methods are based on the concept of interpolation of unequally spaced points, that is, choosing a function, usually a polynomial, that approximates the solution of a differential equation in the range of integration, $x_0 \leq x \leq x_f$, and determining the coefficients of that function from a set of base points.

Let us again consider the set of two differential equations:

$$\frac{dy_1}{dx} = f_1(x, y_1, y_2)$$
$$\frac{dy_2}{dx} = f_2(x, y_1, y_2) \tag{6.12}$$

with split boundary conditions:

$$y_1(x_0) = y_{1,0} \tag{6.13}$$

$$y_2(x_f) = y_{2,f} \tag{6.14}$$

Suppose that the solutions $y_1(x)$ and $y_2(x)$ of Eq. (6.12) can be approximated by the following polynomials which we call *trial functions*:

$$y_1(x) \simeq P_{1,n}(x) = c_{1,0} + c_{1,1}x + c_{1,2}x^2 + \cdots + c_{1,n}x^n \tag{6.40a}$$

$$y_2(x) \simeq P_{2,n}(x) = c_{2,0} + c_{2,1}x + c_{2,2}x^2 + \cdots + c_{2,n}x^n \tag{6.40b}$$

We take the derivatives of both sides of Eq. (6.40) and substitute them in Eq. (6.12):

$$P'_{m,n}(x) \simeq f_m\left(x, P_{1,n}(x), P_{2,n}(x)\right) \quad m = 1, 2 \tag{6.41}$$

We then form the residuals:

$$R_m(x) = P'_{m,n}(x) - f_m\left(x, P_{1,n}(x), P_{2,n}(x)\right) \quad m = 1, 2 \tag{6.42}$$

The objective is to determine the coefficients $\{c_{m,i} \mid i = 0, 1, \ldots, n; \; m = 1, 2\}$ of the polynomials $P_{m,n}(x)$ so that to make the residuals as small as possible over the range of integration of the differential equation. This is accomplished by making the following integral vanish:

$$\int_{x_0}^{x_f} W_k R_m(x)dx = 0 \tag{6.43}$$

where W_k is the weighting functions to be chosen. This technique is called the *method of weighted residuals*.

The collocation method chooses the weighting functions to be the *Dirac delta* (*unit impulse*) function:

$$W_k = \delta(x - x_k) \quad x_0 \leq x_k \leq x_f \tag{6.44}$$

which has the property that:

$$\int_{x_0}^{x_f} a(x)\delta(x - x_k)dx = a(x_k) \tag{6.45}$$

Therefore the integral (6.43) becomes:

$$\int_{x_0}^{x_f} W_k R_m(x) dx = R_m(x_k) = 0 \tag{6.46}$$

Combining Eqs. (6.42) and (6.46) we have:

$$P'_{m,n}(x_k) - f\left(x_k, P_{1,n}(x_k), P_{2,n}(x_k)\right) \qquad m = 1, 2 \tag{6.47}$$

This implies that at a given number of *collocation points*, $\{x_k \mid k = 0, 1, \ldots, n\}$, the coefficients of the polynomials (6.40) are chosen so that Eq. (6.47) is satisfied, that is, the polynomials are *exact* solutions of the differential equations at those collocation points (note that $x_n = x_f$). The larger the number of collocation points, the closer the trial function would resemble the true solution $y_m(x)$ of the differential equations.

Eq. (6.47) contains the $(2n + 2)$ yet-to-be-determined coefficients $\{c_{m,i} \mid i = 0, 1, \ldots, n; m = 1, 2\}$ of the polynomials. These can be calculated by choosing $(2n + 2)$ collocation points. Because it is necessary to satisfy the boundary conditions of the problem, two collocation points are already fixed in this case of boundary value problem. At $x = x_0$:

$$y_1(x_0) = y_{1,0} = c_{1,0} + c_{1,1}x_0 + \cdots + c_{1,n}x_0^n = \sum_{i=0}^{n} c_{1,i}x_0^i \tag{6.48}$$

and at $x = x_f$:

$$y_2(x_f) = y_{2,f} = c_{2,0} + c_{2,1}x_f + \cdots + c_{2,n}x_f^n = \sum_{i=0}^{n} c_{2,i}x_f^i \tag{6.49}$$

Therefore we have the freedom to choose the remaining $(2n)$ internal collocation points and then write Eq. (6.47) for each of these points:

$$\begin{aligned} P'_{1,n}(x_1) - f_1\left(x_1, P_{1,n}(x_1), P_{2,n}(x_1)\right) &= 0 \\ &\vdots \\ P'_{1,n}(x_n) - f_1\left(x_n, P_{1,n}(x_n), P_{2,n}(x_n)\right) &= 0 \end{aligned} \tag{6.50a}$$

$$\begin{aligned} P'_{2,n}(x_0) - f_2\left(x_0, P_{1,n}(x_0), P_{2,n}(x_0)\right) &= 0 \\ &\vdots \\ P'_{2,n}(x_{n-1}) - f_2\left(x_{n-1}, P_{1,n}(x_{n-1}), P_{2,n}(x_{n-1})\right) &= 0 \end{aligned} \tag{6.50b}$$

Note that we have also written Eq. (6.47) for $x = x_f = x_n$ in Eq. (6.50a) and for $x = x_0$ in Eq. (6.50b) because the values $y_{1,f}$ and $y_{2,0}$ are yet unknown. Eqs. (6.48)–(6.50) constitute a complete set of $(2n + 2)$ simultaneous nonlinear equations in $(2n + 2)$ unknowns. The solution to this problem requires the application of Newton's method for simultaneous nonlinear equations (see Chapter 1).

If the collocation points are chosen at equidistant intervals within the interval of integration, then the collocation method is equivalent to the polynomial interpolation of equally spaced points and the finite-difference method. This is not at all surprising because the development of interpolating polynomials and finite differences were all based on expanding the function in the Taylor series (see Chapter 3). It is not necessary, however, to choose the collocation points at equidistant intervals. In fact, it is more advantageous to locate the collocation points at the roots of appropriate orthogonal polynomials, as the following discussion shows.

The *orthogonal collocation method*, which is an extension of the method just described, provides a mechanism for automatically picking the collocation points by making use of orthogonal polynomials. This method chooses the trial functions $y(x)$ and $y(x)$ to be the linear combination:

$$y_m(x) = \sum_{i=0}^{n+1} a_{m,i} P_i(x) \quad m = 1, 2 \tag{6.51}$$

of a series of orthogonal polynomials $P_i(x)$:

$$\begin{aligned}
P_0(x) &= c_{0,0} \\
P_1(x) &= c_{1,0} + c_{1,1}x \\
P_2(x) &= c_{2,0} + c_{2,1}x + c_{2,2}x^2 \\
&\vdots \\
P_i(x) &= c_{i,0} + c_{i,1}x + c_{i,2}x^2 + \cdots + c_{i,i}x^i
\end{aligned} \tag{6.52}$$

This set of polynomials can be written in a condensed form:

$$P_i(x) = \sum_{k=0}^{i} c_{ik}x^k \quad i = 0, 1, \ldots, n+1 \tag{6.53}$$

The coefficients c_{ik} are chosen so that the polynomials obey the orthogonality condition defined in the Appendix:

$$\int_a^b w(x)P_i(x)P_j(x)dx = 0 \quad i \neq j \tag{6.54}$$

When $P_i(x)$ is chosen to be the *Legendre* set of orthogonal polynomials (see Table A.1 in the Appendix), the weight $w(x)$ is unity. The standard interval of integration for Legendre polynomials is $[-1, 1]$. The transformation equation (4.92) is used to transform the Legendre polynomials to the interval $[x_0, x_f]$, which applies to our problem at hand:

$$x = \frac{(x_f - x_0)}{2} z + \frac{(x_f + x_0)}{2} \tag{6.55}$$

Eq. (6.55) relates the variables x and z so that every value of x in the interval $[x_0, x_f]$ corresponds to a value of z in the interval $[-1, 1]$ and vice versa. Therefore, using x or z as independent variables is equivalent. Hereafter, we use z as the independent variable of the Legendre polynomials to stress that the domain under study is the interval $[-1, 1]$. The derivatives with respect to x and z are related to each other by the following relation:

$$\frac{dy_m}{dx} = \frac{2}{(x_f - x_0)} \frac{dy_m}{dz} \tag{6.56}$$

The two-point boundary value problem given by Eqs. (6.12)–(6.14) has $(n + 2)$ collocation points, $\{z_j \mid j = 0, 1, \ldots, n + 1\}$, including the two known boundary values ($z_0 = -1$ and $z_{n+1} = 1$). The location of the n internal collocation points (z_1 to z_n) are determined from the roots of the polynomial $P_n(z) = 0$. The coefficients $a_{m,i}$ in Eq. (6.51) must be determined so that the boundary conditions are satisfied. Eq. (6.51) can be written for the $(n + 2)$ points (z_0 to z_{n+1}) as:

$$y_1(z_j) = \sum_{i=0}^{n+1} d_{1,i} z_j^i \tag{6.57a}$$

$$y_2(z_j) = \sum_{i=0}^{n+1} d_{2,i} z_j^i \tag{6.57b}$$

where the terms of the polynomials have been regrouped. Eqs. (6.57) may be presented in matrix notation as:

$$\mathbf{y}_1 = \mathbf{Q}\mathbf{d}_1 \tag{6.58a}$$

$$\mathbf{y}_2 = \mathbf{Q}\mathbf{d}_2 \tag{6.58b}$$

where $\mathbf{d}_1$ and $\mathbf{d}_2$ are the matrices of coefficients and

$$Q_{j+1,i+1} = z_j^i \qquad \begin{cases} i = 0, 1, \ldots, n+1 \\ j = 0, 1, \ldots, n+1 \end{cases} \tag{6.59}$$

Solving Eqs. (6.58) for $\mathbf{d}_1$ and $\mathbf{d}_2$, we find:

$$\mathbf{d}_1 = \mathbf{Q}^{-1}\mathbf{y}_1 \tag{6.60a}$$

$$\mathbf{d}_2 = \mathbf{Q}^{-1}\mathbf{y}_2 \tag{6.60b}$$

The derivatives of y's are taken as:

$$\frac{dy_1(z_j)}{dz} = \sum_{i=0}^{n+1} d_{1,i} i z_j^{i-1} \tag{6.61a}$$

$$\frac{dy_2(z_j)}{dz} = \sum_{i=0}^{n+1} d_{2,i} i z_j^{i-1} \tag{6.61b}$$

which in matrix form become:

$$\frac{d\mathbf{y}_1}{dz} = \mathbf{C}\mathbf{d}_1 = \mathbf{C}\mathbf{Q}^{-1}\mathbf{y}_1 = \mathbf{A}\mathbf{y}_1 \tag{6.62a}$$

$$\frac{d\mathbf{y}_2}{dz} = \mathbf{C}\mathbf{d}_2 = \mathbf{C}\mathbf{Q}^{-1}\mathbf{y}_2 = \mathbf{A}\mathbf{y}_2 \tag{6.62b}$$

where

$$C_{j+1,i+1} = i z_j^{i-1} \qquad \begin{cases} i = 0, 1, \ldots, n+1 \\ j = 0, 1, \ldots, n+1 \end{cases} \tag{6.63}$$

The two-point boundary value problem of Eq. (6.12) can now be expressed in terms of the orthogonal collocation method as:

or
$$\begin{aligned} \mathbf{A}\mathbf{y}_1 &= \mathbf{f}_1(z, \mathbf{y}_1, \mathbf{y}_2) \\ \mathbf{A}\mathbf{y}_2 &= \mathbf{f}_2(z, \mathbf{y}_1, \mathbf{y}_2) \end{aligned} \tag{6.64}$$

$$\sum_{i=0}^{n+1} A_{ij} y_{1,j} = f_1(z_j, y_{1,j}, y_{2,j}) \tag{6.65a}$$

$$\sum_{i=0}^{n+1} A_{ij} y_{2,j} = f_2(z_j, y_{1,j}, y_{2,j}) \tag{6.65b}$$

with the boundary conditions:

$$y_1(z_0) = y_{1,0} \quad \text{and} \quad y_{2,n+1} = y_2(z_f) = y_{2,f} \tag{6.66}$$

Eqs. (6.65) and (6.66) constitute a set of $(2n + 4)$ simultaneous nonlinear equations whose solution can be obtained using Newton's method for nonlinear equations. It is possible to combine Eqs. (6.65) and present them in matrix form:

$$A_2 Y = F \tag{6.67}$$

where

$$A_2 = \begin{bmatrix} A & 0 \\ 0 & A \end{bmatrix} \tag{6.68}$$

$$Y = \begin{bmatrix} y_1 \\ y_2 \end{bmatrix} = [y_{1,0}, \ldots, y_{1,n+1}, y_{2,0}, \ldots, y_{2,n+1}]' \tag{6.69}$$

$$F = \begin{bmatrix} f_1 \\ f_2 \end{bmatrix} = \begin{bmatrix} f_1(z_0, y_{1,0}, y_{2,0}) \\ \vdots \\ f_1(z_{n+1}, y_{1,n+1}, y_{2,n+1}) \\ f_2(z_0, y_{1,0}, y_{2,0}) \\ \vdots \\ f_2(z_{n+1}, y_{1,n+1}, y_{2,n+1}) \end{bmatrix} \tag{6.70}$$

The bold zeros in Eq. (6.68) are zero matrices of size $(n + 2) \times (n + 2)$; that is the same size as that of matrix A.

It should be noted that Eq. (6.67) is solved for the unknown collocation points which means that we should exclude the equations corresponding to the boundary conditions. In the problem described earlier, the first and the last equations in the set of Eq. (6.67) will not be used because the corresponding dependent values are determined by a boundary condition rather than the collocation method.

The foregoing formulation of the solution for a two-equation boundary value problem can be extended to the solution of m simultaneous fist-order ordinary differential equations. For this purpose, we define the following matrices:

$$A_m = \begin{bmatrix} A & 0 & \cdots & 0 \\ 0 & A & \cdots & 0 \\ \cdots & \cdots & \cdots & \cdots \\ 0 & 0 & \cdots & A \end{bmatrix} \tag{6.71}$$

$$Y = [y_1, y_2, \ldots, y_m]' \tag{6.72}$$

$$F = [f_1, f_2, \ldots, f_m]' \tag{6.73}$$

Note that matrix A in Eq. (6.71) is defined by Eq. (6.62) and appears m times on the diagonal of the matrix A_m. The values of the dependent variables $\{y_{ij} \mid i = 1, 2, \ldots, m; j = 0, 2, \ldots, n + 1\}$ are then evaluated from the simultaneous solution of the following set of nonlinear equations plus boundary conditions:

$$A_m Y - F = 0 \tag{6.74}$$

The equations corresponding to the boundary conditions have to be excluded from Eq. (6.74) at the time of solution.

If the problem to be solved is a second-order two-point boundary value problem in the form:

$$y'' = f(x, y, y') \tag{6.75}$$

with the boundary conditions:

$$y(x_0) = y_0 \quad \text{and} \quad y(x_f) = y_f \tag{6.76}$$

we may follow the similar approach as described earlier and approximate the function $y(x)$ at $(n + 2)$ points, after transforming the independent variable from x to z, as:

$$y(z_j) = \sum_{i=0}^{n+1} d_i z_j^i \tag{6.77}$$

The derivatives of y are then taken as:

$$\frac{dy(z_j)}{dz} = \sum_{i=0}^{n+1} d_i i z_j^{i-1} \tag{6.78}$$

$$\frac{d^2 y(z_j)}{dz^2} = \sum_{i=0}^{n+1} d_i i(i - 1) z_j^{i-2} \tag{6.79}$$

These equations can be written in matrix form:

$$\frac{dy}{dz} = CQ^{-1}y = Ay \tag{6.80}$$

$$\frac{d^2 y}{dz^2} = DQ^{-1}y = By \tag{6.81}$$

where

$$D_{j+1,i+1} = i(i - 1)z_j^{i-2} \quad \begin{cases} i = 0, 1, \ldots, n + 1 \\ j = 0, 1, \ldots, n + 1 \end{cases} \tag{6.82}$$

The two-point boundary value problem of Eq. (6.75) can now be expressed in terms of the orthogonal collocation method as:

$$By = f(z, y, Ay) \tag{6.83}$$

Eq. (6.83) represents a set of $(n + 2)$ simultaneous nonlinear equations, two of which correspond to the boundary conditions (the first and the last equation) and should be neglected when solving the set. The solution of the remaining n nonlinear equations can be obtained using Newton's method for nonlinear equations.

The orthogonal collocation method is more accurate than either the finite-difference method or the collocation method. The choice of collocation points at the roots of the orthogonal polynomials reduces the error considerably. In fact, instead of the user choosing the collocation points, the method locates them automatically such that the best accuracy is achieved.

EXAMPLE 6.4 SOLUTION OF THE OPTIMAL TEMPERATURE PROFILE FOR PENICILLIN FERMENTATION

Apply the orthogonal collocation method to solve the two-point boundary value problem arising from the application of the *maximum principle of Pontryagin* to a batch penicillin fermentation. Obtain the solution to this problem and show the profiles of the state variables, the adjoint variables, and the optimal temperature. The equations which describe the state of the system in a batch penicillin fermentation developed by Constantinides et al. [6] are:

Cell mass production:

$$\frac{dy_1}{dt} = b_1 y_1 - \frac{b_1}{b_2} y_1^2 \quad y_1(0) = 0.03 \tag{1}$$

Penicillin synthesis:

$$\frac{dy_2}{dt} = b_3 y_1 \quad y_2(0) = 0.0 \tag{2}$$

where

y_1 = dimensionless concentration of cell mass
y_2 = dimensionless concentration of penicillin
t = dimensionless time, $0 \le t \le 1$.

The parameters b_i are functions of temperature θ:

$$\begin{aligned}
b_1 &= w_1 \left[\frac{1.0 - w_2(\theta - w_3)^2}{1.0 - w_2(25 - w_3)^2} \right] \\
b_2 &= w_4 \left[\frac{1.0 - w_2(\theta - w_3)^2}{1.0 - w_2(25 - w_3)^2} \right] \\
b_3 &= w_5 \left[\frac{1.0 - w_2(\theta - w_6)^2}{1.0 - w_2(25 - w_6)^2} \right] \\
b_i &\ge 0
\end{aligned} \tag{3}$$

where $w_1 = 13.1$ (value of b_1 at 25°C obtained from fitting the model to experimental data)

$w_2 = 0.005$
$w_3 = 30°C$
$w_4 = 0.94$ (value of b_2 at 25°C)
$w_5 = 1.71$ (value of b_3 at 25°C)
$w_6 = 20°C$
θ = temperature, °C

These parameter-temperature functions are inverted parabolas that reach their peak at 30°C, for b_1 and b_2, and at 20°C, for b_3. The values of the parameters decrease by a factor of 2 over a 10°C change in temperature on either side of the peak. The inequality, $b_i \ge 0$, restricts the

(Continued)

EXAMPLE 6.4 (CONTINUED)

values of the parameters to the positive regime. These functions have shapes typical of those encountered in microbial or enzyme-catalyzed reactions.

The maximum principle has been applied to the foregoing model to determine the optimal temperature profile [7] which maximizes the concentration of penicillin at the final time of the fermentation, $t_f = 1$. The maximum principle algorithm when applied to the state Eqs. (1) and (2) yields the following additional equations:

The adjoint equations:

$$\frac{dy_3}{dt} = -b_1 y_3 + 2\frac{b_1}{b_2} y_1 y_3 - b_3 y_4 \quad y_3(1) = 0 \tag{4}$$

$$\frac{dy_4}{dt} = 0 \quad y_4(1) = 1.0 \tag{5}$$

The Hamiltonian:

$$H = y_3 \left(b_1 y_1 - \frac{b_1}{b_2} y_1^2 \right) + y_4 (b_3 y_1)$$

The necessary condition for maximum:

$$\frac{\partial H}{\partial \theta} = 0 \tag{6}$$

Eqs. (1)−(6) form a two-point boundary value problem. Apply the orthogonal collocation method to obtain the solution to this problem and show the profiles of the state variables, the adjoint variables, and the optimal temperature.

Method of solution

The fundamental numerical problem of optimal control theory is the solution of the two-point boundary value problem, which invariably arises from the application of the maximum principle to determine optimal control profiles. The state and adjoint equations, coupled together through the necessary condition for optimality, constitute a set of simultaneous differential equations that are often unstable. This difficulty is further complicated, in certain problems, when the necessary condition is not solvable explicitly for the control variable θ. Several numerical methods have been developed for the solution of this class of problems.

We first consider the second adjoint equation, Eq. (5), which is independent of the other variables and, therefore, may be integrated directly:

$$y_4 = 1 \quad 0 \le t \le 1 \tag{7}$$

This reduces the number of differential equations to be solved by one. The remaining three differential equations, Eqs. (1), (2), and (4), are solved by Eq. (6.74) where $m = 3$.

(Continued)

EXAMPLE 6.4 (CONTINUED)

Finally, we express the necessary condition (Eq. 6) in terms of the system variables:

$$\frac{\partial H}{\partial \theta} = y_3 \left[y_1 \left(\frac{\partial b_1}{\partial \theta} \right) - y_1^2 \frac{\partial (b_1/b_2)}{\partial \theta} \right] + y_1 y_4 \left(\frac{\partial b_3}{\partial \theta} \right) = 0 \tag{8}$$

The temperature θ can be calculated from Eq. (8) once the system parameters have been determined.

Program description

The programs and functions developed in this example can be found in: https://www.elsevier.com/books-and-journals/book-companion/9780128229613

The MATLAB function *collocation.m* is developed to solve a set of first-order ordinary differential equations in a boundary value problem by the orthogonal collocation method. It starts with checking the input arguments and assigning the default values, if necessary. The number of guessed initial conditions has to be equal to the number of final conditions and also the number of equations should be equal to the total number of boundary conditions (initial and final). If these conditions are not met the function gives a proper error message on the screen and stops execution.

In the next section the function builds the coefficients of the Lagrange polynomial and finds its roots, z_j. The vector of x_j is then calculated from Eq. (6.55). The function applies Newton's method for the solution of the set of nonlinear Eq. (6.74). Therefore the starting values for this technique are generated by the second-order Runge-Kutta method, using the guessed initial conditions. The function continues with building the matrices Q, C, A, A_m, and vectors Y and F.

Just before entering Newton's technique iteration loop, the function keeps track of the equations to be solved, that is, all the equations excluding those corresponding to the boundary conditions. The last part of the function is the solution of the set of Eq. (6.74) by Newton's method. This procedure begins with evaluating the differential equations function values followed by calculating the Jacobian matrix, by differentiating using the forward finite-differences method and, finally, correcting the dependent variables. This procedure is repeated until the convergence is reached at all the collocation points.

It is important to note that the function *collocation.m* requires to receive the values of the set of ordinary differential equations at each point in a column vector with the initial value equations at the top, followed by the final value equations. It is also important to pass the initial and final conditions to the function in the order which corresponds to the order of equations appearing in the file which introduces the ordinary differential equations.

The main program *Example6_4.m* asks the reader to input the parameters required for the solution of the problem. The program then calls the function *collocation* to solve the set of equations. Knowing the system variables, the program calls the function *fzero* to find the temperature at each point. In the end, the program plots the calculated cell

(Continued)

EXAMPLE 6.4 (CONTINUED)

concentration, penicillin concentration, first adjoint variable, and the temperature against time.

The function *Ex6_4_func.m* evaluates the values of the set of Eqs. (1), (2), and (4) at a given point. It is important to note that the first input argument to *Ex6_4_func* is the independent variable although it does not appear in the differential equations in this case. This function also calls the MATLAB function *fzero* to calculate the temperature from Eq. (8) which is introduced in the function *Ex6_4_theta.m*.

Input and results

>> **Example6_4**
Enter w's as a vector: [13.1, 0.005, 30, 0.94, 1.71, 20]
Vector of known initial conditions = [0.03, 0]
Vector of final conditions = 0
Vector of guessed initial conditions = 3
M-file containing the set of differential equations: 'Ex6_4_func'
M-file containing the necessary condition function: 'Ex6_4_theta'
Number of internal collocation points = 10
Relaxation factor = 0.9
Integrating.
Iteration 1
Iteration 2
Iteration 3
Iteration 4
Iteration 5
Iteration 6
Iteration 7
Iteration 8
Iteration 9
Iteration 10
Iteration 11

Discussion of Results

The choice of the value of the missing initial condition for y_3 is an important factor in the convergence of the Newton method because it generates the starting values for the technique. The value of $y_3(0) = 3$ was chosen as the guessed initial condition after some trial and error. The Newton method converged to the correct solution in 11 iterations.

Figs. E6.4a to E6.4d show the profiles of the system variables and the optimal control variable (temperature). For this particular formulation of the penicillin fermentation, the maximum principle indicates that the optimal temperature profile varies from 30°C to 20°C in the pattern shown in Fig. E6.4d.

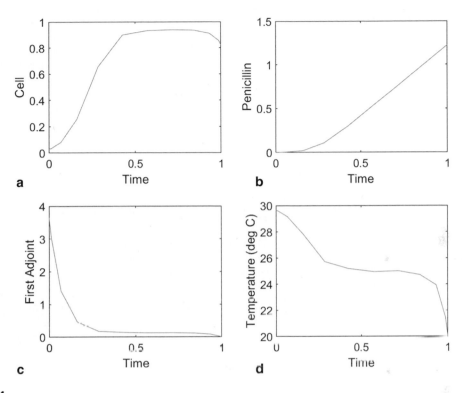

FIG. E6.4

Profiles of the system variables and the optimal control variable for penicillin fermentation.

6.5 USING BUILT-IN MATLAB® FUNCTIONS

MATLAB has two solvers for boundary value problems, *bvp4c* and *bvp5c*. Both these functions are finite-difference codes and use the collocation method to solve the problem. The difference between these two solvers is the accuracy which is fourth-order and fifth-order accurate, respectively. The statement *sol* = *bvp4c*('*y_prime*', '*bc_func*', *guess*) solves the set of ordinary differential equations described in the MATLAB function *y_prime.m*, subject to split boundary conditions given in the function *bc_func.m*. The function returns *sol* that is a structure in which there is the vector *x* containing the values of the independent variable, the matrix *y* which is the matrix of dependent variables at *x*-values and *yp* which provides derivatives of *y*'s. Also, *guess* is a structure containing coordinates of the initial solution guess. It is possible to use *bvpint.m* to generate the initial solution *guess*. The syntax for using *bvp5c* is the same.

6.6 SUMMARY

A boundary value problem is a higher-order equation differential equation or a set of differential equations for which the boundary conditions are given at two or more points of the independent variable. Such a problem is encountered in many chemical engineering applications, such as diffusion and reaction in a catalyst, heat conduction in a fin, and transport phenomena in boundary layers.

A simple algorithm for solving a boundary value problem is the shooting method in which the boundary value problem is transformed into an initial value problem by guessing the unknown initial values. After solving this initial value problem (with a method described in the previous chapter), the final values are compared with those given by the problem and the initial guesses are corrected, if necessary. The iteration continues until all initial values result in the final values determined by the problem.

Another technique for solving a boundary value problem is the finite-difference method. In this method, the derivatives are replaced by their finite-difference approximations. Applying this replacement for all nodes of the grid results in a set of algebraic equations, by the solution of which the values of the dependent variable will be determined at all grid points.

Collocation is another technique for solving boundary value problems. In this method, the function values at the grid points are approximated by a polynomial. Using the method of weighted residuals, the coefficients of this polynomial are determined such that the polynomial becomes the exact solution of the differential equation at the collocation points. In the orthogonal collocation, the collocation points are picked up as the roots of the orthogonal polynomials. The orthogonal collocation method is more accurate than either the finite-difference method or the collocation method.

At the end of the chapter, built-in MATLAB functions which solve boundary value problems are described.

PROBLEMS

6.1 Solve the following boundary value problem:

$$x^2 \frac{d^2y}{dx^2} + (1 - x^2)\frac{dy}{dx} + xy = 0$$
$$x = 1; \quad y = 1$$
$$x = 2; \quad y = 0$$

6.2 A cone-shaped fin, as shown in Fig. P6.2, is in a fluid with the temperature T_a. The temperature of the base of the cone is T_0. The convective heat transfer coefficient between the cone and surrounding is h and the radial profile of the temperature in the cone can be neglected.

Show that the steady-state equation governing the heat transfer in this problem is the following:

$$\frac{1}{S}\frac{d}{dx}\left(S\frac{dT}{dx}\right) - \frac{hP}{KS}(T - T_a) = 0$$

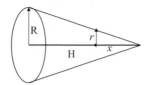

FIG. P6.2

Cone fin.

in which S and P are the cross-sectional area and periphery of the cone at distance x from the tip, respectively. Write the boundary conditions for this problem. Solve the equation using the following values:

$$R = 2 \text{ mm}, \quad H = 6 \text{ cm}, \quad h = 40 \text{ W/m}^2 \cdot \text{K} \quad K = 0.002 \text{ W/m} \cdot \text{K} \quad T_a = 30°\text{C} \quad T_H = 90°\text{C}$$

6.3 Consider the problem of diffusion and reaction in a slab catalyst with second-order kinetics. Show that the mass balance equation written in nondimensional form is:

$$\frac{d^2y}{dx^2} - \varphi^2 y^2 = 0$$

$$x = 0; \quad \frac{dy}{dx} = 0$$

$$x = 1; \quad y = 1$$

where

y = nondimensional concentration (C/C_0)

x = nondimensional distance

$\varphi^2 = kL^2/D_e = 0.5$

k = reaction rate constant

$2L$ = width of the slab

C_0 = concentration of the reactant at the exterior surface of the slab

D_e = effective diffusivity of the reactant in the slab.

Find the dimensionless concentration profile (y) by the shooting method. Use the Euler method in forward integration of the differential equations. Plot the dimensionless concentration against the dimensionless distance (x).

6.4 The mass balance equation for the problem of diffusion and reaction in a catalyst slab is given in nondimensional form as:

$$\frac{d^2y}{dx^2} - \frac{\varphi^2 y}{1 + \delta y + \gamma y^2} = 0$$

$$x = 0; \quad \frac{dy}{dx} = 0$$

$$x = 1; \quad y = 1$$

Find the dimensionless concentration profile (y) by the shooting method. Use the second-order Runge-Kutta method in forward integration of the differential equations. Plot the dimensionless concentration against the dimensionless distance (x).

Additional data: $\varphi^2 = 1.5$, $\delta = 0.2$, $\gamma = 0.01$

6.5 Consider a pressure vessel that is being tested in the laboratory to check its ability to withstand pressure. For a thick pressure vessel of inner radius a and outer radius b, the differential equation for the radial displacement u of a point along the thickness is given by:

$$\frac{d^2u}{dr^2} + \frac{1}{r}\frac{du}{dr} - \frac{u}{r^2} = 0$$

In this problem $u(a)$ and $u(b)$ are known. Solve this boundary value problem by the finite-difference method. Consider $a = 125$ mm, $b = 200$ mm, $u(a) = 0.1$ mm and $u(b) = 0.08$ mm.

6.6 A spherical catalyst is placed in a fluid. Component A diffuses radially into the catalyst and reacts with a second-degree reaction rate. Derive the steady-state mass balance for component A inside the catalyst. Write the boundary conditions for this problem and show that it is a boundary value problem. Consider constant concentration of component A at the surface of the catalyst pellet. Determine the radial distribution of concentration by the collocation method. Diffusivity of A in the catalyst is 3×10^{-6} m^2/s, the concentration of A at the cylinder surface is 0.2 mol/m^3 and the reaction rate constant is 2.5 m^3/mol.s. The radius of the catalyst pellet is 3 mm.

6.7 Steady-state temperature distribution in a metallic rod is described by the following equation:

$$\frac{d^2\theta}{dx^2} - 2\theta^{3/2} = 0$$

$$x = 0, \quad \frac{d\theta}{dx} = 0$$

$$x = 1, \quad \theta = 20$$

Evaluate the temperature along the rod.

6.8 For obtaining the exact velocity profile of fluid in the boundary layer near a semi-infinite flat plate, it is necessary to solve the following ordinary differential equation [1]:

$$f''' = -ff''$$
$$\eta = 0, \quad f = f' = 0$$
$$\eta \to \infty, \quad f' = 1$$

Solve this equation by the shooting method.

6.9 A chemical reaction of order n takes place in a tubular reactor of length L through which the fluid is flowing at a constant velocity u and axial mixing occurs with a dispersion coefficient D. At steady state, the fractional conversion X of the reactant along the reactor can be determined from solving the following dimensionless equation [5]:

$$\frac{1}{Pe}\frac{d^2X}{dz^2} - \frac{dX}{dz} + Da(1-X)^n = 0$$

where Pe is the Peclet number, Da is the Damköhler number and z is the dimensionless length of the reactor. This differential equation can be solved subject to the Danckwerts boundary conditions:

$$z = 0, \quad X = 0$$
$$z = 1, \quad \frac{dX}{dz} = 0$$

What is the converting at the outlet if $Pe = 10$, $Da = 1$ and reaction order is $n = 1.5$? Solve the equation by the orthogonal collocation method.

REFERENCES

[1] Bird, R. B.; Stewart, W. E.; Lightfoot, E. N. *Transport Phenomena*, 2nd ed.; John Wiley & Sons: Hoboken, NJ, 2006.

[2] Kubicek, M.; Hlavacek, V. *Numerical Solution of Nonlinear Boundary Value Problems with Applications;* Dover Publications: Mineola, NY, 2008.

[3] Aziz, A. K., Ed. *Numerical Solutions of Boundary Value Problems for Ordinary Differential Equations;* Academic Press: New York, 1975.

[4] Carreau, P. J.; De Kee, D. C. R.; Chhabra, R. P. *Rheology of Polymeric Systems: Principles and Applications;* Hanser Publishers: Munich, 1998.

[5] Levenspiel, O. *Chemical Reaction Engineering*, 3rd ed.; John Wiley & Sons: New York, 1999.

PARTIAL DIFFERENTIAL EQUATIONS

CHAPTER OUTLINE

MOTIVATION

Differential modeling of physicochemical processes results in a differential equation or a set of differential equations. The model equation involves the derivative(s) of the property of the system with respect to the independent variable for which that property is changing. Because there are four independent variables (three in space and one in time) in a problem, the model equation may include partial derivatives with respect to up to four independent variables. The outcome of the formulation of each of the three fundamental laws (conservation of mass, momentum, and energy) applied to a chemical engineering problem is a second-order partial differential equation. These equations can be classified into three canonical forms (elliptic, parabolic, and hyperbolic). The solution of these partial differential equations based on finite differences is developed in this chapter.

Applied Numerical Methods for Chemical Engineers. DOI: https://doi.org/10.1016/B978-0-12-822961-3.00007-8

7.1 INTRODUCTION

The laws of conservation of mass, momentum, and energy form the basis of the field of transport phenomena. These laws applied to the flow of fluids result in the *equations of change*, which describe the change of velocity, temperature, and concentration with respect to time and position in the system. The dynamics of such systems, which have more than one independent variable, are modeled by *partial differential equations*. These fundamental laws applied to a stationary volume element $\Delta x \Delta y \Delta z$ through which an incompressible fluid is flowing, shown in Fig. 7.1, result in the following equations [1]:
Equation of continuity:

$$\frac{\partial v_x}{\partial x} + \frac{\partial v_y}{\partial y} + \frac{\partial v_z}{\partial z} = 0 \tag{7.1}$$

Equation of continuity for component A:

$$\frac{\partial c_A}{\partial t} + v_x \frac{\partial c_A}{\partial x} + v_y \frac{\partial c_A}{\partial y} + v_z \frac{\partial c_A}{\partial z} = D_{AB}\left(\frac{\partial^2 c_A}{\partial x^2} + \frac{\partial^2 c_A}{\partial y^2} + \frac{\partial^2 c_A}{\partial z^2}\right) + R_A \tag{7.2}$$

Equation of energy:

$$\frac{\partial T}{\partial t} + v_x \frac{\partial T}{\partial x} + v_y \frac{\partial T}{\partial y} + v_z \frac{\partial T}{\partial z} = \alpha\left(\frac{\partial^2 T}{\partial x^2} + \frac{\partial^2 T}{\partial y^2} + \frac{\partial^2 T}{\partial z^2}\right) + \frac{Q}{\rho c} \tag{7.3}$$

Equation of motion:

$$\frac{\partial v_j}{\partial t} + v_x \frac{\partial v_j}{\partial x} + v_y \frac{\partial v_j}{\partial y} + v_z \frac{\partial v_j}{\partial z} = \nu\left(\frac{\partial^2 v_j}{\partial x^2} + \frac{\partial^2 v_j}{\partial y^2} + \frac{\partial^2 v_j}{\partial z^2}\right) - \frac{\partial p}{\partial j} + g_j \qquad j = x, y, \text{ or } z \tag{7.4}$$

where v_x, v_y, and v_z are the velocity components in the three rectangular coordinates, c_A is the molar concentration of A, D_{AB} is the diffusion coefficient, R_A is the molar rate of production of component A, T is the temperature, Q is the rate of production of heat, ρ is the fluid density, c is the heat capacity of the fluid, p is the pressure, and g_j are the components of the gravitational acceleration.

These equations can be simplified based on the assumptions of the problem. For example, the equation of continuity for component A reduces to *Fick's second law of diffusion* when $R_A = 0$ and $v_x = v_y = v_z = 0$:

$$\frac{\partial c_A}{\partial t} = D_{AB}\left(\frac{\partial^2 c_A}{\partial x^2} + \frac{\partial^2 c_A}{\partial y^2} + \frac{\partial^2 c_A}{\partial z^2}\right) \tag{7.5}$$

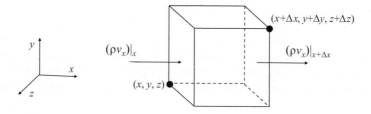

FIG. 7.1

Volume element $\Delta x \Delta y \Delta z$ for fluid flow.

Likewise, for heat conduction in solids, where the velocity terms are zero, Eq. (7.3) simplifies considerably and gives the well-known three-dimensional unsteady state heat conduction equation:

$$\rho C_p \frac{\partial T}{\partial t} = k \left(\frac{\partial^2 T}{\partial x^2} + \frac{\partial^2 T}{\partial y^2} + \frac{\partial^2 T}{\partial z^2} \right) \tag{7.6}$$

This equation has the same form as the respective unsteady state diffusion Eq. (7.5).

The most commonly encountered partial differential equations in chemical engineering are of first and second order. Our discussion in this chapter focuses on these two categories. In the next two sections we attempt to classify these equations and their boundary conditions, and in the remainder of the chapter, we develop the numerical methods, using finite difference and finite element analysis, for the numerical solution of first- and second-order partial differential equations.

7.2 CLASSIFICATION OF PARTIAL DIFFERENTIAL EQUATIONS

Partial differential equations are classified according to their *order, linearity*, and *boundary conditions*. The order of a partial differential equation is determined by the highest-order partial derivative present in that equation. Examples of first-, second-, and third-order partial differential equations are:

$$\text{First order :} \quad \frac{\partial u}{\partial x} - \alpha \frac{\partial u}{\partial y} = 0 \tag{7.7}$$

$$\text{Second order :} \quad \frac{\partial^2 u}{\partial x^2} + u \frac{\partial u}{\partial y} = 0 \tag{7.8}$$

$$\text{Third order :} \quad \left(\frac{\partial^3 u}{\partial x^3} \right)^2 + \frac{\partial^2 u}{\partial x \partial y} + \frac{\partial u}{\partial y} = 0 \tag{7.9}$$

Partial differential equations are categorized into *linear, quasilinear*, and *nonlinear* equations. Consider, for example, the second-order equation:

$$a(.) \frac{\partial^2 u}{\partial y^2} + 2b(.) \frac{\partial^2 u}{\partial x \partial y} + c(.) \frac{\partial^2 u}{\partial x^2} + d(.) = 0 \tag{7.10}$$

If the coefficients are constants or are functions of the independent variables only [$(.) \equiv (x, y)$], then Eq. (7.10) is linear. If the coefficients are functions of the dependent variable and/or any of its derivatives of a lower order than that of the differential equation [$(.) \equiv (x, y, u, \partial u/\partial x, \partial u/\partial y)$], then the equation is quasilinear. Finally, if the coefficients are functions of derivatives of the same order as that of the equation [$(.) \equiv (x, y, u, \partial^2 u/\partial x^2, \partial u^2/\partial y^2, \partial^2 u/\partial x \partial y)$], then the equation is nonlinear. In accordance with these definitions Eq. (7.7) is linear, (7.8) is quasilinear, and (7.9) is nonlinear.

Linear second-order partial differential equations in two independent variables are further classified into three canonical forms: *elliptic, parabolic*, and *hyperbolic*. The general form of this class of equations is:

$$a \frac{\partial^2 u}{\partial x^2} + 2b \frac{\partial^2 u}{\partial x \partial y} + c \frac{\partial^2 u}{\partial y^2} + d \frac{\partial u}{\partial x} + e \frac{\partial u}{\partial y} + fu + g = 0 \tag{7.11}$$

where the coefficients are either constants or functions of the independent variables only. The three canonical forms are determined by the following criterion:

$$b^2 - ac < 0 \quad \text{elliptic} \tag{7.12a}$$

$$b^2 - ac = 0 \quad \text{parabolic} \tag{7.12b}$$

$$b^2 - ac > 0 \quad \text{hyperbolic} \tag{7.12c}$$

If $g = 0$ then Eq. (7.11) is a *homogeneous* differential equation.

The classic examples of second-order partial differential equations that conform to the three canonical forms are:

Laplace's equation (elliptic):

$$\frac{\partial^2 u}{\partial x^2} + \frac{\partial^2 u}{\partial y^2} = 0 \tag{7.13}$$

Heat conduction or diffusion equation (parabolic):

$$\alpha \frac{\partial^2 u}{\partial x^2} = \frac{\partial u}{\partial t} \tag{7.14}$$

Wave equation (hyperbolic):

$$a^2 \frac{\partial^2 u}{\partial x^2} = \frac{\partial^2 u}{\partial t^2} \tag{7.15}$$

Similar classification for second-order partial differential equations with *three* independent variables is given by Tychonov and Samarski [2]. This classification includes elliptic, parabolic, hyperbolic, and ultrahyperbolic. The majority of partial differential equations in engineering and physics are of second order with two, three, or four independent variables. Most of these equations have canonical forms; however, the names elliptic, parabolic, and hyperbolic have been also applied to equations that are not of second order but which possess similar properties [3].

The methods of solution of partial differential equations depend on their canonical form, as will be demonstrated in the rest of this chapter. Because the coefficients of these equations can be functions of the independent variables, it is possible that an equation may shift from one canonical form to another over the range of integration of (x, y).

7.3 INITIAL AND BOUNDARY CONDITIONS

Initial and boundary conditions associated with the partial differential equations must be specified to obtain unique numerical solutions to these equations. In general, boundary conditions for partial differential equations are divided into three categories. These are demonstrated below using the one-dimensional unsteady state heat conduction equation:

$$\alpha \frac{\partial^2 T}{\partial x^2} = \frac{\partial T}{\partial t} \tag{7.16}$$

This is identical (in form) to Eq. (7.14). It is derived from Eq. (7.6) by assuming that the temperature gradients in the y and z dimensions are zero. Eq. (7.20) essentially describes the change in temperature within a solid slab (e.g., the wall of a furnace), where heat transfer takes place in the x-direction (Fig. 7.2).

Following are the three categories of conditions:

Dirichlet conditions (first kind): The values of the dependent variable are given at fixed values of the independent variable. Examples of Dirichlet conditions for the heat conduction equation are:

$$T = f(x) \text{ at } t = 0 \text{ and } 0 \leq x \leq 1$$

and

$$T = T_0 \text{ at } t = 0 \text{ and } 0 \leq x \leq 1$$

These are alternative initial conditions that specify that the initial temperature inside the slab (wall) is a function of position $f(x)$ or a constant T_0 (Fig. 7.2a).

Boundary conditions of the first kind are expressed as

$$T = f(t) \quad \text{at } x = 0 \text{ and } t > 0$$

and

$$T = T_1 \quad \text{at } x = 1 \text{ and } t > 0$$

These boundary conditions specify the value of the independent variable at the left boundary as a function of time $f(t)$ (this may be the condition inside a furnace that is maintained at a prepro-grammed temperature profile) and at the right boundary as a constant T_1 (e.g., the room temperature at the outside of the furnace) (Fig. 7.2a).

Neumann conditions (second kind): The derivative of the dependent variable is given as a constant or as a function of the independent variable. For example

$$\frac{\partial T}{\partial x} = 0 \text{ at } x = 1 \text{ and } t \geq 0$$

This condition specifies that the temperature gradient at the right boundary is zero. In the heat conduction problem, this can be theoretically accomplished by attaching perfect insulation at the right boundary (Fig. 7.2b).

Cauchy conditions: A problem that combines both Dirichlet and Neumann conditions is said to have Cauchy conditions (Fig. 7.2b).

Robin conditions (third kind): The derivative of the dependent variable is given as a function of the dependent variable itself. For the heat conduction problem, the heat flux at the solid-fluid

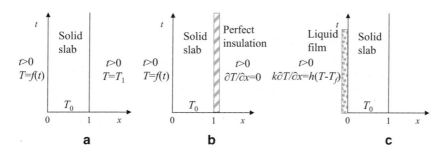

FIG. 7.2 Examples of initial and boundary conditions for the heat conduction process.

(a) Dirichlet conditions; (b) Cauchy conditions (Dirichlet and Neumann); (c) Robin conditions.

interface may be related to the difference between the temperature at the interface and that in the fluid; that is,

$$k\frac{\partial T}{\partial x} = h(T - T_f) \quad \text{at } x = 0 \text{ and } t \geq 0$$

where h is the heat transfer coefficient of the fluid (Fig. 7.2c).

On the basis of their initial and boundary conditions, partial differential equations may be further classified into *initial-value* or *boundary-value* problems. In the first case, at least one of the independent variables has an *open region*. In the unsteady state heat conduction problem, the time variable has the range $0 \leq t \leq \infty$, where no condition has been specified at $t = \infty$; therefore, this is an initial-value problem. When the region is *closed* for all independent variables and conditions are specified at all boundaries, then the problem is of the boundary-value type. An example of this is the two-dimensional steady state heat conduction problem described by the equation:

$$\frac{\partial^2 T}{\partial x^2} + \frac{\partial^2 T}{\partial y^2} = 0 \tag{7.17}$$

with the boundary conditions given at all boundaries:

$$\left.\begin{array}{l} T(0, y) \\ T(L, y) \\ T(x, 0) \\ T(x, H) \end{array}\right\} = \text{specified} \tag{7.18}$$

7.4 SOLUTION OF PARTIAL DIFFERENTIAL EQUATIONS USING FINITE DIFFERENCES

In Chapters 3 and 4 we developed the methods of finite differences and demonstrated that ordinary derivatives can be approximated, with any degree of the desired accuracy, by replacing the differential operators with finite difference operators.

In this section we apply similar procedures in expressing partial derivatives in terms of finite differences. Because partial differential equations involve more than one independent variable, we first establish two-dimensional and three-dimensional grids, in two and three independent variables, respectively, as shown in Fig. 7.3.

The notation (i, j) is used to designate the pivot point for the two-dimensional space and (i, j, k) for the three-dimensional space, where i, j, and k are the counters in the x, y, and z directions, respectively. For unsteady state problems, in which time is one of the independent variables, the counter n is used to designate the time dimension. To keep the notation as simple as possible, we add subscripts only when needed.

The distances between grid points are designated as Δx, Δy, and Δz. When time is one of the independent variables, the time step is shown by Δt.

We now express first, second, and mixed partial derivatives in terms of finite differences. We show the development of these approximations using central differences, and in addition, we summarize in tabular form the formulas obtained from using forward and backward differences.

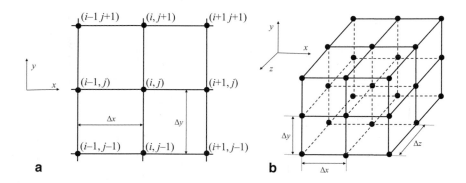

FIG. 7.3

Finite difference grids: (a) two-dimensional grid; (b) three-dimensional grid.

The partial derivative of u with respect to x implies that y and z are held constant; therefore,

$$\left.\frac{\partial u}{\partial x}\right|_{i,j,k} \equiv \left.\frac{du}{dx}\right|_{i,j,k} \tag{7.19}$$

Using Eq. (4.50), which is the approximation of the first-order derivative in terms of central differences, and converting it to the three-dimensional space, we obtain:

$$\left.\frac{\partial u}{\partial x}\right|_{i,j,k} = \frac{1}{2\Delta x}\left(u_{i+1,j,k} - u_{i-1,j,k}\right) + O(\Delta x^2) \tag{7.20}$$

Similarly the first-order partial derivatives in the y- and z-directions are given by the following:

$$\left.\frac{\partial u}{\partial y}\right|_{i,j,k} = \frac{1}{2\Delta y}\left(u_{i,j+1,k} - u_{i,j-1,k}\right) + O(\Delta y^2) \tag{7.21}$$

$$\left.\frac{\partial u}{\partial z}\right|_{i,j,k} = \frac{1}{2\Delta z}\left(u_{i,j,k+1} - u_{i,j,k-1}\right) + O(\Delta z^2) \tag{7.22}$$

In an analogous manner, the second-order partial derivatives are expressed in terms of central differences by using Eq. (4.54):

$$\left.\frac{\partial^2 u}{\partial x^2}\right|_{i,j,k} = \frac{1}{\Delta x^2}\left(u_{i+1,j,k} - 2u_{i,j,k} + u_{i-1,j,k}\right) + O(\Delta x^2) \tag{7.23}$$

$$\left.\frac{\partial^2 u}{\partial y^2}\right|_{i,j,k} = \frac{1}{\Delta y^2}\left(u_{i,j+1,k} - 2u_{i,j,k} + u_{i,j-1,k}\right) + O(\Delta y^2) \tag{7.24}$$

$$\left.\frac{\partial^2 u}{\partial z^2}\right|_{i,j,k} = \frac{1}{\Delta z^2}\left(u_{i,j,k+1} - 2u_{i,j,k} + u_{i,j,k-1}\right) + O(\Delta z^2) \tag{7.25}$$

Finally, the mixed partial derivative is developed as follows:

$$\left.\frac{\partial^2 u}{\partial y \partial x}\right|_{i,j,k} = \frac{\partial}{\partial y}\left[\left.\frac{\partial u}{\partial x}\right|_{i,j,k}\right] \tag{7.26}$$

This is equivalent to applying $\partial u/\partial x$ at points $(i, j + 1, k)$ and $(i, j - 1, k)$, so,

$$\frac{\partial^2 u}{\partial y \partial x}\bigg|_{i,j,k} = \frac{1}{2\Delta y}\left[\frac{1}{2\Delta x}\left(u_{i+1,j+1,k} - u_{i-1,j+1,k}\right) - \frac{1}{2\Delta x}\left(u_{i+1,j-1,k} - u_{i-1,j-1,k}\right)\right] + O\left(\Delta x^2 + \Delta y^2\right)$$

$$= \frac{1}{4\Delta x \Delta y}\left(u_{i+1,j+1,k} - u_{i-1,j+1,k} - u_{i+1,j-1,k} + u_{i-1,j-1,k}\right) + O\left(\Delta x^2 + \Delta y^2\right)$$

The central difference approximations of partial derivatives are summarized in Table 7.1. The corresponding approximations obtained from using forward and backward differences are shown in Tables 7.2 and 7.3, respectively. Equivalent sets of formulas, which are more accurate than those

Table 7.1 Finite difference approximations of partial derivatives using central differences.

Derivative	Central difference	Error
$\dfrac{\partial u}{\partial x}\bigg|_{i,j,k}$	$\dfrac{1}{2\Delta x}\left(u_{i+1,j,k} - u_{i-1,j,k}\right)$	$O(\Delta x^2)$
$\dfrac{\partial u}{\partial y}\bigg|_{i,j,k}$	$\dfrac{1}{2\Delta y}\left(u_{i,j+1,k} - u_{i,j-1,k}\right)$	$O(\Delta y^2)$
$\dfrac{\partial u}{\partial z}\bigg|_{i,j,k}$	$\dfrac{1}{2\Delta z}\left(u_{i,j,k+1} - u_{i,j,k-1}\right)$	$O(\Delta z^2)$
$\dfrac{\partial^2 u}{\partial x^2}\bigg|_{i,j,k}$	$\dfrac{1}{\Delta x^2}\left(u_{i+1,j,k} - 2u_{i,j,k} + u_{i-1,j,k}\right)$	$O(\Delta x^2)$
$\dfrac{\partial^2 u}{\partial y^2}\bigg|_{i,j,k}$	$\dfrac{1}{\Delta y^2}\left(u_{i,j+1,k} - 2u_{i,j,k} + u_{i,j-1,k}\right)$	$O(\Delta y^2)$
$\dfrac{\partial^2 u}{\partial z^2}\bigg|_{i,j,k}$	$\dfrac{1}{\Delta z^2}\left(u_{i,j,k+1} - 2u_{i,j,k} + u_{i,j,k-1}\right)$	$O(\Delta z^2)$
$\dfrac{\partial^2 u}{\partial y \partial x}\bigg|_{i,j,k}$	$\dfrac{1}{14\Delta x \Delta y}\left(u_{i+1,j+1,k} - u_{i-1,j+1,k} - u_{i+1,j-1,k} + u_{i-1,j-1,k}\right)$	$O(\Delta x^2 + \Delta y^2)$

Table 7.2 Finite difference approximations of partial derivatives using forward differences.

Derivative	Forward difference	Error
$\dfrac{\partial u}{\partial x}\bigg|_{i,j,k}$	$\dfrac{1}{\Delta x}\left(u_{i+1,j,k} - u_{i,j,k}\right)$	$O(\Delta x)$
$\dfrac{\partial u}{\partial y}\bigg|_{i,j,k}$	$\dfrac{1}{\Delta y}\left(u_{i,j+1,k} - u_{i,j,k}\right)$	$O(\Delta y)$
$\dfrac{\partial u}{\partial z}\bigg|_{i,j,k}$	$\dfrac{1}{\Delta z}\left(u_{i,j,k+1} - u_{i,j,k}\right)$	$O(\Delta z)$
$\dfrac{\partial^2 u}{\partial x^2}\bigg|_{i,j,k}$	$\dfrac{1}{\Delta x^2}\left(u_{i+2,j,k} - 2u_{i+1,j,k} + u_{i,j,k}\right)$	$O(\Delta x)$
$\dfrac{\partial^2 u}{\partial y^2}\bigg|_{i,j,k}$	$\dfrac{1}{\Delta y^2}\left(u_{i,j+2,k} - 2u_{i,j+1,k} + u_{i,j,k}\right)$	$O(\Delta y)$
$\dfrac{\partial^2 u}{\partial z^2}\bigg|_{i,j,k}$	$\dfrac{1}{\Delta z^2}\left(u_{i,j,k+2} - 2u_{i,j,k+1} + u_{i,j,k}\right)$	$O(\Delta z)$
$\dfrac{\partial^2 u}{\partial y \partial x}\bigg|_{i,j,k}$	$\dfrac{1}{\Delta x \Delta y}\left(u_{i+1,j+1,k} - u_{i,j+1,k} - u_{i+1,j,k} + u_{i,j,k}\right)$	$O(\Delta x + \Delta y)$

Table 7.3 Finite difference approximations of partial derivatives using backward differences.

Derivative	Backward difference	Error
$\left.\dfrac{\partial u}{\partial x}\right|_{i,j,k}$	$\dfrac{1}{\Delta x}\left(u_{i,j,k} - u_{i-1,j,k}\right)$	$O(\Delta x^2)$
$\left.\dfrac{\partial u}{\partial y}\right|_{i,j,k}$	$\dfrac{1}{\Delta y}\left(u_{i,j,k} - u_{i,j-1,k}\right)$	$O(\Delta y^2)$
$\left.\dfrac{\partial u}{\partial z}\right|_{i,j,k}$	$\dfrac{1}{\Delta z}\left(u_{i,j,k} - u_{i,j,k-1}\right)$	$O(\Delta z^2)$
$\left.\dfrac{\partial^2 u}{\partial x^2}\right|_{i,j,k}$	$\dfrac{1}{\Delta x^2}\left(u_{i,j,k} - 2u_{i-1,j,k} + u_{i-2,j,k}\right)$	$O(\Delta x^2)$
$\left.\dfrac{\partial^2 u}{\partial y^2}\right|_{i,j,k}$	$\dfrac{1}{\Delta y^2}\left(u_{i,j,k} - 2u_{i,j-1,k} + u_{i,j-2,k}\right)$	$O(\Delta y^2)$
$\left.\dfrac{\partial^2 u}{\partial z^2}\right|_{i,j,k}$	$\dfrac{1}{\Delta z^2}\left(u_{i,j,k} - 2u_{i,j,k-1} + u_{i,j,k-2}\right)$	$O(\Delta z^2)$
$\left.\dfrac{\partial^2 u}{\partial y \partial x}\right|_{i,j,k}$	$\dfrac{1}{\Delta x \Delta y}\left(u_{i,j,k} - u_{i,j-1,k} - u_{i-1,j,k} + u_{i-1,j-1,k}\right)$	$O(\Delta x^2 + \Delta y^2)$

above, may be developed by using finite difference approximations that have higher accuracies, such as Eqs. (4.59) and (4.64) for central differences, Eqs. (4.41) and (4.46) for forward differences, and Eqs. (4.24) and (4.29) for backward differences. However, more accurate formulas are not commonly used because they involve a larger number of terms and require more extensive computation times.

The use of finite difference approximations is demonstrated in this chapter in setting up the numerical solutions of elliptic, parabolic, and hyperbolic partial differential equations.

7.4.1 ELLIPTIC PARTIAL DIFFERENTIAL EQUATIONS

Elliptic differential equations are often encountered in steady state heat conduction and diffusion operations. For example, in two-dimensional steady state heat conduction in solids, Eq. (7.6) becomes:

$$\frac{\partial^2 T}{\partial x^2} + \frac{\partial^2 T}{\partial y^2} = 0 \tag{7.17}$$

Similarly Fick's second law of diffusion (Eq. 7.5) simplifies to:

$$\frac{\partial^2 c_A}{\partial x^2} + \frac{\partial^2 c_A}{\partial y^2} = 0 \tag{7.28}$$

when a steady state is assumed.

We begin our discussion of numerical solutions of elliptic differential equations by first examining the two-dimensional problem in its general form (Laplace's equation):

$$\frac{\partial^2 u}{\partial x^2} + \frac{\partial^2 u}{\partial y^2} = 0 \tag{7.13}$$

We replace each second-order partial derivative by its approximation in central differences, Eqs. (7.23) and (7.24), to obtain:

$$\frac{1}{\Delta x^2}\left(u_{i+1,j} - 2u_{i,j} + u_{i-1,j}\right) + \frac{1}{\Delta y^2}\left(u_{i,j+1} - 2u_{i,j} + u_{i,j-1}\right) = 0 \tag{7.29}$$

which rearranges to:

$$-2\left(\frac{1}{\Delta x^2} + \frac{1}{\Delta y^2}\right)u_{i,j} + \left(\frac{1}{\Delta x^2}\right)u_{i+1,j} + \left(\frac{1}{\Delta x^2}\right)u_{i-1,j} + \left(\frac{1}{\Delta y^2}\right)u_{i,j+1} + \left(\frac{1}{\Delta y^2}\right)u_{i,j-1} \tag{7.30}$$

This is a linear algebraic equation involving the value of the dependent variable at five adjacent grid points.

A rectangular-shaped object that is divided into p segments in the x-direction and q segments in the y-direction has $(p+1) \times (q+1)$ total grid points and $(p\ 1) \times (q\ 1)$ internal grid points. Eq. (7.30), written for each of the internal points, constitutes a set of $(p-1) \times (q-1)$ simultaneous linear algebraic equations in $(p+1) \times (q+1)-4$ unknowns (the four corner points do not appear in these equations). The boundary conditions provide additional information for the solution of the problem. If the boundary conditions are of the Dirichlet type, the values of the dependent variable are known at all the external grid points. On the other hand, if the boundary conditions at any of the external surfaces are of the Neumann or Robin type, which specify partial derivatives at the boundaries, these conditions must also be replaced by finite difference approximations.

We demonstrate this by specifying a Neumann condition at the left boundary; that is,

$$\frac{\partial u}{\partial x} = \beta \quad \text{at } x = 0 \text{ and all } y \tag{7.31}$$

where β is a constant. Replacing the partial derivative in Eq. (7.31) with a central difference approximation, we obtain:

$$\frac{1}{2\Delta x}\left(u_{i+1,j} - u_{i-1,j}\right) = \beta \tag{7.32}$$

This is valid only at $x = 0$ where $i = 0$; therefore, Eq. (7.32) becomes:

$$u_{-1,j} = u_{1,j} - 2\beta\Delta x \tag{7.33}$$

The points $(-1, j)$ are located outside the object; therefore, $u_{-1,j}$ have fictitious values. Their calculation, however, is necessary for the evaluation of the Neumann boundary condition. Eq. (7.33), written for all y ($j = 0, 1, \ldots, q$), provides $(q+1)$ additional equations but at the same time introduces $(q+1)$ additional variables. To counter this, Eq. (7.30) is also written for $(q+1)$ points along this boundary (at $x = 0$), thus providing the necessary number of independent equations for the solution of the problem.

Replacing the partial derivative in Eq. (7.31) with a forward difference does not require the use of fictitious points. However, it is important to use the forward difference formula with the same accuracy as the other equations. In this case the formula based on forward differences with an error of the order $O(h^2)$, Eq. (4.38), should be used for evaluation of the partial derivative at $x = 0$ ($i = 0$):

$$\frac{1}{2\Delta x}\left(-u_{2,j} + 4u_{1,j} - 3u_{0,j}\right) = \beta \tag{7.34}$$

or

$$-3u_{0,j} + 4u_{1,j} - u_{2,j} = 2\beta\Delta x \tag{7.35}$$

Eq. (7.35) provides $(q+1)$ additional equations without introducing additional variables.

In the case of the Robin condition at the left boundary in the form:

$$\frac{\partial u}{\partial x} = \beta + \gamma u \quad \text{at } x = 0 \text{ and all } y \tag{7.36}$$

where β and γ are constants, a similar derivation as above shows that the following equation should be used at the boundary:

$$-(3 + 2\gamma\Delta x)u_{0,j} + 4u_{1,j} - u_{2,j} = 2\beta\Delta x \tag{7.37}$$

Similarly, for the Robin condition at the right boundary, the formula based on backward differences with an error of the order $O(h^2)$, Eq. (4.21) should be used for evaluation of the partial derivative ($i = p$):

$$u_{p-1,j} - 4u_{p-2,j} + (3 - 2\gamma\Delta x)u_{p,j} = 2\beta\Delta x \tag{7.38}$$

Because Eq. (7.30) and the appropriate boundary conditions constitute a set of linear algebraic equations, the Gauss methods for the solution of such equations may be used. Eq. (7.30) is actually a predominantly diagonal system; therefore, the Gauss-Seidel method (see Section 2.7) is especially suitable for the solution of this problem. Rearranging Eq. (7.30) to solve for $u_{i,j}$, we have

$$u_{i,j} = \frac{\frac{1}{\Delta x^2}\left(u_{i+1,j} + u_{i-1,j}\right) + \frac{1}{\Delta y^2}\left(u_{i,j+1} + u_{i,j-1}\right)}{2\left(\frac{1}{\Delta x^2} + \frac{1}{\Delta y^2}\right)} \tag{7.39}$$

which can be used in the iterative Gauss-Seidel substitution method. An initial estimate of all $u_{i,j}$ is needed, but this can be easily obtained from averaging the Dirichlet boundary conditions.

The Gauss-Seidel method is guaranteed to converge for a predominantly diagonal system of equations. However, its convergence may be quite slow in the solution of elliptic differential equations. The *overrelaxation* method can be used to accelerate the rate of convergence. This technique applies the following weighting algorithm in evaluating the new values of $u_{i,j}$ at each iteration of the Gauss-Seidel method:

$$\left(u_{i,j}\right)_{\text{new}} = w\left(u_{i,j}\right)_{\text{from Eq. (7.43)}} + (1 - w)\left(u_{i,j}\right)_{\text{old}} \tag{7.40}$$

Special care should be taken when processing the nodes at the boundaries if $u_{i,j}$ at these nodes is calculated by a different method of finite differences. In such a case when calculating the new value of $u_{i,j}$, the proper equation should be applied instead of Eq. (7.39). The *relaxation parameter*, w, can be assigned values from the following ranges:

$$0 < w < 1 \quad \text{underrelaxation}$$
$$1 < w \le 2 \quad \text{overrelaxation}$$

When $w = 1$ this method is exactly the same as the unmodified Gauss-Seidel. Methods for estimating the optimal w are given by Lapidus and Pinder [4], who also show that the overrelaxation method is 5 to 100 times faster (depending on step size and convergence criterion) than the Gauss-Seidel method.

In the case when an equidistant grid can be used (i.e., when $\Delta x = \Delta y$), Eq. (7.39) simplifies to:

$$u_{i,j} = \frac{u_{i+1,j} + u_{i-1,j} + u_{i,j+1} + u_{i,j-1}}{4} \tag{7.41}$$

which simply shows that the value of the dependent variable at the pivotal point (i, j) in the Laplace equation is the arithmetic average of the values at the grid points to the right and left of and above and below the pivot point. This is demonstrated by the computational molecule of Fig. 7.4, which is sometimes referred to as a "five-pointed star."

The three-dimensional elliptic partial differential equation:

$$\frac{\partial^2 u}{\partial x^2} + \frac{\partial^2 u}{\partial y^2} + \frac{\partial^2 u}{\partial z^2} = 0 \tag{7.42}$$

can be similarly converted to linear algebraic equations using finite difference approximations in three-dimensional space. Applying Eqs. (7.23)–(7.25) to replace the three partial derivatives of Eq. (7.42), we obtain

$$\frac{1}{\Delta x^2}\left(u_{i+1,j,k} - 2u_{i,j,k} + u_{i-1,j,k}\right) + \frac{1}{\Delta y^2}\left(u_{i,j+1,k} - 2u_{i,j,k} + u_{i,j-1,k}\right)$$
$$+ \frac{1}{\Delta z^2}\left(u_{i,j,k+1} - 2u_{i,j,k} + u_{i,j,k-1}\right) = 0 \tag{7.43}$$

For the equidistant grid ($\Delta x = \Delta y = \Delta z$), the above equation reduces to

$$u_{i,j,k} = \frac{u_{i+1,j,k} + u_{i-1,j,k} + u_{i,j+1,k} + u_{i,j-1,k} + u_{i,j,k+1} + u_{i,j,k-1}}{6} \tag{7.44}$$

In parallel with the two-dimensional case, the value of the dependent variable at the pivot point (i, j, k) is the arithmetic average of the values at the grid points adjacent to the pivot point. The computational molecule for the three-dimensional elliptic equation is shown in Fig. 7.5.

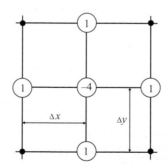

FIG. 7.4 Computational molecule for the Laplace equation using the equidistance grid.

The number in each circle is the coefficient of that point in the difference equation.

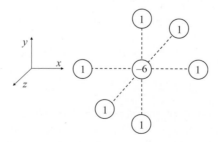

FIG. 7.5

Computational molecule for the three-dimensional elliptic differential equation using the equidistance grid.

The *nonhomogeneous* form of the Laplace equation is the *Poisson* equation:

$$\frac{\partial^2 u}{\partial x^2} + \frac{\partial^2 u}{\partial y^2} = f(x, y) \tag{7.45}$$

which also belongs to the class of elliptic partial differential equations. A form of the Poisson equation,

$$\frac{\partial^2 T}{\partial x^2} + \frac{\partial^2 T}{\partial y^2} = -\frac{Q(x, y)}{k} \tag{7.46}$$

is used to describe heat conduction in a two-dimensional solid plate with an internal heat source. $Q'(x, y)$ is the heat generated per unit volume per time, and k is the thermal conductivity of the material. The finite difference formulation of the Poisson equation is

$$-2\left(\frac{1}{\Delta x^2} + \frac{1}{\Delta y^2}\right)u_{i,j} + \left(\frac{1}{\Delta x^2}\right)u_{i+1,j} + \left(\frac{1}{\Delta x^2}\right)u_{i-1,j} + \left(\frac{1}{\Delta y^2}\right)u_{i,j+1} + \left(\frac{1}{\Delta y^2}\right)u_{i,j-1} = f_{i,j} \tag{7.47}$$

or

$$u_{i,j} = \frac{\frac{1}{\Delta x^2}\left(u_{i+1,j} + u_{i-1,j}\right) + \frac{1}{\Delta y^2}\left(u_{i,j+1} + u_{i,j-1}\right)}{2\left(\frac{1}{\Delta x^2} + \frac{1}{\Delta y^2}\right)} - \frac{f_{i,j}}{2\left(\frac{1}{\Delta x^2} + \frac{1}{\Delta y^2}\right)} \tag{7.48}$$

The numerical solutions of the Laplace and Poisson elliptic partial differential equations are demonstrated in Examples 7.1 and 7.2.

EXAMPLE 7.1 SOLVING THE LAPLACE EQUATION FOR DETERMINING THE TEMPERATURE PROFILE IN A TWO-DIMENSIONAL PLATE

Determine the temperature profile in a 1 m × 1 m plate shown in Fig. E7.1a. The boundary conditions are shown in the figure. No heat is generated in the plate. Use $\Delta x = \Delta y$.

Method of solution

The governing equation for this problem is the Laplace equation:

$$\frac{\partial^2 T}{\partial x^2} + \frac{\partial^2 T}{\partial y^2} = 0$$

This equation should be solved subject to the following boundary conditions:

$$x = 0, \quad T = 250$$

$$x = 1, \quad \frac{\partial T}{\partial x} = 0$$

$$y = 0, \quad \frac{\partial T}{\partial y} = 5(T - 25)$$

$$y = 1, \quad T = 500$$

(Continued)

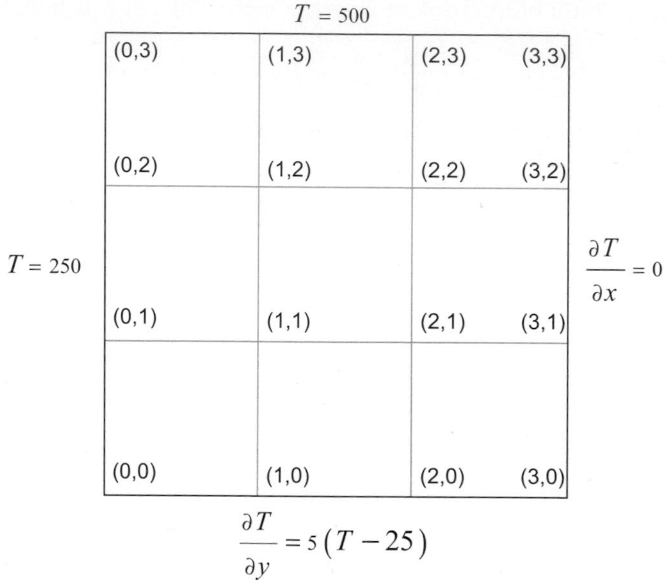

FIG. E7.1a

Flat plate.

EXAMPLE 7.1 (CONTINUED)

Eq. (7.30) can be used for the internal points (1, 1), (2, 1), (1, 2), and (2, 2):

$$T_{i,j-1} + T_{i-1,j} - 4T_{i,j} + T_{i+1,j} + T_{i,j+1} = 0$$

At the left boundary, points (0, 1) and (0, 2), there is a Dirichlet condition:

$$T_{0,j} = 250$$

At the right boundary on points (3, 1) and (3, 2), for which a Neumann condition exists, Eq. (7.38) should be used ($\beta = \gamma = 0$):

$$T_{1,j} - 4T_{2,j} + 3T_{3,j} = 0$$

Eq. (7.37) for the Robin condition should be used for points (1, 0) and (2, 0) on the lower boundary ($\beta = -5 \times 25$ and $\gamma = 5$):

$$-\frac{19}{3}T_{i,0} + 4T_{i,1} - T_{i,2} = -\frac{250}{3}$$

At the top boundary, points (1, 3) and (2, 3), there is a Dirichlet condition:

$$T_{0,j} = 500$$

(Continued)

EXAMPLE 7.1 (CONTINUED)

Therefore, the complete set of equations for calculating the temperature profile in this example is:

Point (1, 0): $-\frac{19}{3}T_{1,0} + 4T_{1,1} - T_{1,2} = -\frac{250}{3}$

Point (2, 0): $-\frac{19}{3}T_{2,0} + 4T_{2,1} - T_{2,2} = -\frac{250}{3}$

Point (0, 1): $T_{0,1} = 250$

Point (1, 1): $T_{1,0} + T_{0,1} - 4T_{1,1} + T_{2,1} + T_{1,2} = 0$

Point (2, 1): $T_{2,0} + T_{1,1} - 4T_{2,1} + T_{3,1} + T_{2,2} = 0$

Point (3, 1): $T_{1,1} - 4T_{2,1} + 3T_{3,1} = 0$

Point (0, 2): $T_{0,2} = 250$

Point (1, 2): $T_{1,1} + T_{0,2} - 4T_{1,2} + T_{2,2} + T_{1,3} = 0$

Point (2, 2): $T_{2,1} + T_{1,2} - 4T_{2,2} + T_{3,2} + T_{2,3} = 0$

Point (3, 2): $T_{1,2} - 4T_{2,2} + 3T_{3,2} = 0$

Point (1, 3): $T_{1,3} = 500$

Point (2, 3): $T_{2,3} = 500$

This set of linear algebraic equations can be shown in matrix form as:

$$
\begin{bmatrix}
-\frac{19}{3} & 0 & 0 & 4 & 0 & 0 & 0 & -1 & 0 & 0 & 0 & 0 \\
0 & -\frac{19}{3} & 0 & 0 & 4 & 0 & 0 & 0 & -1 & 0 & 0 & 0 \\
0 & 0 & 1 & 0 & 0 & 0 & 0 & 0 & 0 & 0 & 0 & 0 \\
1 & 0 & 1 & -4 & 1 & 0 & 0 & 1 & 0 & 0 & 0 & 0 \\
0 & 1 & 0 & 1 & -4 & 1 & 0 & 0 & 1 & 0 & 0 & 0 \\
0 & 0 & 0 & 1 & -4 & 3 & 0 & 0 & 0 & 0 & 0 & 0 \\
0 & 0 & 0 & 0 & 0 & 1 & 0 & 0 & 0 & 0 & 0 & 0 \\
0 & 0 & 1 & 0 & 0 & 1 & -4 & 1 & 1 & 0 & 0 & 0 \\
0 & 0 & 0 & 1 & 0 & 0 & 1 & -4 & 1 & 0 & 1 & 0 \\
0 & 0 & 0 & 0 & 0 & 0 & 1 & -4 & 3 & 0 & 0 & 0 \\
0 & 0 & 0 & 0 & 0 & 0 & 0 & 0 & 0 & 1 & 0 & 0 \\
0 & 0 & 0 & 0 & 0 & 0 & 0 & 0 & 0 & 0 & 0 & 1 \\
\end{bmatrix}
\begin{bmatrix}
T_{1,0} \\ T_{2,0} \\ T_{0,1} \\ T_{1,1} \\ T_{2,1} \\ T_{3,1} \\ T_{0,2} \\ T_{1,2} \\ T_{2,2} \\ T_{3,2} \\ T_{1,3} \\ T_{2,3}
\end{bmatrix}
=
\begin{bmatrix}
-\frac{250}{3} \\ -\frac{250}{3} \\ 250 \\ 0 \\ 0 \\ 0 \\ 250 \\ 0 \\ 0 \\ 0 \\ 500 \\ 500
\end{bmatrix}
$$

Also, if this set of equations is to be solved by the Gauss-Seidel or Jacobi method, it should be rearranged as:

$T_{1,0} \quad = -\frac{3}{19}\left(-\frac{250}{3} - 4T_{1,1} + T_{1,2}\right)$

$T_{2,0} \quad = -\frac{3}{19}\left(-\frac{250}{3} - 4T_{2,1} + T_{2,2}\right)$

(Continued)

EXAMPLE 7.1 (CONTINUED)

$$T_{0,1} = 250$$
$$T_{1,1} = \tfrac{1}{4}\left(T_{1,0} + T_{0,1} + T_{2,1} + T_{1,2}\right)$$
$$T_{2,1} = \tfrac{1}{4}\left(T_{2,0} + T_{1,1} + T_{3,1} + T_{2,2}\right)$$
$$T_{3,1} = \tfrac{1}{3}\left(-T_{1,1} + 4T_{2,1}\right)$$
$$T_{0,2} = 250$$
$$T_{1,2} = \tfrac{1}{4}\left(T_{1,1} + T_{0,2} + T_{2,2} + T_{1,3}\right)$$
$$T_{2,2} = \tfrac{1}{4}\left(T_{2,1} + T_{1,2} + T_{3,2} + T_{2,3}\right)$$
$$T_{3,2} = \tfrac{1}{3}\left(-T_{1,2} + 4T_{2,2}\right)$$
$$T_{1,3} = 500$$
$$T_{2,3} = 500$$

We do not present the details of the solution of this set of equations but show the final results in Fig. E7.1b

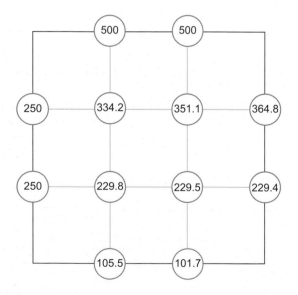

FIG. E7.1b

Calculated temperature profile.

EXAMPLE 7.2 SOLUTION OF THE LAPLACE AND POISSON EQUATIONS

Write a general MATLAB® function to determine the numerical solution of a two-dimensional elliptic partial differential equation of the general form:

$$\frac{\partial^2 u}{\partial x^2} + \frac{\partial^2 u}{\partial y^2} = f$$

for a rectangular object of variable width and height. The object could have Dirichlet, Neumann, or Robin boundary conditions. The value of f should be assumed to be constant. Use this function to find the solution to the following problems ($u = T$):

(a) A thin square metal plate of dimensions 1 m × 1.5 m is subjected to four heat sources that maintain the temperature on its four edges as follows:

$$T(0, y) = 250°C$$
$$T(1, y) = 100°C$$
$$T(x, 0) = 500°C$$
$$T(x, 1.5) = 25°C$$

The flat sides of the plate are insulated so that no heat is transferred through these sides. Calculate the temperature profiles within the plate.

(b) Perfect insulation is installed on two edges (right and top) of the plate of part (a). The other two edges are maintained at constant temperatures. The set of Dirichlet and Neumann boundary conditions is:

$$T(0, y) = 250°C$$
$$\left.\frac{\partial T}{\partial x}\right|_{1,y} = 0$$
$$T(x, 0) = 500°C$$
$$\left.\frac{\partial T}{\partial x}\right|_{x,1.5} = 0$$

Calculate the temperature profiles within the plate and compare these with the results of part (a).

(c) The thin metal plate of part (a) is made of an alloy that has a melting point of 800°C and thermal conductivity of 16 W/m.K. The plate is subject to an electric current that creates a uniform heat source within the plate. The amount of heat generated is $Q = 100$ kW/m^3. All four edges of the plate are in contact with fluid at 25°C. The set of Robin boundary conditions is:

$$\left.\frac{\partial T}{\partial x}\right|_{0,y} = 5[T(0, y) - 25]$$
$$\left.\frac{\partial T}{\partial x}\right|_{1,y} = 5[25 - T(1, y)]$$
$$\left.\frac{\partial T}{\partial y}\right|_{x,0} = 5[T(x, 0) - 25]$$
$$\left.\frac{\partial T}{\partial y}\right|_{x,1.5} = 5[25 - T(x, 1)]$$

(Continued)

EXAMPLE 7.2 (CONTINUED)

Examine the temperature profiles within the plate to ascertain whether the alloy will begin to melt under these conditions.

Method of solution

We solve this problem by matrix inversion because matrix operations are much faster than element-by-element operations in MATLAB, especially when a large number of equations are to be solved. To solve the set of equations in a matrix format, $u_{i,j}$ values have to be rearranged as a column vector. Therefore, we put in order all the dependent variables in a vector and renumber them from 1 to $(p+1)(q+1)$, as illustrated in Fig. E7.2a. Using this single numbering system, Eq. (7.47) can be written as:

$$-2\left(\frac{1}{\Delta x^2} + \frac{1}{\Delta y^2}\right)u_n + \left(\frac{1}{\Delta x^2}\right)u_{n+1} + \left(\frac{1}{\Delta x^2}\right)u_{n-1} + \left(\frac{1}{\Delta y^2}\right)u_{n+p+1} + \left(\frac{1}{\Delta y^2}\right)u_{n-p-1} = f \quad (1)$$

Note that for the internal points,

$$n = j(p+1) + i + 1, \quad i = 1,\dots,p-1, \quad j = 1,\dots,q-1$$

When the Laplace equation is being solved, f is zero, which makes Eq. (7.47) equivalent to Eq. (7.30). For the Poisson equation, the value of f is assumed constant throughout the plate.

(Continued)

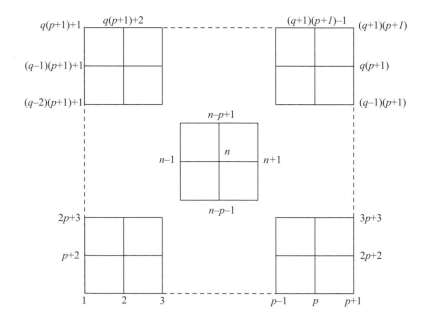

FIG. E7.2a

Scheme of renumbering values of $u_{i,j}$ to convert to a column vector.

EXAMPLE 7.2 (CONTINUED)

At each boundary, if the condition is of the Dirichlet type, the values of u remain unchanged, provided that the related equation at this point is:

$$u_N = \text{constant}$$

where N is a node at the boundary with the Dirichlet condition. However, if the condition is of the Neumann or Robin type, the forward or backward difference is used for evaluating the first-order derivative of the order $O(h^2)$ at the boundaries:

Lower x boundary (forward differences):

$$\frac{\partial u}{\partial x}\bigg|_{x=0} = \frac{1}{2\Delta x}(-3u_N + 4u_{N+1} - u_{N+2})$$

where N is a node on the line $x = 0$.

Upper x boundary (backward differences):

$$\frac{\partial u}{\partial x}\bigg|_{x=L} = \frac{1}{2\Delta x}(3u_N - 4u_{N-1} + u_{N-2})$$

where N is a node on the line $x = L$.

Lower y boundary (forward differences):

$$\frac{\partial u}{\partial y}\bigg|_{y=0} = \frac{1}{2\Delta y}\left(-3u_N + 4u_{N+p+1} - u_{N+2p+2}\right)$$

where N is a node on the line $y = 0$.

Upper y boundary (backward differences):

$$\frac{\partial u}{\partial y}\bigg|_{y=W} = \frac{1}{2\Delta y}\left(3u_N - 4u_{N-p-1} - u_{N-2p-2}\right)$$

where N is a node on the line $y = W$.

The value of the dependent variable at the four corner grid points cannot be calculated by this method. The arithmetic average of the values of the dependent variables at the three adjacent points on the boundaries is assigned to the dependent variable at each corner point. The flowchart of solving Poisson's equation based on the algorithm used in this example is illustrated in Fig. E7.2b.

Program description

The programs and functions developed in this example can be found in: https://www.elsevier.com/books-and-journals/book-companion/9780128229613

The MATLAB function *elliptic.m* is written for the solution of the Laplace equation, or the Poisson equation with constant right-hand side value, for a rectangular plate. The first part of the function is the initialization and checking of the input arguments. If the last input argument, which introduces the constant of the right-hand side of the elliptic equation to the function, is omitted, it is assumed to be zero, and the function solves Laplace's equation. The input argument that carries boundary conditions should consist of four rows, one for each boundary.

(Continued)

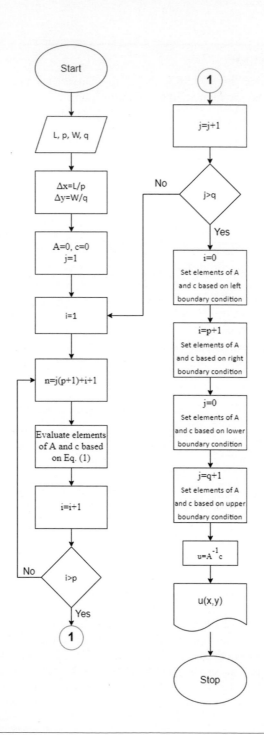

FIG. E7.2b

Flowchart of solving Poisson's equation.

EXAMPLE 7.2 (CONTINUED)

In each row, the first element is a flag that indicates the type of condition (1 for Dirichlet, 2 for Neumann, and 3 for Robin), followed by the boundary condition value or parameters.

Next in the function are several sections dealing with building the matrix of coefficients and the vector of constants according to what is discussed in the method of solution. Finally, the function calculates all u values by the matrix left division operator ($A\backslash c$). The outputs of *elliptic.m* are the vectors of x and y and the matrix of $u_{i,j}$ values.

The main program *Example7_2.m* asks the user to input all the necessary parameters for solving the elliptic equation from the keyboard. It then calls the function *elliptic.m* to solve the equation and finally plots the results in a three-dimensional graph.

Input and results
Part (a)

>> **Example7_2**
Solution of elliptic partial differential equation.
Length of the plate (x-direction) (m) = 1
Width of the plate (y-direction) (m) = 1.5
Number of divisions in x-direction = 20
Number of divisions in y-direction = 20
Right hand side of the equation (f) = 0
Boundary conditions:
Lower x boundary condition:
1 - Dirichlet
2 - Neumann
3 - Robin
Enter your choice: 1
Value = 250
Upper x boundary condition:
1 - Dirichlet
2 - Neumann
3 - Robin
Enter your choice: 1
Value = 100
Lower y boundary condition:
1 - Dirichlet
2 - Neumann
3 - Robin
Enter your choice: 1
Value = 500
Upper y boundary condition:
1 - Dirichlet
2 - Neumann

(Continued)

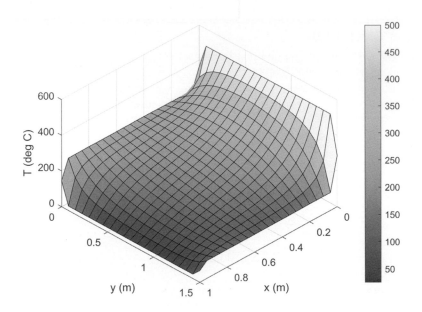

FIG. E7.2c

Solution of the Laplace equation with Dirichlet conditions.

EXAMPLE 7.2 (CONTINUED)

3 - Robin
Enter your choice: 1
Value = 25

Part (b)

>> **Example7_2**
Length of the plate (x-direction) (m) = 1
Width of the plate (y-direction) (m) = 1.5
Number of divisions in x-direction = 20
Number of divisions in y-direction = 20
Right hand side of the equation (f) = 0
Boundary conditions:
Lower x boundary condition:
1 - Dirichlet
2 - Neumann
3 - Robin
Enter your choice: 1
Value = 250

(Continued)

EXAMPLE 7.2 (CONTINUED)

Upper x boundary condition:
1 - Dirichlet
2 - Neumann
3 - Robin
Enter your choice: 2
Value = 0
Lower y boundary condition:
1 - Dirichlet
2 - Neumann
3 - Robin
Enter your choice: 1
Value = 500
Upper y boundary condition:
1 - Dirichlet
2 - Neumann
3 - Robin
Enter your choice: 2
Value = 0

(Continued)

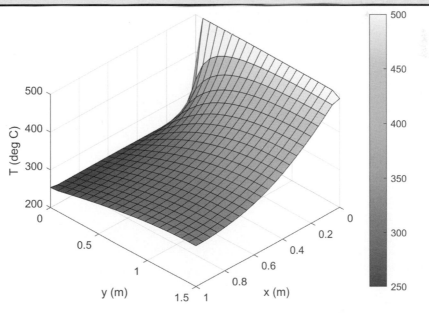

FIG. E7.2d

Solution of the Laplace equation with Neumann conditions.

EXAMPLE 7.2 (CONTINUED)

Part (c)

>> **Example 7_2**
Length of the plate (x-direction) (m) = 1
Width of the plate (y-direction) (m) = 1.5
Number of divisions in x-direction = 20
Number of divisions in y-direction = 20
Right hand side of the equation (f) = -100e3/16
Boundary conditions:
Lower x boundary condition:
1 - Dirichlet
2 - Neumann
3 - Robin
Enter your choice: 3
u' = (beta) + (gamma)*u
Constant (beta) = -5*25
Coefficient (gamma) = 5
Upper x boundary condition:
1 - Dirichlet
2 - Neumann
3 - Robin
Enter your choice: 3
u' = (beta) + (gamma)*u
Constant (beta) = 5*25
Coefficient (gamma) = -5
Lower y boundary condition:
1 - Dirichlet
2 - Neumann
3 - Robin
Enter your choice: 3
u' = (beta) + (gamma)*u
Constant (beta) = -5*25
Coefficient (gamma) = 5
Upper y boundary condition:
1 - Dirichlet
2 - Neumann
3 - Robin
Enter your choice: 3
u' = (beta) + (gamma)*u
Constant (beta) = 5*25
Coefficient (gamma) = -5

(Continued)

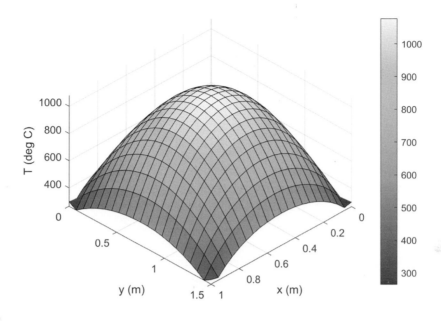

FIG. E7.2e

Solution of the Poisson equation with Robin conditions.

EXAMPLE 7.2 (CONTINUED)

Discussion of results

Part (*a*): By entering $f = 0$ as input to the program, the Laplace equation with Dirichlet boundary conditions is solved, and the graphical result is shown in Fig. E7.2c.

Part (*b*): The result of this part is shown in Fig. E7.2d. The effect of insulation on the right and top edges of the plate is evident. The gradient of the temperature near these boundaries approaches zero to satisfy the imposed boundary conditions. Because the insulation stops the flow of heat through these boundaries, the temperature along the insulated edges is higher than that of part (*a*).

Part (*c*): The Poisson equation is solved with a Poisson constant determined from Eq. (7.46):

$$f = -\frac{Q}{k} = -\frac{100000}{16}$$

The solution is shown in Fig. E7.2e. It can be seen from this figure that the temperature within the plate rises sharply to its highest value of 1078.3°C at the center point. Under these circumstances, the metal will begin to melt at the center core. Increasing the heat removed from the edges by convection, either by lowering the fluid temperature or increasing the heat transfer coefficient, lowers the internal temperature and can prevent melting of the plate.

7.4.2 PARABOLIC PARTIAL DIFFERENTIAL EQUATIONS

Classic examples of parabolic differential equations are the unsteady state heat conduction equation:

$$\frac{\partial T}{\partial t} = \alpha \left(\frac{\partial^2 T}{\partial x^2} + \frac{\partial^2 T}{\partial y^2} + \frac{\partial^2 T}{\partial z^2} \right) \tag{7.6}$$

and Fick's second law of diffusion:

$$\frac{\partial c_A}{\partial t} = D_{AB} \left(\frac{\partial^2 c_A}{\partial x^2} + \frac{\partial^2 c_A}{\partial y^2} + \frac{\partial^2 c_A}{\partial z^2} \right) \tag{7.5}$$

with Dirichlet, Neumann, or Cauchy boundary conditions.

Let us consider this class of equations in the general one-dimensional form:

$$\frac{\partial u}{\partial t} = \alpha \frac{\partial^2 u}{\partial x^2} \tag{7.14}$$

In this section we develop several methods of solution of Eq. (7.14) using finite differences.

Explicit methods: We express the derivatives in terms of central differences around the point (i, n), using the counter i for the x-direction and n for the t-direction:

$$\left. \frac{\partial^2 u}{\partial x^2} \right|_{i,n} = \frac{1}{\Delta x^2} \left(u_{i+1,n} - 2u_{i,n} + u_{i-1,n} \right) + O(\Delta x^2) \tag{7.49}$$

$$\left. \frac{\partial u}{\partial t} \right|_{i,n} = \frac{1}{2\Delta t} \left(u_{i,n+1} - u_{i,n-1} \right) + O(\Delta t^2) \tag{7.50}$$

Combining Eqs. (7.18), (7.54), and (7.55) and rearranging, we have:

$$u_{i,n+1} = u_{i,n-1} + \frac{2\alpha\Delta t}{\Delta x^2} \left(u_{i+1,n} - 2u_{i,n} + u_{i-1,n} \right) + O(\Delta x^2 + \Delta t^2) \tag{7.51}$$

This is an *explicit* algebraic formula that calculates the value of the dependent variable at the next time step $(u_{j,n+1})$ from values at the current and earlier time steps. Once the initial and boundary conditions of the problem are specified, the solution of an explicit formula is usually straightforward. However, this particular explicit formula is *unstable*, because it contains negative terms on the right side. A rigorous discussion of stability analysis is given in Section 7.6, but as a rule of thumb, when all the known values are arranged on the right side of the finite difference formulation, if there are any negative coefficients, the solution is unstable. This is stated more precisely by the positivity rule [5]: "For:

$$u_{i,n+1} = Au_{i+1,n} + Bu_{i,n} + Cu_{i-1,n} \tag{7.52}$$

if A, B, C are positive and $A + B + C \leq 1$, then the numerical scheme is stable."

To eliminate the instability problem, we replace the first-order derivative in Eq. (7.14) with the forward difference:

$$\left. \frac{\partial u}{\partial t} \right|_{i,n} = \frac{1}{\Delta t} \left(u_{i,n+1} - u_{i,n} \right) + O(\Delta t) \tag{7.53}$$

Combining Eqs. (7.14), (7.49), and (7.53) we obtain the explicit formula:

$$u_{i,n+1} = \left(\frac{\alpha\Delta t}{\Delta x^2} \right) u_{i+1,n} + \left(1 - 2\frac{\alpha\Delta t}{\Delta x^2} \right) u_{i,n} + \left(\frac{\alpha\Delta t}{\Delta x^2} \right) u_{i-1,n} + O(\Delta x^2 + \Delta t) \tag{7.54}$$

For a stable solution, the positivity rule requires that:

$$1 - 2\frac{\alpha \Delta t}{\Delta x^2} \geq 0 \tag{7.55}$$

Rearranging Eq. (7.55), we get:

$$\frac{\alpha \Delta t}{\Delta x^2} \leq \frac{1}{2} \tag{7.56}$$

This inequality determines the relationship between the two integration steps, Δx in the x-direction and Δt in the t-direction. As Δx gets smaller, Δt becomes much smaller, thus requiring longer computation times.

If we choose to work with the equality part of (7.55) or (7.56); that is,

$$\frac{\alpha \Delta t}{\Delta x^2} = \frac{1}{2} \tag{7.57}$$

then Eq. (7.54) simplifies to:

$$u_{i,n+1} = \frac{1}{2}\left(u_{i+1,n} + u_{i-1,n}\right) + O(\Delta x^2 + \Delta t) \tag{7.58}$$

This explicit formula calculates the value of the dependent variable at position i of the next time step $(n + 1)$ from values to the right and left of i at the present time step n. The computational molecule for this equation is shown in Fig. 7.6.

It should be emphasized that using the forward difference for the first-order derivative introduces the error of order $O(\Delta t)$. Therefore, Eq. (7.53) is of order $O(\Delta t)$ in the time direction and $O(\Delta x^2)$ in the x-direction. However, the advantage of gaining stability outweighs the loss of accuracy in this case.

The finite difference solution to the nonhomogeneous parabolic equation:

$$\frac{\partial u}{\partial t} = \alpha \frac{\partial^2 u}{\partial x^2} + f(x, t) \tag{7.59}$$

is given by the following explicit formula:

$$u_{i,n+1} = \left(\frac{\alpha \Delta t}{\Delta x^2}\right) u_{i+1,n} + \left(1 - 2\frac{\alpha \Delta t}{\Delta x^2}\right) u_{i,n} + \left(\frac{\alpha \Delta t}{\Delta x^2}\right) u_{i-1,n} + (\Delta t)f_{i,n} \tag{7.60}$$

We encounter equations of the type (7.59) when there is a source or sink in the physical problem.

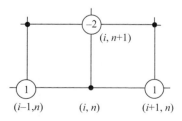

FIG. 7.6

Computational molecule of the one-dimensional parabolic differential equation.

The same treatment for the two-dimensional parabolic formula:

$$\frac{\partial u}{\partial t} = \alpha \left(\frac{\partial^2 u}{\partial x^2} + \frac{\partial^2 u}{\partial y^2} \right) + f(x, y, t) \tag{7.61}$$

results in:

$$u_{i,j,n+1} = \left(\frac{\alpha \Delta t}{\Delta x^2} \right) (u_{i+1,j,n} + u_{i-1,j,n}) + \left(\frac{\alpha \Delta t}{\Delta y^2} \right) (u_{i,j+1,n} + u_{i,j-1,n})$$
$$+ \left(1 - 2\frac{\alpha \Delta t}{\Delta x^2} - 2\frac{\alpha \Delta t}{\Delta y^2} \right) u_{i,j,n} + (\Delta t) f_{i,j,n} \tag{7.62}$$

The stability condition is obtained from the positivity rule is:

$$\frac{\alpha \Delta t}{\Delta x^2 + \Delta y^2} \leq \frac{1}{8} \tag{7.63}$$

The formula for the three-dimensional parabolic equation can be derived by adding to Eq. (7.62) the terms that come from $\partial^2 u/\partial z^2$. The right-hand side of the stability condition, in this case, is 1/18.

Parabolic partial differential equations can have initial and boundary conditions of the Dirichlet, Neumann, Cauchy, or Robin type. These boundary and initial conditions were discussed in Section 7.3. Examples of these conditions for the heat conduction problem are demonstrated in Fig. 7.2. The boundary conditions must be discretized using the same finite difference grid as used for the differential equation. For Dirichlet conditions, this simply involves setting the values of the dependent variable along the appropriate boundary equal to the given boundary condition. For Neumann and Robin conditions, the gradient at the boundaries must be replaced by finite difference approximations, resulting in additional algebraic equations that must be incorporated into the overall scheme of the solution of the resulting set of algebraic equations. In the case of Neumann or Robin conditions at the left boundary in the form:

$$\frac{\partial u}{\partial x} = \beta + \gamma u \quad \text{at } x = 0 \tag{7.64}$$

where $\gamma = 0$ for the Neumann condition, the following equation, based on forward differences with an error of the order $O(h^2)$, should be used at the boundary:

$$u_{0,n} = \frac{4u_{1,n} - u_{2,n} - 2\beta \Delta x}{3 + 2\gamma \Delta x} \tag{7.65}$$

In the case of Neumann or Robin condition at the right boundary, the formula based on backward differences with an error of the order $O(h^2)$ should be used ($i = N$):

$$u_{N,n} = \frac{-u_{N-1,n} + 4u_{N-2,n} + 2\beta \Delta x}{3 - 2\gamma \Delta x} \tag{7.66}$$

Eqs. (7.65) and (7.66) should be used (if and where applicable) for calculating the dependent variables at start and endpoints, respectively, after evaluating the dependent variable for the internal points from Eq. (7.54).

The flowchart of solving the parabolic partial differential equation by the explicit method is illustrated in Fig. 7.7.

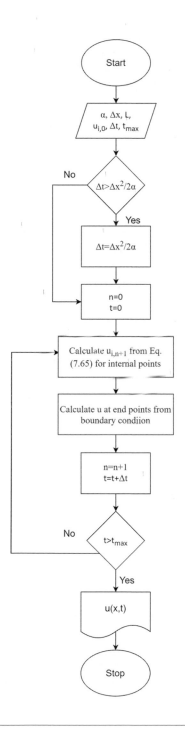

FIG. 7.7

Flowchart of solving the parabolic differential equation by the explicit method.

EXAMPLE 7.3 SOLVING PARABOLIC PARTIAL DIFFERENTIAL EQUATION BY THE EXPLICIT METHOD FOR DETERMINING THE UNSTEADY STATE VELOCITY PROFILE BETWEEN TWO FLAT PLATES

Water is placed between two horizontal flat plates that are 4 mm apart. Initially, the fluid and the plates are at rest. At time $t = 0$, the lower plate is set in motion with a velocity of 1 m/s while the upper plate remains motionless, as shown in Fig. E7.3a. Find the velocity of water as a function of x and t.

The unsteady state flow of water between two horizontal flat plates.

Assuming laminar flow and neglecting the end effects, the equation of motion for this problem is:

$$\frac{\partial V}{\partial t} = \nu \frac{\partial^2 V}{\partial x^2}$$

Initial and boundary values in this problem are:

$$\begin{aligned} &I.C. \quad t = 0, \quad V = 0 \\ &B.C. \quad x = 0, \quad V = 1 \\ &B.C. \quad x = 4, \quad V = 0 \end{aligned}$$

Use the explicit method for solving this parabolic differential equation and determine the velocity profile of water between the two plates. Consider $\Delta x = 1$ mm. The dynamic viscosity of water, ν, is 1×10^{-6} m²/s.

Method of solution

From the stability condition (7.55), we have:

$$\Delta t \leq \frac{\Delta x^2}{2\nu} = \frac{0.001^2}{2 \times 10^{-6}} = 0.5 \ s$$

The segments and initial values are illustrated in Fig. E7.3b.

Considering the maximum time step, Eq. (7.58) can be used for solving this parabolic partial differential equation.

(Continued)

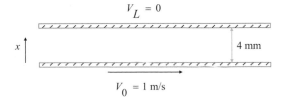

FIG. E7.3a

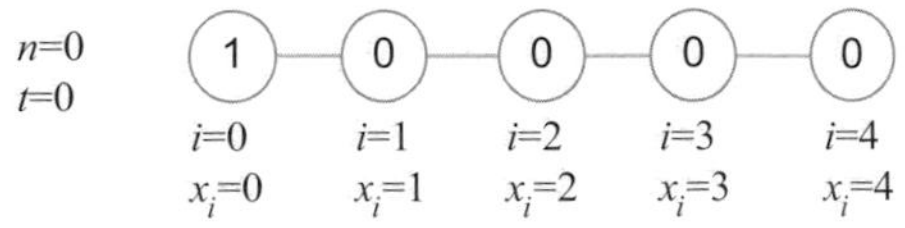

FIG. E7.3b

Initial condition.

EXAMPLE 7.3 (CONTINUED)

$$V_{i,n+1} = \frac{1}{2}\left(V_{i+1,n} + V_{i-1,n}\right)$$

Therefore the velocity profile in the next time step ($n+1=1$, $t=\Delta t = 0.5$ s) is calculated as follows:

$$V_{1,1} = \frac{1}{2}\left(V_{2,0} + V_{0,0}\right) = \frac{1}{2}(0+1) = 0.5$$

$$V_{2,1} = \frac{1}{2}\left(V_{3,0} + V_{1,0}\right) = \frac{1}{2}(0+0) = 0$$

$$V_{3,1} = \frac{1}{2}\left(V_{4,0} + V_{2,0}\right) = \frac{1}{2}(0+0) = 0$$

Similarly the velocity profile in the next time step ($n+1=2$, $t=2\Delta t = 1$ s) is:

$$V_{1,2} = \frac{1}{2}\left(V_{2,1} + V_{0,1}\right) = \frac{1}{2}(0+1) = 0.5$$

$$V_{2,2} = \frac{1}{2}\left(V_{3,1} + V_{1,1}\right) = \frac{1}{2}(0+0.5) = 0.25$$

$$V_{3,2} = \frac{1}{2}\left(V_{4,1} + V_{2,1}\right) = \frac{1}{2}(0+0) = 0$$

For the next time step ($n+1=3$, $t=3\Delta t = 1.5$ s), we have:

$$V_{1,3} = \frac{1}{2}\left(V_{2,2} + V_{0,2}\right) = \frac{1}{2}(0.25+1) = 0.625$$

$$V_{2,3} = \frac{1}{2}\left(V_{3,2} + V_{1,2}\right) = \frac{1}{2}(0+0.5) = 0.25$$

$$V_{3,3} = \frac{1}{2}\left(V_{4,2} + V_{2,2}\right) = \frac{1}{2}(0+0.25) = 0.125$$

Calculations can be continued in the same manner until the steady state condition (a linear velocity profile) is reached. Results of calculations for the first four time steps are shown in Fig. E7.3c.

Implicit methods

Let us now consider some implicit methods for the solution of parabolic equations. We use the grid of Fig. 7.8, in which the half point in the t-direction $(i, n + \frac{1}{2})$ is shown. Instead of expressing $\partial u / \partial t$ in terms of forward difference around (i, n) as it was done in the explicit form, we express this partial derivative in terms of central difference around the half-point:

$$\left. \frac{\partial u}{\partial t} \right|_{i,n+1/2} = \frac{1}{\Delta t} \left(u_{i,n+1} - u_{i,n} \right) \qquad (7.67)$$

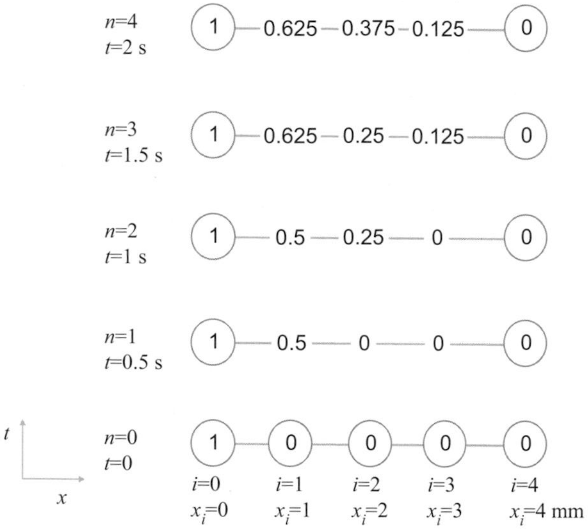

$n=4$
$t=2$ s $\boxed{1}$ — 0.625 — 0.375 — 0.125 — $\boxed{0}$

$n=3$
$t=1.5$ s $\boxed{1}$ — 0.625 — 0.25 — 0.125 — $\boxed{0}$

$n=2$
$t=1$ s $\boxed{1}$ — 0.5 — 0.25 — 0 — $\boxed{0}$

$n=1$
$t=0.5$ s $\boxed{1}$ — 0.5 — 0 — 0 — $\boxed{0}$

t $n=0$
$t=0$ $\boxed{1}$ — $\boxed{0}$ — $\boxed{0}$ — $\boxed{0}$ — $\boxed{0}$

x $\quad i=0 \qquad i=1 \qquad i=2 \qquad i=3 \qquad i=4$
$\quad x_i=0 \qquad x_i=1 \qquad x_i=2 \qquad x_i=3 \qquad x_i=4$ mm

FIG. E7.3c

The unsteady state velocity profile in the first four time steps, calculated by the explicit formula.

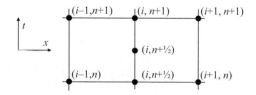

FIG. 7.8

Finite difference grid for the derivation of implicit formulas.

In addition, the second-order partial derivative is expressed at the halfway point as a weighted average of the central differences at points $(i, n + 1)$ and (i, n):

$$\frac{\partial^2 u}{\partial x^2}\bigg|_{i,n+1/2} = \theta\frac{\partial^2 u}{\partial x^2}\bigg|_{i,n+1} + (1 - \theta)\frac{\partial^2 u}{\partial x^2}\bigg|_{i,n} = \theta\left[\frac{1}{\Delta x^2}\left(u_{i+1,n+1} - 2u_{i,n+1} + u_{i-1,n+1}\right)\right]$$

$$+ (1 - \theta)\left[\frac{1}{\Delta x^2}\left(u_{i+1,n} - 2u_{i,n} + u_{i-1,n}\right)\right] \tag{7.68}$$

where θ is in the range $0 \leq \theta \leq 1$. A combination of Eqs. (7.14), (7.67), and (7.68) results in the *variable-weighted implicit* approximation of the parabolic partial differential equation:

$$\alpha\theta\left[\frac{1}{\Delta x^2}\left(u_{i+1,n+1} - 2u_{i,n+1} + u_{i-1,n+1}\right)\right] - \frac{1}{\Delta t}u_{i,n+1}$$

$$= -\alpha(1 - \theta)\left[\frac{1}{\Delta x^2}\left(u_{i+1,n} - 2u_{i,n} + u_{i-1,n}\right)\right] - \frac{1}{\Delta t}u_{i,n} \tag{7.69}$$

This formula is implicit because the left-hand side involves more than one value at the $(n + 1)$ position of the difference grid (i.e., more than one unknown at any step in the time domain).

When $\theta = 0$, Eq. (7.69) becomes identical to the classic explicit formula (7.60). When $\theta = 1$, Eq. (7.69) becomes:

$$-\left(\frac{\alpha\Delta t}{\Delta x^2}\right)u_{i-1,n+1} + \left(1 + 2\frac{\alpha\Delta t}{\Delta x^2}\right)u_{i,n+1} - \left(\frac{\alpha\Delta t}{\Delta x^2}\right)u_{i+1,n+1} = u_{i,n} \tag{7.70}$$

This is called the *backward implicit* approximation, which can also be obtained by approximating the first-order partial derivative using the backward difference at $(i, n + 1)$ and the second-order partial derivative by the central difference at $(i, n + 1)$.

Finally, when $\theta = \frac{1}{2}$, Eq. (7.69) yields the well-known *Crank-Nicolson implicit formula*:

$$-\left(\frac{\alpha\Delta t}{\Delta x^2}\right)u_{i-1,n+1} + 2\left(1 + \frac{\alpha\Delta t}{\Delta x^2}\right)u_{i,n+1} - \left(\frac{\alpha\Delta t}{\Delta x^2}\right)u_{i+1,n+1}$$

$$= \left(\frac{\alpha\Delta t}{\Delta x^2}\right)u_{i-1,n} + 2\left(1 - \frac{\alpha\Delta t}{\Delta x^2}\right)u_{i,n} + \left(\frac{\alpha\Delta t}{\Delta x^2}\right)u_{i+1,n} \tag{7.71}$$

For an implicit solution to the nonhomogeneous parabolic equation:

$$\frac{\partial u}{\partial t} = \alpha\frac{\partial^2 u}{\partial x^2} + f(x,t) \tag{7.59}$$

by the above method, we also need to calculate the value of f at the midpoint $(i, n + \frac{1}{2})$, which we take as the average of the value of f at grid points $(i, n + 1)$ and (i, n):

$$f_{i,n+1/2} = \frac{1}{2}\left(f_{i,n+1} + f_{i,n}\right) \tag{7.72}$$

Putting Eqs. (7.67), (7.68) (considering $\theta = \frac{1}{2}$), and (7.72) into Eq. (7.59) results in:

$$-\left(\frac{\alpha \Delta t}{\Delta x^2}\right)u_{i-1,n+1} + 2\left(1 + \frac{\alpha \Delta t}{\Delta x^2}\right)u_{i,n+1} - \left(\frac{\alpha \Delta t}{\Delta x^2}\right)u_{i+1,n+1} - (\Delta t)f_{i,n+1}$$
$$= \left(\frac{\alpha \Delta t}{\Delta x^2}\right)u_{i-1,n} + 2\left(1 - \frac{\alpha \Delta t}{\Delta x^2}\right)u_{i,n} + \left(\frac{\alpha \Delta t}{\Delta x^2}\right)u_{i+1,n} + (\Delta t)f_{i,n} \tag{7.73}$$

Eq. (7.73) is the Crank-Nicolson implicit formula for the solution of the nonhomogeneous parabolic partial differential Eq. (7.59).

When written for the entire difference grid, implicit formulas generate sets of simultaneous linear algebraic equations whose matrix of coefficients is usually tridiagonal. This type of problem may be solved using a Gauss elimination procedure or more efficiently using the Thomas algorithm [4], which is a variation of Gauss elimination.

Implicit formulas of the type described above are unconditionally stable. It can be generalized that most explicit finite difference approximations are conditionally stable, whereas most implicit approximations are unconditionally stable. The explicit methods, however, are computationally easier to solve than the implicit techniques.

EXAMPLE 7.4 SOLVING PARABOLIC PARTIAL DIFFERENTIAL EQUATION BY THE CRANK-NICOLSON IMPLICIT METHOD FOR DETERMINING THE UNSTEADY STATE VELOCITY PROFILE BETWEEN TWO FLAT PLATES

Repeat Example 7.3 and employ the Crank-Nicolson implicit formula for calculating the velocity profile.

Method of solution

Eq. (7.71) is used for solving this problem. Considering $\Delta t = 0.5$ s, the set of equation in the first time step ($n + 1 = 1$) is:

$$i = 1, \quad -0.5V_{0,1} + 3V_{1,1} - 0.5V_{2,1} = 0.5V_{0,0} + V_{1,0} + 0.5V_{2,0}$$
$$i = 2, \quad -0.5V_{1,1} + 3V_{2,1} - 0.5V_{3,1} = 0.5V_{1,0} + V_{2,0} + 0.5V_{3,0}$$
$$i = 3, \quad -0.5V_{2,1} + 3V_{3,1} - 0.5V_{4,1} = 0.5V_{2,0} + V_{3,0} + 0.5V_{4,0}$$

Inserting initial and boundary conditions into this set of equations and rearranging results in:

$$3V_{1,1} - 0.5V_{2,1} = 1$$
$$-0.5V_{1,1} + 3V_{2,1} - 0.5V_{3,1} = 0$$
$$-0.5V_{2,1} + 3V_{3,1} = 0$$

This set of linear algebraic equations can be solved by one of the methods described in Chapter 2, from which we get:

$$V_{1,1} = 0.3431, \quad V_{2,1} = 0.0588, \quad V_{3,1} = 0.0098$$

(Continued)

EXAMPLE 7.4 (CONTINUED)

In the next time step ($n + 1 = 2$), the set of equations becomes:

$$i = 1, \quad -0.5V_{0,2} + 3V_{1,2} - 0.5V_{2,2} = 0.5V_{0,1} + V_{1,1} + 0.5V_{2,1}$$
$$i = 2, \quad -0.5V_{1,2} + 3V_{2,2} - 0.5V_{3,2} = 0.5V_{1,1} + V_{2,1} + 0.5V_{3,1}$$
$$i = 3, \quad -0.5V_{2,2} + 3V_{3,2} - 0.5V_{4,2} = 0.5V_{2,1} + V_{3,1} + 0.5V_{4,1}$$

or

$$3V_{1,1} - 0.5V_{2,1} = 1.4019$$
$$-0.5V_{1,1} + 3V_{2,1} - 0.5V_{3,1} = 0.23525$$
$$-0.5V_{2,1} + 3V_{3,1} = 0.0392$$

and

$$V_{1,1} = 0.4953, \quad V_{2,1} = 0.1678, \quad V_{3,1} = 0.0410$$

Calculations can be continued in the same manner for the next time steps. Results of calculations for the first four time steps are shown in Fig. E7.4.

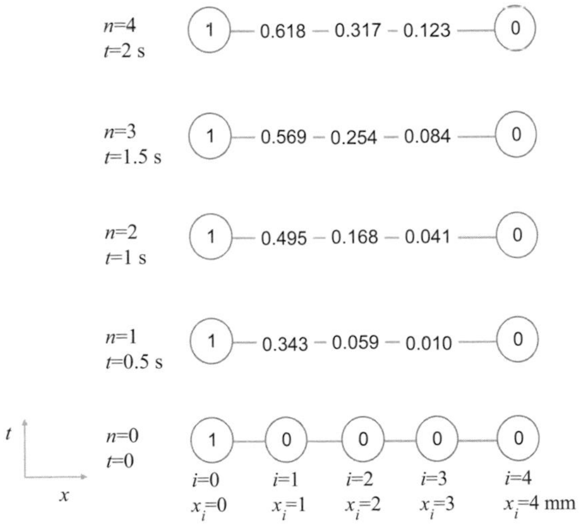

FIG. E7.4

The unsteady state velocity profile in the first four time steps, calculated by the Crank-Nicolson formula.

Method of lines

Another technique for the solution of parabolic partial differential equations is the *method of lines*. This is based on the concept of converting the partial differential equation into a set of ordinary differential equations by discretizing only the spatial derivatives using finite differences and leaving the time derivatives unchanged. This concept, applied to Eq. (7.14), results in:

$$\frac{du_i}{dt} = \frac{\alpha}{\Delta x^2}(u_{i+1} - 2u_i + u_{i-1}) \tag{7.74}$$

There will be as many of these ordinary differential equations as there are grid points in the *x*-direction (Fig. 7.9). The complete set of differential equations for $0 \le i \le N$ would be:

$$\frac{du_0}{dt} = \frac{\alpha}{\Delta x^2}(u_{-1} - 2u_0 + u_1) \tag{7.75a}$$

$$\frac{du_i}{dt} = \frac{\alpha}{\Delta x^2}(u_{i-1} - 2u_i + u_{i+1}) \tag{7.75b}$$

$$\frac{du_N}{dt} = \frac{\alpha}{\Delta x^2}(u_{N-1} - 2u_N + u_{N+1}) \tag{7.75c}$$

The two equations at the boundaries, (7.75a) and (7.75c), would have to be modified according to the boundary conditions that are specified in the particular problem. For example, if a Dirichlet condition is given at $x = 0$ and $t > 0$, that is,

$$u_0 = \beta \text{ for } t > 0 \tag{7.76}$$

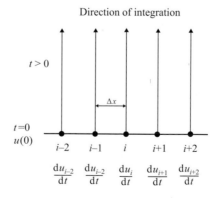

FIG. 7.9

Method of lines.

then Eq. (7.75a) is modified to:

$$\frac{du_0}{dt} = 0, \qquad u_0(0) = \beta \tag{7.77}$$

On the other hand, if a Neumann condition is given at this boundary, that is,

$$\frac{\partial u}{\partial x}\bigg|_{0,t} = \beta \quad \text{at } x = 0 \text{ and } t > 0 \tag{7.78}$$

the partial derivative is replaced by a central difference approximation:

$$\frac{\partial u}{\partial x}\bigg|_{0,t} = \frac{u_1 - u_{-1}}{2\Delta x} = \beta \tag{7.79}$$

We then extract u_1 from this equation and replace it in Eq. (7.75a). Thus, the first equation in the set of differential equations (7.75) becomes:

$$\frac{du_0}{dt} = \frac{2\alpha}{\Delta x^2}(u_1 - u_0 - \beta\Delta x) \tag{7.80}$$

For the Robin conditions at the left boundary,

$$\frac{\partial u}{\partial x} = \beta + \gamma u \quad \text{at } x = 0 \tag{7.81}$$

the same procedure can be followed, and Eq. (7.75a) in this case becomes:

$$\frac{du_0}{dt} = \frac{2\alpha}{\Delta x^2}[u_1 - (1 + \gamma\Delta x)u_0 - \beta\Delta x] \tag{7.82}$$

The complete set of simultaneous differential equations must be integrated forward in time (the *n*-direction) starting with the initial conditions of the problem. This method gives stable solutions for parabolic partial differential equations.

EXAMPLE 7.5 SOLVING PARABOLIC PARTIAL DIFFERENTIAL EQUATION BY THE METHOD OF LINES FOR DETERMINING THE UNSTEADY STATE VELOCITY PROFILE BETWEEN TWO FLAT PLATES

Repeat Example 7.3 and use the method of lines to write the system of ordinary differential equations for determining the unsteady state velocity profile.

Method of solution

Let us use the same number of divisions in the *x*-direction. There is a Dirichlet condition at both ends of the domain.

(Continued)

EXAMPLE 7.5 (CONTINUED)

$$\frac{dV_0}{dt} = 0 \qquad\qquad V_0(0) = 1$$

$$\frac{dV_1}{dt} = V_0 - 2V_1 + V_2 \qquad\qquad V_1(0) = 0$$

$$\frac{dV_2}{dt} = V_1 - 2V_2 + V_3 \qquad\qquad V_2(0) = 0$$

$$\frac{dV_3}{dt} = V_2 - 2V_3 + V_4 \qquad\qquad V_3(0) = 0$$

$$\frac{dV_4}{dt} = 0 \qquad\qquad V_4(0) = 0$$

This is a set of linear ordinary differential equations that can be presented in the matrix form as follows:

(Continued)

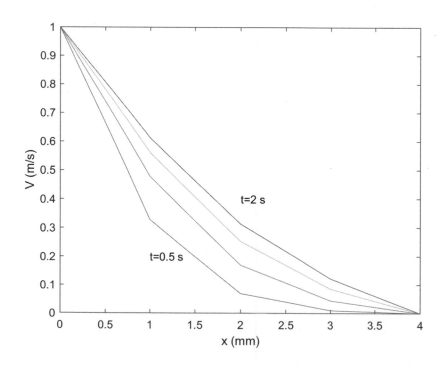

FIG. E7.5

Unsteady state profile of the velocity water flowing between two flat plates.

EXAMPLE 7.5 (CONTINUED)

$$\frac{dV}{dt} = AV \quad V(0) = V_0$$

where

$$A = \begin{bmatrix} 0 & 0 & 0 & 0 & 0 \\ 1 & -2 & 1 & 0 & 0 \\ 0 & 1 & -2 & 1 & 0 \\ 0 & 0 & 1 & -2 & 1 \\ 0 & 0 & 0 & 0 & 0 \end{bmatrix} \quad V = \begin{bmatrix} V_0 \\ V_1 \\ V_2 \\ V_3 \\ V_4 \end{bmatrix} \quad V_0 = \begin{bmatrix} 1 \\ 0 \\ 0 \\ 0 \\ 0 \end{bmatrix}$$

This set of equations can be solved by the function *LinearODE.m* developed in Example 5.2 (see Chapter 5). The results are shown graphically in Fig. E7.5.

EXAMPLE 7.6 SOLUTION OF THE PARABOLIC PARTIAL DIFFERENTIAL EQUATION FOR DIFFUSION

Write a general MATLAB function to determine the numerical solution of the parabolic partial differential equation

$$\frac{\partial u}{\partial t} = \alpha \frac{\partial^2 u}{\partial x^2} + f(x, t) \tag{7.59}$$

using the Crank-Nicolson implicit formula. The function f may be a constant value or linear with respect to u. Apply this MATLAB function to solve the following problems (in this problem, $u \equiv c_A$, and z is used instead of x to indicate the length).

(a) The stagnant liquid B in a container that is 10 cm high ($L = 10$ cm) is exposed to the nonreactant gas A at time $t = 0$. The concentration of A, dissolved physically in B, reaches $c_{A0} = 0.01$ mol/m^3 at the interface instantly and remains constant. The diffusion coefficient of A in B is $D_{AB} = 2 \times 10^{-9}$ m^2/s. Determine the evolution of the concentration of A within the container. Plot the flux of A dissolved in B against time.

(b) Repeat part (a), but this time consider that A reacts with B according to the following reaction:

$$A + B \rightarrow C \quad -r_A = 2 \times 10^{-7} c_A \text{ mol/s.m}^3$$

Method of solution

The physical problem is sketched in Fig. E7.6a. The mole balance of A for part (a) leads to:

$$\frac{\partial c_A}{\partial t} = D_{AB} \frac{\partial^2 c_A}{\partial z^2} \tag{1}$$

The boundary and initial conditions for Eq. (1) are:

(Continued)

Gas A

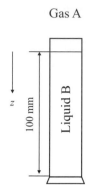

FIG. E7.6a

Diffusion of A through B.

EXAMPLE 7.6 (CONTINUED)

$$\text{I.C.} \quad c_A(z,0) = 0 \quad \text{for } z > 0 \tag{2}$$

$$\text{B.C. 1} \quad c_A(0,t) = c_{A0} \quad \text{for } t \geq 0 \tag{3}$$

$$\text{B.C. 2} \quad \frac{\partial c_A}{\partial z}\bigg|_{z=L} = 0 \quad \text{for } t \geq 0 \tag{4}$$

For part (b), moles of A are consumed by the liquid B while diffusing in it. Assuming that the concentration of the product C is negligible, so that the diffusion coefficient remains unchanged, the mole balance of A results in

$$\frac{\partial c_A}{\partial t} = D_{AB}\frac{\partial^2 c_A}{\partial z^2} + kc_A \tag{5}$$

Initial and boundary conditions remain the same (Eqs. 2–4).

Once the concentration profile of A is known, the molar flux of A entering the liquid B can be calculated from Fick's law for both parts (a) and (b):

$$N_{Az}(t) = -D_{AB}\frac{\partial c_A}{\partial z}\bigg|_{z=0} \tag{6}$$

The Crank-Nicolson implicit formula (7.73) is used for the solution of this problem:

$$-\left(\frac{\alpha\Delta t}{\Delta x^2}\right)u_{i-1,n+1} + 2\left(1 + \frac{\alpha\Delta t}{\Delta x^2}\right)u_{i,n+1} - \left(\frac{\alpha\Delta t}{\Delta x^2}\right)u_{i+1,n+1} - (\Delta t)f_{i,n+1}$$

$$= \left(\frac{\alpha\Delta t}{\Delta x^2}\right)u_{i-1,n} + 2\left(1 - \frac{\alpha\Delta t}{\Delta x^2}\right)u_{i,n} + \left(\frac{\alpha\Delta t}{\Delta x^2}\right)u_{i+1,n} + (\Delta t)f_{i,n}$$

(Continued)

EXAMPLE 7.6 (CONTINUED)

Because the function f is linear with respect to u, this equation represents a set of linear algebraic equations that may be solved by the matrix inversion method, at each time step.

When a Neumann or Robin condition is specified, a forward or backward finite difference approximation of the first derivative of order $O(h^2)$ is applied at the start or endpoint of the x-direction, respectively.

Program description

The programs and functions developed in this example can be found in: https://www.elsevier.com/books-and-journals/book-companion/9780128229613

The MATLAB function *parabolic1D.m* is developed to solve the parabolic partial differential equation in an unsteady state one-dimensional problem. The boundary conditions are passed to the function in the same format as that of Example 7.2, with the exception that they are given in only the x-direction. The function also needs the initial condition, u_0, which is a vector containing the values of the dependent variable for all x at time $t = 0$.

The first part of the function is initialization, which checks the inputs and sets the values required in the calculations. The solution of the equation follows next and consists of an outer loop on the time interval. At each time interval, the matrix of coefficients and the vector of constants of the set of Eq. (7.74) is formed. The function then solves the set of linear algebraic equations that gives the value of the dependent variable in this time interval. This procedure continues until the limit time is reached. If the problem at hand is nonhomogeneous, the name of the MATLAB function containing the function f should be given as the eighth input argument. Because this function is assumed to be linear with respect to u, the set of algebraic Eq. (7.74) remain linear. The function corrects the matrix of coefficients and the vector of constant in this case, accordingly.

The main program *Example7_6.m* asks the user to input all the necessary parameters for solving the problem from the keyboard. It then calls the function *parabolic1D.m* to solve the partial differential equation and finally plots the contour line graph of concentration profiles and the plot of the molar flux of A entered the container versus time.

The function *Ex7_6_func.m* contains the rate law equation for the reaction of part (b). It is important to note that the first to third input arguments to this function have to be u, x, and t, respectively, even if one of them is not used in the function f.

Input and results
Part (a).

>> **Example7_6**
Solution of parabolic partial differential equation.
Depth of the container (m) = 0.1
Maximum time (s) = 70*3600*24
Number of divisions in z-direction = 10

(Continued)

EXAMPLE 7.6 (CONTINUED)

Number of divisions in t-direction = 500
Diffusion coefficient of A in B = 2e-9
1 - No reaction between A and B
2 - A reacts with B
Enter your choice: 1
Boundary conditions:
Concentration of A at interface (mol/m^3) = 0.01
Condition at Bottom of the container:
   1 - Dirichlet
   2 - Neumann
   3 - Robin
Enter your choice: 2
Value = 0

Part (b).

>> **Example7_6**
Solution of parabolic partial differential equation.
Depth of the container (m) = 0.1
Maximum time (s) = 70*3600*24
Number of divisions in z-direction = 10
Number of divisions in t-direction = 500
Diffusion coefficient of A in B = 2e-9
1 - No reaction between A and B
2 - A reacts with B
Enter your choice: 2
Rate constant = 2e-7
Name of the file containing the rate law = 'Ex7_6_func'
Boundary conditions:
Concentration of A at interface (mol/m^3) = 0.01
Condition at Bottom of the container:
1 - Dirichlet
2 - Neumann
3 - Robin
Enter your choice: 2
Value = 0

Discussion of results

Part (*a*): The unsteady state concentration profile is plotted in Fig. E7.6b. The steady state concentration profile is $c_A = 0.01$ mol/m^3 at all levels. The unsteady state mole flux of *A*

(Continued)

EXAMPLE 7.6 (CONTINUED)

entering the container is shown in Fig. E7.6c. This flux decreases with time and reaches zero at steady state.

Part (b): The unsteady state concentration profile is plotted in Fig. E7.6d. The steady state concentration profile is:

$$\frac{c_A}{c_{A_0}} = \frac{\cosh\beta\left[1 - (z/L)\right]}{\cosh\beta} \quad \text{where} \quad \beta = \sqrt{kL^2/D_{AB}}$$

The unsteady state mole flux of A entering the container is shown in Fig. E7.6e. This flux decreases with time in the beginning. However, it reaches the steady state value of 1.3×10^{-5} mol/m^{3}day. This happens because A is constantly consumed by B in the container. In fact, the steady state flux at top of the container is equal to the consumption of A in the container by the reaction:

$$-D_{AB}\frac{\partial c_A}{\partial z}\bigg|_{z=0} = \int_0^L (-r_A)dz = \frac{D_{AB}c_{A_0}}{L}\beta\tanh\beta \quad t \to \infty$$

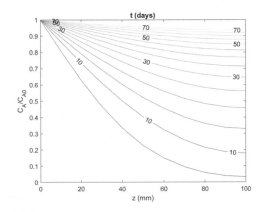

FIG. E7.6b

Concertation profile of A with no reaction.

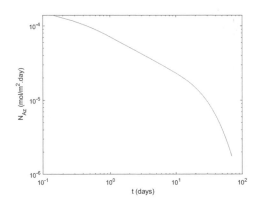

FIG. E7.6c

Flux profile of A with no reaction.

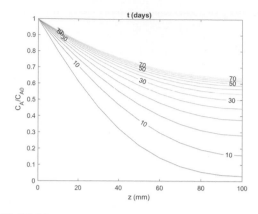

FIG. E7.6d

Concertation profile of *A* with reaction.

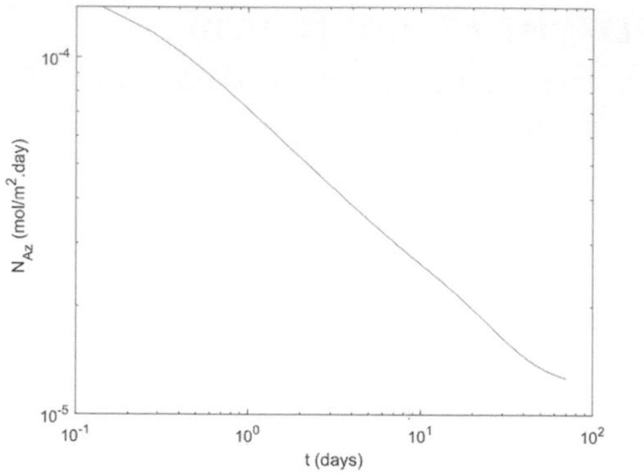

FIG. E7.6e

Flux profile of *A* with reaction.

EXAMPLE 7.7 TWO-DIMENSIONAL PARABOLIC PARTIAL DIFFERENTIAL EQUATION FOR HEAT TRANSFER

Write a general MATLAB function to determine the numerical solution of the parabolic partial differential equation:

$$\frac{\partial u}{\partial t} = \alpha \left(\frac{\partial^2 u}{\partial x^2} + \frac{\partial^2 u}{\partial y^2} \right) + f(x, y, t) \tag{7.61}$$

by explicit method. Apply this function to solve the following problems ($u \equiv T$):

(a) The wall of a furnace is 20 cm thick (*x*-direction) and 50 cm long (*y*-direction) and is made of brick, which has a thermal diffusivity of 2×10^{-7} m²/s. The temperature of the wall is 25°C when the furnace is off. When the furnace is fired, the temperature on the inside face of the wall ($x = 0$) reaches 500°C quite rapidly. The temperature of the outside face of the wall is maintained at 25°C. The other two faces of the wall (*y*-direction) are assumed to be perfectly insulated. Determine the evolution of temperature profiles within the brick wall.

(b) Insulation is placed on the outside surface of the wall. Assume this is also perfect insulation and show the evolution of the temperature profiles within the wall when the furnace is fired to 500°C.

(c) The furnace wall of part (a) is initially at a uniform temperature of 500°C. Both sides of the wall are exposed to forced air circulation at 25°C and the heat transfer coefficient is 20 W/m²°C. The faces of the wall in the *y*-direction are assumed to be perfectly insulated. Show the temperature profiles within the wall.

Method of solution

The explicit formula (7.62) is used for the solution of this problem:

$$\begin{aligned} u_{i,j,n+1} = {} & \left(\frac{\alpha \Delta t}{\Delta x^2} \right) \left(u_{i+1,j,n} + u_{i-1,j,n} \right) + \left(\frac{\alpha \Delta t}{\Delta y^2} \right) \left(u_{i,j+1,n} + u_{i,j-1,n} \right) \\ & + \left(1 - 2 \frac{\alpha \Delta t}{\Delta x^2} - 2 \frac{\alpha \Delta t}{\Delta y^2} \right) u_{i,j,n} + (\Delta t) f_{i,j,n} \end{aligned} \tag{7.62}$$

(Continued)

EXAMPLE 7.7 (CONTINUED)

The value of the time increment Δt for a stable solution is calculated from Eq. (7.63):

$$\frac{\alpha \Delta t}{\Delta x^2 + \Delta y^2} \leq \frac{1}{8} \tag{7.63}$$

When Neumann or Robin conditions are specified (e.g., for Neumann condition $\gamma = 0$),

$$\frac{\partial u}{\partial x}\Big|_{0,y,t} = \beta + \gamma u(0, y, t)$$

a forward difference approximation of the condition is used at this boundary:

$$\frac{1}{2\Delta x}(-u_2 + 4u_1 - 3u_0) = \beta + \gamma u_0$$

from which the dependent variable at the boundary, u_0, can be obtained:

$$u_0 = \frac{-2\beta\Delta x - u_2 + 4u_1}{3 + 2\gamma\Delta x}$$

Similarly if the Neumann or Robin condition is at the upper boundary, the dependent variable can be calculated from:

$$u_{p+1} = \frac{-2\beta\Delta x + u_{p-1} + 4u_p}{-3 + 2\gamma\Delta x}$$

The same discussion applies to y-direction boundaries.

Program description

The programs and functions developed in this example can be found in: https://www.elsevier.com/books-and-journals/book-companion/9780128229613

The MATLAB function *parabolic2D.m* is written for the solution of the parabolic partial differential equation in an unsteady state two-dimensional problem. The boundary conditions are passed to the function in the same format as that of Example 7.2. The initial condition, u_0, is a matrix of the values of the dependent variable for all x and y at time $t = 0$. If the problem at hand is nonhomogeneous, the name of the MATLAB function containing the function f should be given as the 10th input argument.

The function starts with the initialization section, which checks the inputs and sets the values required in the calculations. The solution of the equation follows next and consists of an outer loop on the time interval. At each time interval, values of the dependent variable for inner grid points are being calculated based on Eq. (7.61), followed by the calculation of the grid points on the boundaries according to the formula developed in the previous section. The values of the dependent variable at corner points are assumed to be the average of their adjacent points on the converging boundaries.

(Continued)

EXAMPLE 7.7 (CONTINUED)

In the main program *Example7_7.m*, all the necessary parameters for solving the problem are introduced from the keyboard. The program then asks for initial and boundary conditions, builds the matrix of initial conditions, and calls the function *parabolic2D.m* to solve the partial differential equation. It is possible to repeat the same problem with different initial and boundary conditions.

The last part of the program is the visualization of the results. There are two ways to look at the results. One way is dynamic visualization, which is an animation of the temperature profile evolution of the wall. This method may be time-consuming, because it makes individual frames of the temperature profiles at each time interval and then shows them one after another using the *movie* command. Instead, the user may select the other option, which is to see a summary of the results in nine succeeding chronological frames.

Input and results
Part (a).

>> **Example7_7**
Solution of two-dimensional parabolic
partial differential equation.
Width of the plate (x-direction) (m) = 0.2
Length of the plate (y-direction) (m) = 0.5
Maximum time (hr) = 12
Number of divisions in x-direction = 8
Number of divisions in y-direction = 8
Number of divisions in t-direction = 30
Thermal diffusivity of the wall = 2e-7
Initial temperature of the wall (deg C) = 25
Boundary conditions:
Lower x boundary condition:
1 - Dirichlet
2 - Neumann
3 - Robin
Enter your choice: 1
Value = 500
Upper x boundary condition:
1 - Dirichlet
2 - Neumann
3 - Robin
Enter your choice: 1
Value = 25
Lower y boundary condition:
1 - Dirichlet

(Continued)

EXAMPLE 7.7 (CONTINUED)

2 - Neumann
3 - Robin
Enter your choice: 2
Value = 0
Upper y boundary condition:
1 - Dirichlet
2 - Neumann
3 - Robin
Enter your choice: 2
Value = 0
Create a movie of temperature profile evolution (0/1)? 0

Part (b).

>> **Example7_7**
Solution of two-dimensional parabolic
partial differential equation.
Width of the plate (x-direction) (m) = 0.2
Length of the plate (y-direction) (m) = 0.5
Maximum time (hr) = 12
Number of divisions in x-direction = 8
Number of divisions in y-direction = 8
Number of divisions in t-direction = 30
Thermal diffusivity of the wall = 2e-7
Initial temperature of the wall (deg C) = 25
Boundary conditions:
Lower x boundary condition:
1 - Dirichlet
2 - Neumann
3 - Robin
Enter your choice: 1
Value = 500
Upper x boundary condition:
1 - Dirichlet
2 - Neumann
3 - Robin
Enter your choice: 2
Value = 0
Lower y boundary condition:
1 - Dirichlet
2 - Neumann

(Continued)

EXAMPLE 7.7 (CONTINUED)

3 - Robin
Enter your choice: 2
Value = 0
Upper y boundary condition:
1 - Dirichlet
2 - Neumann
3 - Robin
Enter your choice: 2
Value = 0
Create a movie of temperature profile evolution (0/1)? 0

Part (c).

>> **Example7_7**
Solution of two-dimensional parabolic
partial differential equation.
Width of the plate (x-direction) (m) = 0.2
Length of the plate (y-direction) (m) = 0.5
Maximum time (hr) = 12
Number of divisions in x-direction = 8
Number of divisions in y-direction = 8
Number of divisions in t-direction = 30
Thermal diffusivity of the wall = 2e-7
Initial temperature of the wall (deg C) = 500
Boundary conditions:
Lower x boundary condition:
1 - Dirichlet
2 - Neumann
3 - Robin
Enter your choice: 3
u' = (beta) + (gamma)*u
Constant (beta) = −25*20
Coefficient (gamma) = 20
Upper x boundary condition:
1 - Dirichlet
2 - Neumann
3 - Robin
Enter your choice: 3
u' = (beta) + (gamma)*u
Constant (beta) = 25*20
Coefficient (gamma) = -20

(Continued)

EXAMPLE 7.7 (CONTINUED)

Lower y boundary condition:

1 - Dirichlet

2 - Neumann

3 - Robin

Enter your choice: 2

Value = 0

Upper y boundary condition:

1 - Dirichlet

2 - Neumann

3 - Robin

Enter your choice: 2

Value = 0

Create a movie of temperature profile evolution (0/1)? 0

Discussion of results

Part (a): Heat transfers from the inside of the furnace (left boundary), where the temperature is at 500°C, toward the outside (right boundary), where the temperature is maintained at 25°C. Therefore, the temperature profile progresses as shown in Fig. E7.7a. If the integration is continued for a sufficiently long time, the profile will reach the steady state, which in this case is a straight plane connecting the two Dirichlet conditions. This is easily verified from the analytical solution of the steady state problem,

$$\frac{\partial^2 T}{\partial x^2} + \frac{\partial^2 T}{\partial y^2} = 0$$

Because the two faces of the wall in the y-direction are insulated, this becomes essentially a one-dimensional problem that yields the equation of a straight line,

$$T = -2375x + 500$$

which is calculated using the Dirichlet conditions.

Part (b): In this case, the insulation installed on the outside surface of the furnace wall causes the temperature within the wall to continue rising, as shown in Fig. E7.7b. The steady state temperature profile would be $T = 500$°C throughout the solid wall. This is also verifiable from the analytical solution of the steady state problem in conjunction with the imposed boundary conditions.

Part (c): The cooling of the wall occurs from both sides, and the temperature profile moves, symmetrically, as shown in Fig. E7.7c. The final temperature would be 25°C.

The reader is encouraged to repeat this example and choose the movie option to see the temperature profile evolution dynamically. It should be noted that the rate of evolution of the temperature profile on the screen is not the same as that of the process itself.

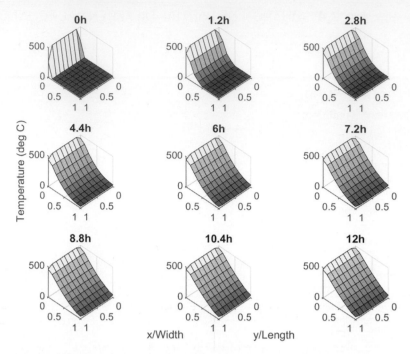

FIG. E7.7a Evolution of temperature in the wall of the furnace with no insulation.

The length and width have been normalized to be in the range of (0, 1).

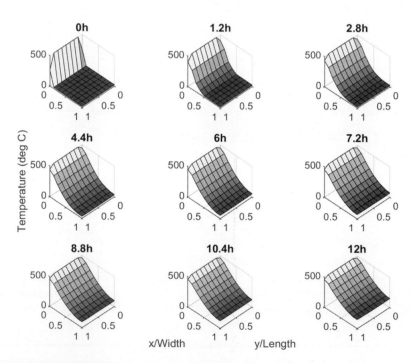

FIG. E7.7b Evolution of temperature in the wall of the furnace with insulation.

The length and width have been normalized to be in the range of (0, 1).

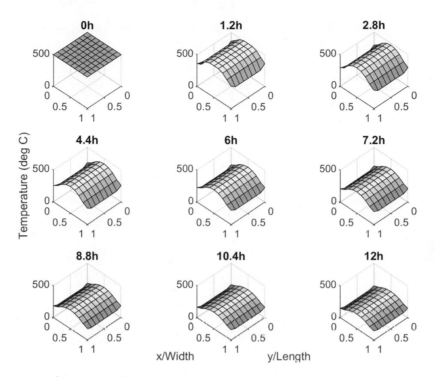

FIG. E7.7c **Evolution of temperature in the wall of the furnace with cooling from both sides.**

The length and width have been normalized to be in the range of (0, 1).

7.4.3 HYPERBOLIC PARTIAL DIFFERENTIAL EQUATIONS

Second-order partial differential equations of the hyperbolic type occur principally in physical problems connected with vibration processes. For example, the one-dimensional wave equation:

$$\rho \frac{\partial^2 u}{\partial t^2} = T_0 \frac{\partial^2 u}{\partial x^2} + f(x, t) \tag{7.83}$$

describes the transverse motion of a vibrating string that is subjected to tension T_0 and external force $f(x, t)$. In the case of constant density ρ, the equation is written in the form:

$$\frac{\partial^2 u}{\partial t^2} = a^2 \frac{\partial^2 u}{\partial x^2} + F(x, t) \tag{7.84}$$

where

$$a^2 = \frac{T_0}{\rho} \quad \text{and} \quad F(x, t) = \frac{1}{\rho} f(x, t) \tag{7.85}$$

If no external force acts on the string, Eq. (7.82) becomes a homogeneous equation:

$$\frac{\partial^2 u}{\partial t^2} = a^2 \frac{\partial^2 u}{\partial x^2} \tag{7.86}$$

The two-dimensional extension of Eq. (7.84) is:

$$\frac{\partial^2 u}{\partial t^2} = a^2 \left(\frac{\partial^2 u}{\partial x^2} + \frac{\partial^2 u}{\partial y^2} \right) + F(x, y, t) \tag{7.87}$$

which describes the vibration of a membrane subjected to tension T_0 and external force $f(x, y, t)$.

To find the numerical solution of Eq. (7.86), we expand each second-order derivative in terms of central finite differences to obtain:

$$\frac{u_{i,n+1} - 2u_{i,n} + u_{i,n-1}}{\Delta t^2} = a^2 \left(\frac{u_{i+1,n} - 2u_{i,n} + u_{i-1,n}}{\Delta x^2} \right) + O(\Delta x^2 + \Delta t^2) \tag{7.88}$$

Rearranging to solve for $u_{i,n+1}$,

$$u_{i,n+1} = 2\left(1 - \frac{a^2 \Delta t^2}{\Delta x^2} \right) u_{i,n} + \frac{a^2 \Delta t^2}{\Delta x^2} \left(u_{i+1,n} + u_{i-1,n} \right) - u_{i,n-1} + O(\Delta x^2 + \Delta t^2) \tag{7.89}$$

This is an *explicit* numerical solution of the hyperbolic Eq. (7.86).

The positivity rule (Eq. 7.52) applied to Eq. (7.89) shows that this solution is stable if the following inequality limit is obeyed:

$$\frac{a^2 \Delta t^2}{\Delta x^2} \leq 1 \tag{7.90}$$

Similarly the homogeneous form of the two-dimensional hyperbolic equation:

$$\frac{\partial^2 u}{\partial t^2} = a^2 \left(\frac{\partial^2 u}{\partial x^2} + \frac{\partial^2 u}{\partial y^2} \right) \tag{7.91}$$

is expanded using central finite difference approximation to yield:

$$\frac{u_{i,j,n+1} - 2u_{i,j,n} + u_{i,j,n-1}}{\Delta t^2} = a^2 \left(\frac{u_{i+1,j,n} - 2u_{i,j,n} + u_{i-1,j,n}}{\Delta x^2} \right) + a^2 \left(\frac{u_{i,j+1,n} - 2u_{i,j,n} + u_{i,j-1,n}}{\Delta y^2} \right) + O(\Delta x^2 + \Delta y^2 + \Delta t^2) \tag{7.92}$$

Rearranging this equation to the explicit form, using an equidistant grid in x- and y-directions, results in:

$$u_{i,j,n+1} = 2\left[1 - 2\left(\frac{a^2 \Delta t^2}{\Delta x^2} \right) \right] u_{i,j,n} - u_{i,j,n-1} + \frac{a^2 \Delta t^2}{\Delta x^2} \left(u_{i+1,j,n} + u_{i-1,j,n} + u_{i,j+1,n} + u_{i,j-1,n} \right) \tag{7.93}$$

This solution is stable when:

$$\frac{a^2 \Delta t^2}{\Delta x^2} \leq 1 \tag{7.94}$$

The flowchart of solving the hyperbolic partial differential equation by the explicit method is illustrated in Fig. 7.10.

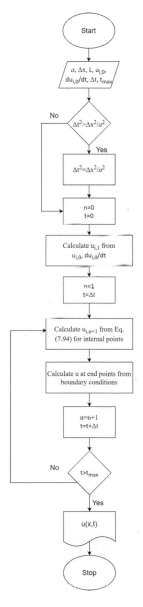

FIG. 7.10

Flowchart of solving the hyperbolic differential equation by the explicit method.

Implicit methods for the solution of hyperbolic partial differential equations can be developed using the *variable-weight* approach, where space partial derivatives are weighted at $(n + 1)$, n, and $(n - 1)$. The implicit formulation of Eq. (7.86) is:

$$\frac{u_{i,n+1} - 2u_{i,n} + u_{i,n-1}}{\Delta t^2} = \frac{a^2}{\Delta x^2} \left[\theta \left(u_{i+1,n+1} - 2u_{i,n+1} + u_{i-1,n+1} \right) \right.$$

$$\left. + (1 - 2\theta) \left(u_{i+1,n} - 2u_{i,n} + u_{i-1,n} \right) + \theta \left(u_{i+1,n-1} - 2u_{i,n-1} + u_{i-1,n-1} \right) \right] \quad (7.95)$$

where $0 \leq \theta \leq 1$. When $\theta = 0$, Eq. (7.95) reverts back to the explicit method (7.89). When $\theta = \frac{1}{2}$, Eq. (7.95) is a Crank-Nicolson-type approximation. Implicit methods yield tridiagonal sets of linear algebraic equations whose solutions can be obtained using Gauss elimination methods.

EXAMPLE 7.8 SOLVING ONE-DIMENSIONAL HYPERBOLIC PARTIAL DIFFERENTIAL EQUATION FOR VIBRATING STRING BY THE EXPLICIT METHOD

Use the explicit formula to solve the following wave equation for computing the displacement of a vibrating string,

$$\frac{\partial^2 u}{\partial t^2} = \frac{\partial^2 u}{\partial x^2} \quad (1)$$

subject to the boundary conditions:

$$\text{B.C. 1} \quad u = 0 \quad \text{for } x = 0 \text{ and } t \geq 0$$

$$\text{B.C. 2} \quad u = 0 \quad \text{for } x = 0.5 \text{ and } t \geq 0$$

and the initial position and velocity:

$$\text{I.C. 1} \quad u = p(x) = \sin 2\pi x \quad \text{for } t = 0 \text{ and } 0 \leq x \leq 0.5$$

$$\text{I.C. 2} \quad \partial u / \partial t = q(x) = 0 \quad \text{for } t = 0 \text{ and } 0 \leq x \leq 0.5$$

Consider $\Delta x = 0.1$ and use the maximum Δt.

Method of solution
The explicit formula (7.86) is used for the solution of this problem:

$$u_{i,n+1} = 2 \left(1 - \frac{a^2 \Delta t^2}{\Delta x^2} \right) u_{i,n} + \frac{a^2 \Delta t^2}{\Delta x^2} \left(u_{i+1,n} + u_{i-1,n} \right) - u_{i,n-1} \quad (2)$$

The maximum time step can be obtained from Eq. (7.90):

$$\Delta t = \frac{\Delta x}{a} = 0.1$$

Note that $a^2 = 1$ in Eq. (1). With this time step, Eq. (2) simplifies to:

(Continued)

EXAMPLE 7.8 (CONTINUED)

$$u_{i,n+1} = u_{i+1,n} + u_{i-1,n} - u_{i,n-1} \tag{3}$$

Before using Eq. (2), the dependent variable at two starting times ($n = 0$ and $n = 1$) should be known. The first one ($n = 0$) is supplied by the initial condition 1. We can use the Taylor series for computing the dependent variable at time Δt ($n = 1$):

$$u_{i,1} = u_{i,0} + \Delta t \frac{\partial u}{\partial t}\bigg|_{i,0} + \frac{\Delta t}{2!} \frac{\partial^2 u}{\partial t^2}\bigg|_{i,0} + \cdots \tag{4}$$

The first derivative at $n = 0$ is available from the initial condition 2 and is equal to $q(x)$. The second derivative in Eq. (4) can be evaluated from the differential equation itself:

$$\frac{\partial^2 u}{\partial t^2}\bigg|_{i,0} = \frac{\partial^2 u}{\partial x^2}\bigg|_{i,0} = \frac{d^2 p}{dx^2} = -4\pi^2 \sin 2\pi x \tag{5}$$

Therefore Eq. (4) becomes:

$$u_{i,1} = \sin 2\pi x - 2\pi^2 \Delta t^2 \sin 2\pi x \tag{6}$$

We can now start solving the equation. The first two steps are calculated from I.C. 1 and Eq. (6), respectively, and are shown in Table E7.8.

Now Eq. (3) can be used to determine the values in the next time steps.

For $n + 1 = 2$ ($t = 2\Delta t$):

$i = 0 \qquad u_{0,2} = 0$
$i = 1 \qquad u_{1,2} = u_{2,1} + u_{0,1} - u_{1,0} = 0.1755$
$i = 2 \qquad u_{2,2} = u_{3,1} + u_{1,1} - u_{2,0} = 0.2840$
$i = 3 \qquad u_{3,2} = u_{4,1} + u_{2,1} - u_{3,0} = 0.2840$
$i = 4 \qquad u_{4,2} = u_{5,1} + u_{3,1} - u_{4,0} = 0.1755$
$i = 5 \qquad u_{5,2} = 0$

For $n + 1 = 3$ ($t = 3\Delta t$):

$i = 0 \qquad u_{0,3} = 0$
$i = 1 \qquad u_{1,3} = u_{2,2} + u_{0,2} - u_{1,1} = -0.1877$

(Continued)

Table E7.8 Values of $u_{i,n}$ for the first two steps.

	$x = 0$ $i = 0$	$x = 0.1$ $i = 1$	$x = 0.2$ $i = 2$	$x = 0.3$ $i = 3$	$x = 0.4$ $i = 4$	$x = 0.5$ $i = 5$
$n = 0$	0	0.5878	0.9511	0.9511	0.5878	0
$n = 1$	0	0.4718	0.7633	0.7633	0.4718	0

EXAMPLE 7.8 (CONTINUED)

$i = 2$ $u_{2,3} = u_{3,2} + u_{1,2} - u_{2,1} = -0.3038$

$i = 3$ $u_{3,3} = u_{4,2} + u_{2,2} - u_{3,1} = -0.3038$

$i = 4$ $u_{4,3} = u_{5,2} + u_{3,2} - u_{4,1} = -0.1877$

$i = 5$ $u_{5,2} = 0$

    Calculations can be continued in the same manner for the next time steps. Results of calculations for the first four time steps are shown in Fig. E7.8.

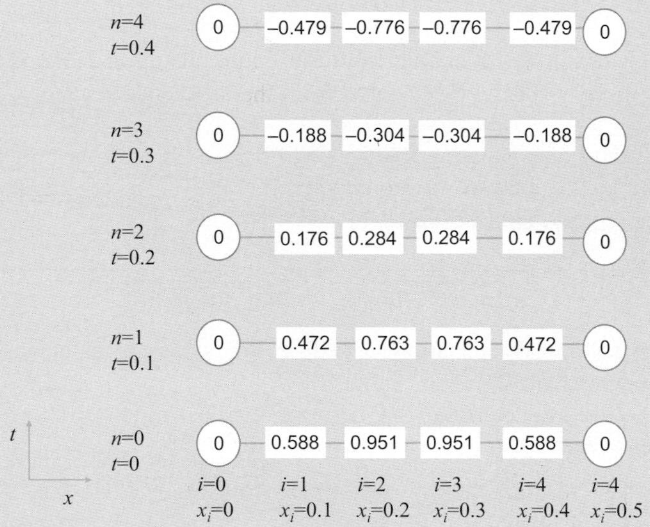

FIG. E7.8

The first five steps of calculations in the solution of the wave equation by the explicit method.

7.4.4 IRREGULAR BOUNDARIES AND POLAR COORDINATE SYSTEMS

The finite difference approximations of partial differential equations developed in this chapter so far were based on regular Cartesian coordinate systems. Quite often, however, the objects, whose properties are being modeled by the partial differential equations, may have circular, cylindrical, or spherical shapes or may have altogether irregular boundaries. The finite difference approximations may be modified to handle such cases.

    Let us first consider an object that is well described by Cartesian coordinates everywhere except near the boundary, which is of irregular shape, as shown in Fig. 7.11. There are two methods of treating the curved boundary. One simple method is to reshape the boundary to pass through the grid point closest to it. For example, in Fig. 7.11, the point (i, j) can be assumed to be on the boundary instead of the point $(i + 1, j)$ on the original boundary. Although it is a simple method, the approximation of the boundary introduces an error in the calculations, especially at the boundary itself.

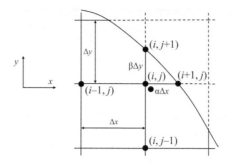

FIG. 7.11

Finite difference grid for irregular boundaries.

A more precise method of expressing the finite difference equation at the irregular boundary is to modify it accordingly. We may use a Taylor series expansion of the dependent variable at the point (i, j) in the x-direction to get (see Fig. 7.11):

$$u_{i+1,j} = u_{i,j} + (\alpha \Delta x)\frac{\partial u}{\partial x}\Big|_{i,j} + \frac{(\alpha^2 \Delta x^2)}{2!}\frac{\partial^2 u}{\partial x^2}\Big|_{i,j} + O(\Delta x^3) \tag{7.96}$$

$$u_{i-1,j} = u_{i,j} - (\Delta x)\frac{\partial u}{\partial x}\Big|_{i,j} + \frac{(\Delta x^2)}{2!}\frac{\partial^2 u}{\partial x^2}\Big|_{i,j} + O(\Delta x^3) \tag{7.97}$$

Eliminating $(\partial^2 u/\partial x^2)$ from Eqs. (7.96) and (7.97) gives:

$$\frac{\partial u}{\partial x}\Big|_{i,j} = \frac{1}{\Delta x}\left[\frac{1}{\alpha(1+\alpha)}\right]\left[u_{i+1,j} - (1-\alpha^2)u_{i,j} - \alpha^2 u_{i-1,j}\right] \tag{7.98}$$

and eliminating $(\partial u/\partial x)$ from Eqs. (7.96) and (7.97) gives:

$$\frac{\partial^2 u}{\partial x^2}\Big|_{i,j} = \frac{1}{\Delta x^2}\left[\frac{2}{\alpha(1+\alpha)}\right]\left[u_{i+1,j} - (1+\alpha)u_{i,j} + \alpha u_{i-1,j}\right] \tag{7.99}$$

Eqs. (7.98) and (7.99) were derived for the configuration shown in Fig. 7.11 in which we encounter the irregular boundary on the right side of the domain. For the irregular boundary on the left side (the point $(i-1, j)$ is on the boundary), the same approach results in:

$$\frac{\partial u}{\partial x}\Big|_{i,j} = \frac{1}{\Delta x}\left[\frac{1}{\alpha(1+\alpha)}\right]\left[\alpha^2 u_{i+1,j} + (1-\alpha^2)u_{i,j} - u_{i-1,j}\right] \tag{7.100}$$

$$\frac{\partial^2 u}{\partial x^2}\Big|_{i,j} = \frac{1}{\Delta x^2}\left[\frac{2}{\alpha(1+\alpha)}\right]\left[\alpha u_{i+1,j} - (1+\alpha)u_{i,j} + u_{i-1,j}\right] \tag{7.101}$$

Similarly in the y-direction, for the upper irregular boundary:

$$\frac{\partial u}{\partial y}\Big|_{i,j} = \frac{1}{\Delta y}\left[\frac{1}{\beta(1+\beta)}\right]\left[u_{i,j+1} - (1-\beta^2)u_{i,j} - \beta^2 u_{i,j-1}\right] \tag{7.102}$$

$$\frac{\partial^2 u}{\partial y^2}\Big|_{i,j} = \frac{1}{\Delta y^2}\left[\frac{2}{\beta(1+\beta)}\right]\left[u_{i,j+1} - (1+\beta)u_{i,j} + \beta u_{i,j-1}\right] \tag{7.103}$$

and for the lower irregular boundary (the point $(i, j - 1)$ is on the boundary):

$$\frac{\partial u}{\partial y}\bigg|_{i,j} = \frac{1}{\Delta y}\left[\frac{1}{\beta(1+\beta)}\right]\left[\beta^2 u_{i,j+1} - \left(1 - \beta^2\right)u_{i,j} - u_{i,j-1}\right] \tag{7.104}$$

$$\frac{\partial^2 u}{\partial y^2}\bigg|_{i,j} = \frac{1}{\Delta y^2}\left[\frac{2}{\beta(1+\beta)}\right]\left[\beta u_{i,j+1} - (1 + \beta)u_{i,j} + u_{i,j-1}\right] \tag{7.105}$$

When $\alpha = \beta = 1$, Eqs. (7.98) to (7.105) become identical to those developed earlier in this chapter for regular Cartesian coordinate systems. Therefore, for objects with irregular boundaries, the partial differential equations would be converted to algebraic equations using Eqs. (7.98) to (7.105). For points adjacent to the boundary, the parameters α and β would assume values that reflect the irregular shape of the boundary, and for internal points away from the boundary, the value of α and β would be unity.

Eqs. (7.98)–(7.105) can be used at the boundaries with the Dirichlet condition where the dependent variable at the boundary is known. Treatment of Neumann and Robin conditions where the normal derivative at the curved or irregular boundary is specified is more complicated. Considering again Fig. 7.11, the normal derivative of the dependent variable at the boundary can be expressed as:

$$\frac{\partial u}{\partial \mathbf{n}}\bigg|_{i+1,j} = \frac{\partial u}{\partial x}\bigg|_{i+1,j}\cos\gamma + \frac{\partial u}{\partial y}\bigg|_{i+1,j}\sin\gamma \tag{7.106}$$

where $\mathbf{n}$ is the unit vector normal to the boundary and γ is the angle between the vector $\mathbf{n}$ and x-axis. The derivatives with respect to x and y in Eq. (7.106) can be approximated by Taylor series expansions:

$$\frac{\partial u}{\partial x}\bigg|_{i+1,j} = \frac{\partial u}{\partial x}\bigg|_{i,j} + (\alpha\Delta x)\frac{\partial^2 u}{\partial x^2}\bigg|_{i,j} \tag{7.107}$$

$$\frac{\partial u}{\partial y}\bigg|_{i+1,j} = \frac{\partial u}{\partial y}\bigg|_{i,j} + (\alpha\Delta x)\frac{\partial^2 u}{\partial x\partial y}\bigg|_{i,j} \tag{7.108}$$

The derivatives at the grid point (i, j) should be known to calculate the normal derivative at the boundary. For the particular configuration of Fig. 7.11, we may use backward finite differences to evaluate the derivatives in Eqs. (7.107) and (7.108) (see Table 7.3):

$$\frac{\partial u}{\partial x}\bigg|_{i,j} = \frac{1}{\Delta x}\left(u_{i,j} - u_{i-1,j}\right) \tag{7.109}$$

$$\frac{\partial^2 u}{\partial x^2}\bigg|_{i,j} = \frac{1}{\Delta x^2}\left(u_{i,j} - 2u_{i-1,j} + u_{i-2,j}\right) \tag{7.110}$$

$$\frac{\partial u}{\partial y}\bigg|_{i,j} = \frac{1}{\Delta y}\left(u_{i,j} - u_{i,j-1}\right) \tag{7.111}$$

$$\frac{\partial^2 u}{\partial x\partial y} = \frac{1}{\Delta x\Delta y}\left(u_{i,j} - u_{i-1,j} - u_{i,j-1} + u_{i-1,j-1}\right) \tag{7.112}$$

Combining Eqs. (7.107), (7.109), and (7.110) gives:

$$\frac{\partial u}{\partial x}\bigg|_{i+1,j} = \frac{1}{\Delta x}\left[(1 + \alpha)u_{i,j} - (1 + 2\alpha)u_{i-1,j} + \alpha u_{i-2,j}\right] \tag{7.113}$$

and combining Eqs. (7.108), (7.111), and (7.112) gives:

$$\frac{\partial u}{\partial y}\bigg|_{i+1,j} = \frac{1}{\Delta y}\left[(1+\alpha)u_{i,j} - \alpha u_{i-1,j} - (1+\alpha)u_{i,j-1} + \alpha u_{i-1,j-1}\right] \tag{7.114}$$

Replacing Eqs. (7.113) and (7.114) into Eq. (7.106) provides the normal derivative that can be used when dealing with Neumann or Robin conditions at an irregular boundary.

Similarly in the *y*-direction,

$$\frac{\partial u}{\partial n}\bigg|_{i,j+1} = \frac{\partial u}{\partial x}\bigg|_{i,j+1}\cos\gamma + \frac{\partial u}{\partial y}\bigg|_{i,j+1}\sin\gamma \tag{7.115}$$

where

$$\frac{\partial u}{\partial x}\bigg|_{i,j+1} = \frac{1}{\Delta x}\left[(1+\beta)u_{i,j} - \beta u_{i,j-1} - (1+\beta)u_{i-1,j} + \beta u_{i-1,j-1}\right] \tag{7.116}$$

$$\frac{\partial u}{\partial y}\bigg|_{i,j+1} = \frac{1}{\Delta y}\left[(1+\beta)u_{i,j} - (1+2\alpha)u_{i,j-1} + \beta u_{i,j-2}\right] \tag{7.117}$$

It is important to remember that Eqs. (7.113), (7.114), (7.116), and (7.117) are specific to the configuration shown in Fig. 7.11. For other possible configurations, forward differences, or a combination of forward and backward differences (in different directions), may be used to treat the derivative boundary condition. For example, in the third quadrant (lower-left boundary), we have:

$$\frac{\partial u}{\partial x}\bigg|_{i-1,j} = \frac{1}{\Delta x}\left[\alpha u_{i+2,j} + (1+2\alpha)u_{i+1,j} - (1+\alpha)u_{i,j}\right] \tag{7.118}$$

$$\frac{\partial u}{\partial y}\bigg|_{i-1,j} = \frac{1}{\Delta y}\left[-\alpha u_{i+1,j+1} + (1+\alpha)u_{i,j+1} + \alpha u_{i+1,j} - (1+\alpha)u_{i,j}\right] \tag{7.119}$$

$$\frac{\partial u}{\partial x}\bigg|_{i,j-1} = \frac{1}{\Delta x}\left[-\beta u_{i+1,j+1} + (1+\beta)u_{i+1,j} + \beta u_{i,j+1} - (1+\beta)u_{i,j}\right] \tag{7.120}$$

$$\frac{\partial u}{\partial y}\bigg|_{i,j+1} = \frac{1}{\Delta y}\left[\beta u_{i,j+2} + (1+2\beta)u_{i,j+1} - (1+\beta)u_{i,j}\right] \tag{7.121}$$

Cylindrical shaped objects are more conveniently expressed in polar coordinates. The transformation from Cartesian coordinate to polar coordinate systems is performed using the following relationships, which are based on Fig. 7.12:

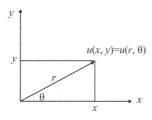

FIG. 7.12

Transformation to polar coordinates.

$$x = r\cos\theta \quad y = r\sin\theta$$
$$r = \sqrt{x^2 + y^2} \quad \theta = \tan^{-1}\frac{y}{x} \tag{7.122}$$

The Laplacian operator in polar coordinates becomes:

$$\frac{\partial^2 u}{\partial x^2} + \frac{\partial^2 u}{\partial y^2} = \frac{\partial^2 u}{\partial r^2} + \frac{1}{r}\frac{\partial u}{\partial r} + \frac{1}{r^2}\frac{\partial^2 u}{\partial \theta^2} \tag{7.123}$$

Fick's second law of diffusion (Eq. 7.10) in polar coordinates is:

$$\frac{\partial c_A}{\partial t} = D_{AB}\left(\frac{\partial^2 c_A}{\partial r^2}\frac{1}{r}\frac{\partial c_A}{\partial r} + \frac{1}{r^2}\frac{\partial^2 c_A}{\partial \theta^2} + \frac{\partial^2 c_A}{\partial z^2}\right) \tag{7.124}$$

Using the finite difference grid for polar coordinates shown in Fig. 7.13, the partial derivatives are approximated by:

$$\left.\frac{\partial^2 u}{\partial r^2}\right|_{i,j} = \frac{1}{\Delta r^2}\left(u_{i,j+1} - 2u_{i,j} + u_{i,j-1}\right) \tag{7.125}$$

$$\left.\frac{\partial^2 u}{\partial \theta^2}\right|_{i,j} = \frac{1}{\Delta \theta^2}\left(u_{i+1,j} - 2u_{i,j} + u_{i-1,j}\right) \tag{7.126}$$

$$\left.\frac{\partial u}{\partial r}\right|_{i,j} = \frac{1}{2\Delta r}\left(u_{i,j+1} - u_{i,j-1}\right) \tag{7.127}$$

where j and i are counters in r- and θ-directions, respectively. Partial derivatives in z- and t-directions (not shown in Fig. 7.13) would be similarly expressed, with the use of additional subscripts.

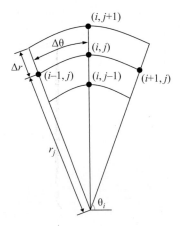

FIG. 7.13

Finite difference grid for polar coordinates.

Table 7.4 Partial differential equation solvers in MATLAB's PDE Toolbox.

Solver	Description
adaptmesh	Adapts mesh generation and solves elliptic partial differential equation
assempde	Assembles and solves the elliptic partial differential equation
hyperbolic	Solves hyperbolic partial differential equation
parabolic	Solves parabolic partial differential equation
pdenonlin	Solves nonlinear elliptic partial differential equation
poisolv	Solves the Poisson equation on a rectangular grid

7.5 USING BUILT-IN MATLAB® FUNCTIONS

To solve a partial differential equation using the PDE toolbox, one may simply use the *graphical user interface* by employing the *pdetool* command. In this separate environment, the user is able to define the two-dimensional geometry, introduce the boundary conditions, solve the partial differential equation, and visualize the results. In special cases where the problem is complicated or nonstandard, the user may wish to solve it using command-line functions. Some of these functions (solvers only) are listed in Table 7.4.

7.6 STABILITY ANALYSIS

In this section, we discuss the stability of finite difference approximations using the well-known von Neumann procedure. This method introduces an initial error represented by a finite Fourier series and examines how this error propagates during the solution. The von Neumann method applies to initial-value problems. For this reason, it is used to analyze the stability of the explicit method for parabolic equations developed in Section 7.4.2 and the explicit method for hyperbolic equations developed in Section 7.4.3.

Define the error $\varepsilon_{m,n}$ as the difference between the solution $u_{m,n}$ of the finite difference approximation and the exact solution $\bar{u}_{m,n}$ of the differential equation at step (m, n):

$$\varepsilon_{m,n} = u_{m,n} - \bar{u}_{m,n} \tag{7.128}$$

The explicit finite difference solution (7.54) of the parabolic partial differential Eq. (7.14) can be written for $u_{m,n+1}$ and $\bar{u}_{m,n+1}$ as follows:

$$u_{m,n+1} = \left(\frac{\alpha\Delta t}{\partial x^2}\right)u_{m+1,n} + \left(1 - 2\frac{\alpha\Delta t}{\partial x^2}\right)u_{m,n} + \left(\frac{\alpha\Delta t}{\partial x^2}\right)u_{m-1,n} + R_{E_{m,n+1}} \tag{7.129}$$

$$\bar{u}_{m,n+1} = \left(\frac{\alpha\Delta t}{\partial x^2}\right)\bar{u}_{m+1,n} + \left(1 - 2\frac{\alpha\Delta t}{\partial x^2}\right)\bar{u}_{m,n} + \left(\frac{\alpha\Delta t}{\partial x^2}\right)\bar{u}_{m-1,n} + T_{E_{m,n+1}} \tag{7.130}$$

where $R_{Em,n+1}$ and $T_{Em,n+1}$ are the roundoff and truncation errors, respectively, at step $(m, n + 1)$.

Combining Eqs. (7.128) to (7.130), we obtain:

$$\varepsilon_{m,n+1} - \left(\frac{\alpha\Delta t}{\partial x^2}\right)\varepsilon_{m+1,n} - \left(1 - 2\frac{\alpha\Delta t}{\partial x^2}\right)\varepsilon_{m,n} - \left(\frac{\alpha\Delta t}{\partial x^2}\right)\varepsilon_{m-1,n} = R_{E_{m,n+1}} - T_{E_{m,n+1}} \qquad (7.131)$$

This is a *nonhomogeneous finite difference equation in two dimensions*, representing the propagation of error during the numerical solution of the parabolic partial differential Eq. (7.14). The solution to this finite difference equation is rather difficult to obtain. For this reason, the von Neumann analysis considers the *homogeneous* part of Eq. (7.131):

$$\varepsilon_{m,n+1} - \left(\frac{\alpha\Delta t}{\partial x^2}\right)\varepsilon_{m+1,n} - \left(1 - 2\frac{\alpha\Delta t}{\partial x^2}\right)\varepsilon_{m,n} - \left(\frac{\alpha\Delta t}{\partial x^2}\right)\varepsilon_{m-1,n} = 0 \qquad (7.132)$$

which represents the propagation of the error introduced at the initial point ($n = 0$) only and ignores truncation and roundoff errors that enter the solution at $n > 0$.

The solution of the homogeneous finite difference equation may be written in the separable form

$$\varepsilon_{m,n} = ce^{\gamma n\Delta t}e^{i\beta m\Delta x} \qquad (7.133)$$

where $i = \sqrt{-1}$ and c, γ, and β are constants. At $n = 0$,

$$\varepsilon_{m,0} = ce^{i\beta m\Delta x} \qquad (7.134)$$

which is the error at the initial point. Therefore, the term $e^{\gamma\Delta t}$ is the *amplification factor* of the initial error. In order for the original error not to grow as n increases, the amplification factor must satisfy the *von Neumann condition for stability*:

$$\left|e^{\gamma\Delta t}\right| \leq 1 \qquad (7.135)$$

The amplification factor can have complex values. In that case the modulus of the complex numbers must satisfy the above inequality, that is,

$$|r| \leq 1 \qquad (7.136)$$

Therefore, the stability region in the complex plane is a circle of radius $= 1$, as shown in Fig. 7.14.

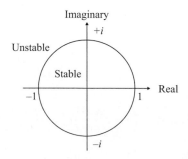

FIG. 7.14

Stability region in the complex plane.

The amplification factor is determined by substituting Eq. (7.133) into Eq. (7.132) and rearranging to obtain:

$$e^{\gamma \Delta t} = \left(1 - 2\frac{\alpha \Delta t}{\partial x^2}\right) + \frac{\alpha \Delta t}{\partial x^2}\left(e^{i\beta \Delta x} + e^{-i\beta \Delta x}\right) \tag{7.137}$$

Using trigonometric identities,

$$\frac{e^{i\beta \Delta x} + e^{-i\beta \Delta x}}{2} = \cos\beta \Delta x \tag{7.138}$$

and

$$1 - \cos\beta \Delta x = 2\sin^2\frac{\beta \Delta x}{2} \tag{7.139}$$

Eq. (7.137) becomes:

$$e^{\gamma \Delta t} = 1 - \left(4\frac{\alpha \Delta t}{\partial x^2}\right)\left(\sin^2\frac{\beta \Delta x}{2}\right) \tag{7.140}$$

Combining this with the von Neumann condition for stability, we obtain the stability bound:

$$0 \leq \left(\frac{\alpha \Delta t}{\partial x^2}\right)\left(\sin^2\frac{\beta \Delta x}{2}\right) \leq \frac{1}{2} \tag{7.141}$$

The term $\sin^2(\beta \Delta x/2)$ has its highest value equal to unity; therefore,

$$0 < \frac{\alpha \Delta t}{\partial x^2} \leq \frac{1}{2} \tag{7.142}$$

is the limit for conditional stability for this method. It should be noted that this limit is identical to that obtained by using the positivity rule (Section 7.4.2).

The stability of the explicit solution (7.89) of the hyperbolic Eq. (7.86) can be similarly analyzed using the von Neumann method. The homogeneous equation for the error propagation of that solution is:

$$\varepsilon_{m,n+1} - 2\left(1 - 2\frac{\alpha^2 \Delta t^2}{\partial x^2}\right)\varepsilon_{m,n} - \frac{\alpha^2 \Delta t^2}{\partial x^2}\left(\varepsilon_{m+1,n} + \varepsilon_{m-1,n}\right) + \varepsilon_{m,n-1} = 0 \tag{7.143}$$

Substitution of the solution (7.133) into (7.143) and use of the trigonometric identities (7.138) and (7.139) give the amplification factor as:

$$e^{\gamma \Delta t} = \left(1 - 2\frac{\alpha^2 \Delta t^2}{\partial x^2}\sin^2\frac{\beta \Delta x}{2}\right) \pm \sqrt{\left(1 - 2\frac{\alpha^2 \Delta t^2}{\partial x^2}\sin^2\frac{\beta \Delta x}{2}\right)^2 - 1} \tag{7.144}$$

The above amplification factor satisfies inequality (7.136) in the complex plane, that is, when

$$\left(1 - 2\frac{\alpha^2 \Delta t^2}{\partial x^2}\sin^2\frac{\beta \Delta x}{2}\right)^2 - 1 \leq 0 \tag{7.145}$$

which converts to the inequality,

$$\frac{\alpha^2 \Delta t^2}{\partial x^2} \leq \frac{1}{\sin^2(\beta \Delta x/2)} \tag{7.146}$$

The term $\sin^2(\beta\Delta x/2)$ has its highest value equal to unity; therefore,

$$\frac{\alpha^2\Delta t^2}{\partial x^2} \le 1 \tag{7.147}$$

is the conditional stability limit for this method.

In a similar manner, the stability of other explicit and implicit finite difference methods may be examined. This has been done by Lapidus and Pinder [4], who conclude that "most explicit finite difference approximations are conditionally stable, whereas most implicit approximations are unconditionally stable."

7.7 SUMMARY

Second-order partial differential equations can be classified into elliptic, parabolic, and hyperbolic forms. Elliptic equations are encountered in steady state diffusion problems. Unsteady state diffusion problems are expressed by parabolic equations. Wave equations are of the hyperbolic form. In the finite difference method, the derivatives in the partial differential equation are substituted by their numerical formulas derived in Chapter 4.

Replacing the second-order derivatives with their approximation using central differences results in a set of linear algebraic equations. For solving this set, additional equations should be added to this set of equations by discretizing equations of boundary conditions. Attention should be made to use the finite difference formulas with the same order of error for both the differential equation and boundary conditions.

Two general methods exist for solving a parabolic partial differential equation: explicit and implicit. In the explicit method presented in this chapter, the first-order derivative is replaced with the forward difference formula, whereas the central difference formula is employed for the second-order derivative with respect to the other independent variable.

In the implicit method, the derivatives are evaluated at the halfway point of the grid. In this way, more than one unknown at any step in the time domain can be found in the discretized equation. Writing this equation for the entire difference grid, a set of simultaneous linear algebraic equations, whose matrix of coefficients is usually tridiagonal, is obtained.

The method of lines is another method for solving parabolic equations. In this method, the second-order special derivative is discretized, whereas the time derivative remains unchanged. This is repeated for all grid points and a set of linear ordinary differential equations is obtained by solving which the solution of the parabolic partial differential equation is achieved.

Both explicit and implicit methods for solving hyperbolic partial differential equations are developed. The explicit method is easy to implement, although it is conditionally stable (i.e., there is an upper limit for the step size). The implicit method, on the other hand, is unconditionally stable, although is more difficult to implement.

A simple technique for solving partial differential equations with irregular boundaries is described. Some partial differential equation solvers in the PDE Toolbox of MATLAB are also introduced.

PROBLEMS

7.1 Modify *elliptic.m* in Example 7.2 to solve the three-dimensional problem:

$$\frac{\partial^2 u}{\partial x^2} + \frac{\partial^2 u}{\partial y^2} + \frac{\partial^2 u}{\partial z^2} = 0$$

Apply this function to calculate the distribution of the dependent variable within a solid body that is subject to the following boundary conditions:

$$
\begin{aligned}
u(0, y, z) &= 100 \quad u(1, y, z) = 100 \\
u(x, 0, z) &= 0 \quad\;\;\; u(x, 1, z) = 0 \\
u(x, y, 0) &= 50 \quad\;\; u(x, y, 1) = 50
\end{aligned}
$$

7.2 Solve Laplace's equation with the following boundary conditions and discuss the results:

$$u(0, y) = 100 \quad \left.\frac{\partial u}{\partial x}\right|_{10,y} = 10 \quad \left.\frac{\partial u}{\partial y}\right|_{x,0} = 0 \quad \left.\frac{\partial u}{\partial y}\right|_{x,1} = 0$$

7.3 The ambient temperature surrounding a house is 50°F. The heat in the house had been turned off; therefore, the internal temperature is also at 50°F at $t = 0$. The heating system is turned on and raises the internal temperature to 70°F at the rate of 4°F/h. The ambient temperature remains at 50°F. The wall of the house is 0.5 ft thick and is made of a material that has an average thermal diffusivity $\alpha = 0.01$ ft²/h and a thermal conductivity $k = 0.2$ Btu/ (h.ft².°F). The heat transfer coefficient on the inside of the wall is $h_{in} = 1.0$ Btu/(h.ft².°F), and the heat transfer coefficient on the outside is $h_{out} = 2.0$ Btu/(h.ft².°F). Estimate how long it will take to reach a steady state temperature distribution across the wall.

7.4 Develop the finite difference approximation of Fick's second law of diffusion in polar coordinates. Write a MATLAB program that can be used to solve the following problem [6]:

A wet cylinder of agar gel at 278 K with a uniform concentration of urea of 0.1 kg mol/ m³ has a diameter of 30.48 mm and is 38.1 mm long with flat, parallel ends. The diffusivity is 4.72×10^{-10} m²/s. Calculate the concentration at the midpoint of the cylinder after 100 h for the following cases if the cylinder is suddenly immersed in turbulent pure water:
(a) For radial diffusion only
(b) Diffusion that occurs radially and axially

7.5 Express the two-dimensional parabolic partial differential equation:

$$\frac{\partial u}{\partial t} = \alpha \left(\frac{\partial^2 u}{\partial x^2} + \frac{\partial^2 u}{\partial y^2} \right)$$

in an explicit finite difference formulation. Determine the limits of conditional stability for this method using:
(a) The von Neumann stability.
(b) The positivity rule.

7.6 Show that the unsteady state heat transfer in a thin, triangular surface is governed by the equation:

$$\frac{\partial T}{\partial t} = \alpha \left(\frac{\partial^2 T}{\partial x^2} + \frac{\partial^2 T}{\partial y^2} \right)$$

(a) Assuming that $\Delta x = \Delta y = 1/3$, determine the maximum value of the time step (Δt) to obtain a stable numerical solution of this equation by employing the positivity rule.

(b) Solve the equation for the initial and boundary conditions shown in Fig. P7.6. Use the time step evaluated in (a).

Additional data: $L = H = 1$, $\alpha = 0.1$

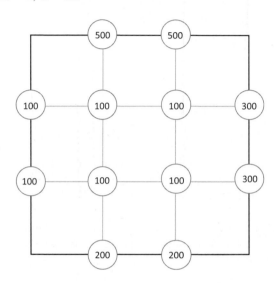

FIG. P7.6

Initial and boundary conditions.

7.7 Determine the steady state temperature profile of a rectangular plate as shown in Fig. P7.7. Assume constant thermal diffusivity k and constant heat generation in the plate q'. Assume $\Delta x = \Delta y = 0.25$ m and $q'/k = 0.2$ K/m².

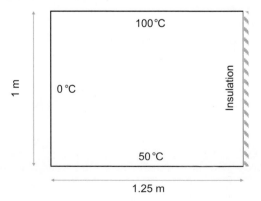

FIG. P7.7

Geometry and boundary conditions.

7.8 Unsteady state heat transfer in a rod with heat generation along the rod is governed by the equation:

$$\frac{\partial T}{\partial t} = \alpha \frac{\partial^2 T}{\partial x^2} + \beta T$$
$$T(x,0) = 0$$
$$T(0,t) = 1$$
$$\frac{\partial T(1,t)}{\partial x} = 0$$

(a) Assuming that the length of the rod is 1 and $\Delta x = 0.2$, determine the maximum value of the time step to obtain a stable numerical solution.

(b) Solve this equation by the explicit method.

Additional data: $\alpha = 0.1$, $\beta = 1$

7.9 Unsteady state mass transfer in a slab is given by the equation:

$$\frac{\partial c}{\partial t} = 10^{-4} \frac{\partial^2 c}{\partial y^2}$$
$$t = 0, c = 0$$
$$y = 0, c = 1$$
$$y = 1, \frac{dc}{dy} = 0$$

Solve this equation numerically and show the evolution of the concentration profile in the slab for 3 consecutive time steps. Use $\Delta t = 500$ and $\Delta y = 0.25$.

7.10 Fig. P7.10 illustrates a rectangular box. The depth of this box is relatively small such that the changes inside the box can be considered to be two-dimensional, as shown in the figure. In corner A, steam is injected into the box with a relative concentration equal to 1. Side B of the box is open to the ambient environment whose relative moisture concentration is 0.

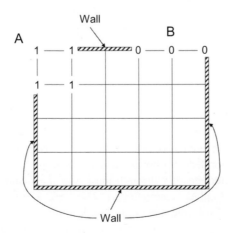

FIG. P7.10

Top view of the box with steam injection at corner A and outlet at side B.

(a) Derive the steady state mass balance equation for steam in the box.

(b) Solve the equation subject to boundary conditions shown in Fig. P7.10, and determine the steady state concentration profile of steam inside the box ($\Delta x = \Delta y$).

7.11 A 12-in-square membrane (no bending or shear stresses) with a 4-in-square hole in the middle is fastened at the outside and inside boundaries, as shown in Fig. P7.11 [7]. If a highly stretched membrane is subject to a pressure p, the partial differential equation for the deflection w in the z-direction is:

$$\frac{\partial^2 w}{\partial x^2} + \frac{\partial^2 w}{\partial y^2} = -\frac{p}{T}$$

where T is the tension (pounds per linear inch). For a highly stretched membrane, the tension T may be assumed constant for small deflections. Using the following values of pressure and tension,

$$p = 5 \text{ psi(uniformly distributed)}$$
$$T = 100 \text{ lb/in}$$

(a) List all the boundary conditions needed for the numerical solution of the problem.

(b) Express the differential equation in finite difference form to obtain the deflection w of the membrane.

(c) Solve the equation numerically.

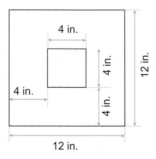

FIG. P7.11

Stretched membrane.

7.12 Fig. P7.12 shows a cross section of a long cooling fin of width W, thickness t, and thermal conductivity k that is bonded to a hot wall, maintaining its base (at $x = 0$) at a temperature T_w [8]. Heat is conducted steadily through the fin in the plane of Fig. P7.12 so that the fin temperature T obeys Laplace's equation, $\partial^2 T/\partial x^2 + \partial^2 T/\partial y^2 = 0$. (Temperature variations along the length of the fin in the z-direction are ignored.)

Heat is lost from the sides and tip of the fin by convection to the surrounding air (radiation is neglected at sufficiently low temperatures) at a local rate $q = h(T_s - T_a)$ Btu/ (h.ft^2). Here, T_s and T_a, in degrees Fahrenheit, are the temperatures at a point on the fin surface and of the air, respectively. If the surface of the fin is vertical, the heat transfer coefficient h obeys the dimensional correlation $h = 0.21(T_s - T_a)^{1/3}$.

(a) Set up the equations for a numerical solution of this problem to determine the temperature at a finite number of points within the fin and the surface.

(b) Describe in detail the step-by-step procedure for solving the equation of part (a) and evaluating the temperature within the fin and at the surface.

(c) Solve the problem numerically using the following quantities:

$$T_w = 200\ °F \qquad T_a = 70\ °F$$
$$t = 0.25 \quad \text{in} \quad k = 25.9\ \text{Btu/(h.ft.}°\text{F)}$$
$$w = 0.5\ \text{in}$$

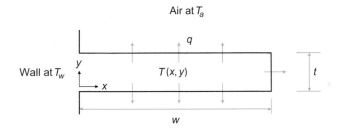

FIG. P7.12

Cooling fin.

7.13 Consider a steady state plug flow reactor of length z through which a substrate is flowing with a constant velocity v with no dispersion effects. The reactor is made up of a series of collagen membranes, each impregnated with two enzymes catalyzing the sequential reaction [9]:

$$A \xrightarrow{\text{Enzyme 1}} B \xrightarrow{\text{Enzyme 2}} C$$

The membranes in the reactor are arranged in parallel, as shown in Fig. P7.13. The nomenclature for this problem is shown in Table P7.13.

For a substrate molecule to encounter the immobilized enzymes, it must diffuse across a *Nernst* diffusion layer on the surface of the support and then some distance into the membrane. The coupled reaction takes place in the membrane, and the product, the unreacted intermediate, and substrate diffuse back into the bulk fluid phase. No inactivation of the enzymes occurs, and it is assumed that the enzymes behave independently of each other.

Because the membrane can accommodate only a finite number of enzyme molecules per unit weight, it becomes necessary to introduce a control parameter ε that measures the ratio of the molar concentration of enzyme 1 to the molar concentration of enzymes 1 plus 2. It is implicitly assumed that the binding sites on collagen do not discriminate between the enzymes. Thus, when both enzymes are present, the maximum reaction velocities reduce to εV_1 and $(1 - \varepsilon)V_2$. The control ε is constrained between the bounds of 0 (only enzyme 2 present) and 1 (only enzyme 1 present).

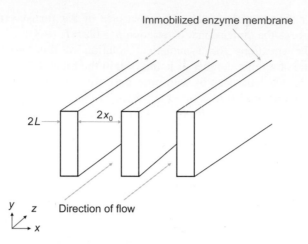

FIG. P7.13

Biocatalytic reactor.

Table P7.13 Nomenclature for problem 7.13.	
$C_{Af} =$	concentration of A in feed, mol/L
$D =$	molecular diffusivity of reactants or products in the membrane, cm^2/s
$k_L =$	overall mass transfer coefficient in the fluid phase, cm/s
$K_{M1}, K_{M2} =$	Michaelis-Menten constant for enzymes 1 and 2, mol/L
$L =$	the half thickness of membrane, mils
$v =$	superficial fluid velocity in reactor, cm/s
$V_1, V_2 =$	maximum reaction velocity for enzymes 1 and 2, mol/(L.s)
$X =$	variable axial distance from the center of the membrane to surface, cm
$x_0 =$	half distance between two consecutive membranes, cm
$X =$	x/L, dimensionless distance
$Y_{Ab}, Y_{Bb}, Y_{Cb} =$	the bulk concentration of species A, B, or C divided by the feed concentration of A (C_{Af}), dimensionless
$Y_{Am}, Y_{Bm}, Y_{Cm} =$	membrane concentration of species A, B, or C divided by the feed concentration of A (C_{Af}), dimensionless
$Y_{As}, Y_{Bs}, Y_{Cs} =$	the surface concentration of species A, B, or C divided by the feed concentration of A (C_{Af}), dimensionless
$z =$	variable longitudinal distance from the entrance of the reactor, cm
$\varepsilon =$	control, the ratio of the molar concentration of enzyme 1 to total molar concentration of enzymes 1 plus 2, dimensionless
$\theta =$	z/v, space time, s

The reaction rates for the two sequential reactions are given by the Michaelis-Menten relationship:

$$R_1 = \frac{\varepsilon V_1 Y_{Am}}{K_{M1} + C_{Af} Y_{Am}}$$

$$R_2 = \frac{(1 - \varepsilon) V_2 Y_{Bm}}{K_{M2} + C_{Af} Y_{Bm}}$$

Material balance for the species A, B, and C in the membrane yield the following differential equations:

$$\frac{D}{L^2} \frac{\partial^2 Y_{Am}}{\partial X^2} - R_1 = 0$$

$$\frac{D}{L^2} \frac{\partial^2 Y_{Bm}}{\partial X^2} + R_2 - R_1 = 0$$

$$\frac{D}{L^2} \frac{\partial^2 Y_{Cm}}{\partial X^2} + R_2 = 0$$

In the bulk fluid phase, the material balances for species A, B, and C can be defined as:

$$\frac{dY_{Ab}}{d\theta} + \frac{k_L}{x_0}(Y_{Ab} - Y_{As}) = 0$$

$$\frac{dY_{Bb}}{d\theta} + \frac{k_L}{x_0}(Y_{Bb} - Y_{Bs}) = 0$$

$$\frac{dY_{Cb}}{d\theta} + \frac{k_L}{x_0}(Y_{Cb} - Y_{Cs}) = 0$$

Because each membrane is symmetric about $X = 0$, the boundary conditions at $X = 0$ and $X = 1$ become:

$$\frac{\partial Y_{Am}}{\partial X} = \frac{\partial Y_{Bm}}{\partial X} = \frac{\partial Y_{Cm}}{\partial X} = 0 \quad \text{at} \quad X = 0$$

$$\left.\begin{array}{c} Y_{Am} = Y_{As} \\ Y_{Bm} = Y_{Bs} \\ Y_{Cm} = Y_{Cs} \end{array}\right\} \quad \text{at} \quad X = 1$$

The surface concentrations are determined by equating the surface flux to the bulk transport flux, that is,

$$\frac{D}{L} \left(\frac{\partial Y_{Am}}{\partial X}\right)_{X=1} = k_L(Y_{Ab} - Y_{As})$$

$$\frac{D}{L} \left(\frac{\partial Y_{Bm}}{\partial X}\right)_{X=1} = k_L(Y_{Bb} - Y_{Bs})$$

$$\frac{D}{L} \left(\frac{\partial Y_{Cm}}{\partial X}\right)_{X=1} = k_L(Y_{Cb} - Y_{Cs})$$

Finally at the entrance of the reactor, that is, at $\theta = 0$,

$$Y_{Ab} = 1 \quad Y_{Bb} = 0 \quad Y_C = 0$$

Develop a numerical procedure for solving the above set of equations and write a computer program to calculate the concentration profiles in the membranes and in the bulk fluid for the following set of kinetic and transport parameters:

$$V_1 = 4.4 \times 10^{-3} \text{ mol/L.s} \quad V_2 = 12.0 \times 10^{-3} \text{ mol/L.s}$$
$$K_{M1} = 0.022 \text{ mol/L} \quad K_{M2} = 0.010 \text{ mol/L}$$
$$D = 5.7 \times 10^{-8} \text{ cm}^2/\text{s} \quad C_{Af} = 1.0 \text{ mol/L}$$
$$k_L = 1.2 \times 10^{-4} \text{ cm/s} \quad x_0 = 23 \text{ mils}$$
$$\varepsilon = 0.75 \quad 2L = 3 \text{ mils}$$

7.14 Coulet et al. [10] have developed a glucose sensor that has glucose oxidase enzyme immobilized as a surface layer on a highly polymerized collagen membrane. In this system, glucose (analyte) is converted to hydrogen peroxide, which is subsequently detected on the membrane face (which is not exposed to the analyte solution) by an amperometric electrode. The hydrogen peroxide flux is a direct measure of the sensor response [11].

The physical model and coordinate system are shown in Fig. P7.14. The local analyte concentration at the enzyme surface is low, so that the reaction kinetics are adequately described by a first-order law. This latter assumption ensures that the electrode response is proportional to the analyte concentration.

The governing dimensionless equation describing analyte transport within the membrane is:

$$\frac{\partial C}{\partial \xi} = \frac{\partial^2 C}{\partial \zeta^2}$$

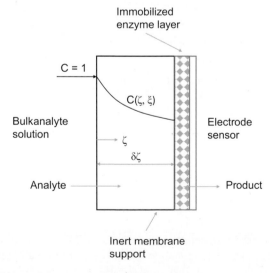

FIG. P7.14 Schematic description of an anisotropic enzyme reactor.

The membrane (exaggerated) has an active enzyme deposited as a surface layer at the electrode sensor interface. The product flux is the result of the reaction involving analyte diffusion through the membrane.

where the dimensionless time ξ and penetration ζ variables are defined as:

$$\xi = \frac{Dt}{\delta^2}$$
$$\zeta = \frac{x}{\delta}$$

where δ is the membrane thickness and D the diffusion coefficient. The initial and boundary conditions are:

$$
\begin{array}{lll}
C = 0 & \xi = 0 & 0 \le \zeta \le 1 \\
C = 1 & \xi > 0 & \zeta = 0 \\
\dfrac{\partial C}{\partial \zeta} = -\mu C & \xi > 0 & \zeta = 1
\end{array}
$$

where μ is the *Damköhler* number, defined as:

$$\mu = \frac{k''\delta}{D}$$

The surface rate constant k'' is related to the surface concentration of the enzyme $[E'']$, the turnover number k_{cat}, and the intrinsic Michaelis-Menten constant K_m by:

$$k'' = \frac{k_{cat}[E'']}{K_m}$$

(a) Predict the electrode response as a function of the dimensionless time ξ for a 0.3-mm-thick membrane with the analyte diffusion coefficient $D = 2 \times 10^{-6}$ cm²/s and immobilized enzyme with the surface rate constant $k'' = 0.24$ cm/h.

(b) Repeat part (a) for the reaction kinetics defined by the Michaelis-Menten law.

7.15 The radial dispersion coefficient of solids in a fluidized bed can be evaluated by the injection of tracer particles at the center of the fluidized bed and monitoring the unsteady state dispersion of these particles [12]. Assuming instantaneous axial mixing of solids and radial mixing occurring by dispersion, the governing partial differential equation of the model, in cylindrical coordinates, is:

$$\frac{\partial C}{\partial t} = D_{sr} \frac{1}{r} \frac{\partial}{\partial r}\left(r \frac{\partial C}{\partial r}\right)$$

where C is the concentration of the tracer, t is time, r is the radial position, and D_{sr} is the radial solid dispersion coefficient. The appropriate initial and boundary conditions are:

$$
\begin{array}{lll}
t = 0 & 0 \le r \le a & C = 100\% \\
t > 0 & r = 0 & \dfrac{\partial C}{\partial r} = 0 \\
t > 0 & r = R & \dfrac{\partial C}{\partial r} = 0
\end{array}
$$

where a is the radius of the tracer injection tube and R is the radius of the column. The analytical solution of the dispersion equation, subject to these conditions, is:

$$\frac{C}{C_\infty} = 1 + \frac{2}{a} \sum_{i=0}^{\infty} \frac{J_1(\lambda_i a) J_0(\lambda_i r)}{\lambda_i [J_0(\lambda_i R)]^2} \exp\left[-D_{sr}\lambda_i^2 t\right]$$

where C_∞ is the concentration of the tracer at the steady state condition, J is the Bessel function of the first kind, and λ_i is calculated from:

$$J_1(\lambda_i R) = 0$$

(a) Use the analytical solution of the dispersion equation to plot the unsteady state concentration profiles of the tracer.

(b) Solve the dispersion equation numerically and compare it with the exact solution.

Additional data: $2R = 0.27$ m, $2a = 19$ mm, $D_{sr} = 2 \times 10^{-4}$ m²/s

7.16 Consider a first-order chemical reaction carried out under isothermal steady state conditions in a tubular-flow reactor. On the assumptions of laminar flow and negligible axial diffusion, the material balance equation is:

$$-v_0\left[1 - \left(\frac{r}{R}\right)^2\right]\frac{\partial c}{\partial z} + D\left(\frac{\partial^2 c}{\partial r^2} + \frac{1}{r}\frac{\partial c}{\partial r}\right) - kc = 0$$

where

v_0 = velocity of central streamline
R = tube radius
k = reaction rate constant
c = concentration of reactant
D = radial diffusion constant
z = axial distance along the length of the tube
r = radial distance from the center of the tube

Upon defining the following dimensionless variables,

$$\lambda = \frac{kz}{v_0} \quad C = \frac{c}{c_0} \quad \alpha = \frac{D}{kR^2} \quad U = \frac{r}{R}$$

the equation becomes:

$$(1 - U^2)\frac{\partial C}{\partial \lambda} = \alpha\left(\frac{\partial^2 C}{\partial U^2} + \frac{1}{U}\frac{\partial C}{\partial U}\right) - C$$

where c_0 is the entering concentration of the reactant to the reactor.

(a) Choose a set of appropriate boundary conditions for this problem. Explain your choice.

(b) What class of PDE is the above equation (hyperbolic, parabolic, or elliptic)?

(c) Set up the equation for a numerical solution using finite difference approximations.

(d) Does your choice of finite differences result in an explicit or implicit set of equations? Give the details of the procedure for the solution of this set of equations.

(e) Discuss stability considerations with respect to the method you have chosen.

REFERENCES

[1] Bird, R. B.; Stewart, W. E.; Lightfoot, E. N. *Transport Phenomena*, 2nd ed.; John Wiley & Sons: New York, 2006.

[2] Tychonov, A. N.; Samarski, A. A. *Partial Differential Equations of Mathematical Physics;* Holden-Day: San Francisco, 1964.

[3] Vichnevetsky, R. *Computer Methods for Partial Differential Equations*, vol. I. Prentice-Hall: Englewood Cliffs, NJ, 1981.

[4] Lapidus, L.; Pinder, G. F. *Numerical Solution of Partial Differential Equations in Science and Engineering;* Wiley: New York, 2011.

[5] Finlayson, B. A. *Nonlinear Analysis in Chemical Engineering;* McGraw-Hill: New York, 1980.

[6] Geankoplis, C. J.; Hersel, A.; Lepek, D. H. *Transport Processes and Separation Process Principles*, 5th ed.; Pearson: Boston, 2018.

[7] James, M. L.; Smith, G. M.; Wolford, J. C. *Applied Numerical Methods for Digital Computation with FORTRAN and CSMP*, 2nd ed.; Harper & Row: New York, 1977.

[8] Carnahan, B.; Luther, H. A.; Wilkes, J. O. *Applied Numerical Methods;* Wiley: New York, 1969.

[9] Fernandes, P. M.; Constantinides, A.; Vieth, W. R.; Venkatasubramanian, K. Enzyme Engineering: Part V. Modeling and Optimizing Multi-Enzyme Reactor Systems. *Chemtech* **1975,** *July*, 438−445.

[10] Coulet, P. R.; Sternberg, R.; Thevenot, D. R. Electrochemical Study of Reactions at Interfaces of Glucose Oxidase Collagen Membranes. *Biochimica et Biophysica Acta* **1980,** *612*, 317−327.

[11] Pedersen, H.; Chotani, G. K. Analysis of a Theoretical Model for Anisotropic Enzyme Membranes: Application to Enzyme Electrodes. *Applied Biochemistry and Biotechnology* **1981,** *6*, 309−317.

[12] Berruti, F.; Scott, D. S.; Rhodes, E. Measurement and Modelling Lateral Solid Mixing in a Three-Dimensional Batch Gas-Solid Fluidized Bed Reactor. *Canadian Journal of Chemical Engineering* **1986,** *64*, 48−56.

LINEAR AND NONLINEAR REGRESSION ANALYSIS

CHAPTER OUTLINE

MOTIVATION

Regression analysis is the application of mathematical and statistical methods for the analysis of the experimental data and the fitting of the mathematical models to these data by the estimation of the unknown parameters of the models. The series of statistical tests, which normally accompany regression analysis, serve in model identification, model verification, and efficient design of the experimental program. Most mathematical models encountered in engineering and science are non-linear in the parameters. These parameters can be determined by the implementation of linear and

Applied Numerical Methods for Chemical Engineers. DOI: https://doi.org/10.1016/B978-0-12-822961-3.00008-X

nonlinear regression methods. In this chapter after giving a brief review of statistical terminology, we develop the basic algorithm of linear regression and then show how this is extended to nonlinear regression. We develop the methods in matrix notation so that the algorithms are equally applicable to fitting single or multiple variables and to using single or multiple sets of experimental data.

8.1 PROCESS ANALYSIS, MATHEMATICAL MODELING, AND REGRESSION ANALYSIS

Engineers and scientists are often required to analyze complex physical or chemical systems and to develop mathematical models that simulate the behavior of such systems. *Process analysis* is a term commonly used by chemical engineers to describe the study of complex chemical, biochemical, or petrochemical processes. Moreover, phrases such as *systems engineering* and *systems analysis* are used by electrical engineers and computer scientists to refer to the analysis of electric networks and computer systems. No matter what the phraseology is, the principles applied are the same.

In the process analysis, scientific methods are used for the recognition and definition of problems and the development of procedures for their solution. The process analysis includes (1) mathematical specification of the problem for the given physical solution, (2) detailed analysis to obtain mathematical models, and (3) synthesis and presentation of results to ensure full comprehension.

At the heart of successful process analysis is the step of *mathematical modeling*. The objective of modeling is to construct, from theoretical and empirical knowledge of a process, a mathematical formulation that can be used to predict the behavior of this process. Complete understanding of the mechanism of the chemical, physical, or biological aspects of the process under investigation is not usually possible. However, some information on the mechanism of the system may be available; therefore, a combination of empirical and theoretical methods can be used. It is important to note that no model can present a precise description of reality. Nevertheless, a theoretical model that works provides information on the system under study over important ranges of the variables by means of equations that reflect at least the major features of the mechanism.

The engineer in the process industries is usually concerned with the operation of existing plants and the development of new processes. In the first case the control, improvement, and optimization of the operation are the engineer's main objectives. In order to achieve this a quantitative representation of the process—a model—is needed that would give the relationship between the various parts of the system. In the design of new processes, the engineer draws information from theory and the literature to construct mathematical models that may be used to simulate the process (Fig. 8.1). The development of mathematical models often requires the implementation of an experimental program in order to obtain the necessary information for the verification of the models. The experimental program is originally designed based on the theoretical considerations coupled with *a priori* knowledge of the process and is subsequently modified based on the results of regression analysis.

Regression analysis is the application of mathematical and statistical methods for the analysis of the experimental data and the fitting of the mathematical models to these data by the estimation

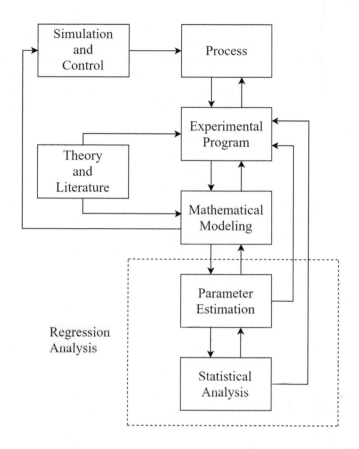

FIG. 8.1

Mathematical modeling and regression analysis.

of the unknown parameters of the models. The series of statistical tests, which normally accompany regression analysis, serve in model identification, model verification, and efficient design of the experimental program.

Strictly speaking, a mathematical model of a dynamic system is a set of equations that can be used to calculate how the *state* of the system evolves through time under the action of the *control variables*, given the state of the system at some initial time. The state of the system is described by a set of variables known as *state variables*. The first stage in the development of a mathematical model is to identify the state and control variables.

The control variables are those that can be directly controlled by the experimenter and that influence the way the system changes from its initial state to that of any later time. Examples of control variables in a chemical reaction system may be the temperature, pressure, and/or concentration of some of the components. The state variables are those that describe the state of the system and that are not under direct control. The concentrations of reactants and products are state variables in chemical systems. The distinction between state and control variables is not always fixed, but can change when the method of operating the system changes. For example, if the temperature is not directly controlled, it becomes a state variable.

The equations comprising the mathematical model of the process are called the *performance equations*. These equations should show the effect of the control variables on the evolution of the state variables. The performance equation may be a set of differential equations and/or a set of algebraic equations. For example, a set of ordinary differential equations describing the dynamics of a process may have the general form:

$$\frac{dy}{dx} = g(x, y, \theta, b) \tag{8.1}$$

where x = independent variable
y = vector of state (dependent) variables
θ = vector of control variables
b = vector of parameters whose values must be determined

In this chapter we concern ourselves with the methods of estimating the parameter vector b using regression analysis. For this purpose, we assume that the vector of control variables θ is fixed; therefore, the mathematical model simplifies to:

$$\frac{dy}{dx} = g(x, y, b) \tag{8.2}$$

In their integrated form, the previous set of performance equations converts to:

$$y = f(x, b) \tag{8.3}$$

For regression analysis, mathematical models are classified as *linear* or *nonlinear* with respect to the *unknown parameters*. For example, the following differential equation,

$$\frac{dy}{dt} = ky \tag{8.4}$$

which we classified earlier as linear with respect to the dependent variable (see Chapter 5), is *nonlinear* with respect to the parameter k. This is clearly shown by the integrated form of Eq. (8.4):

$$y = y_0 e^{kt} \tag{8.5}$$

where y is highly nonlinear with respect to k.

Most mathematical models encountered in engineering and science are nonlinear in the parameters. Attempts at linearizing these models—by rearranging the equations and regrouping the variables—were common practice in the pre-computer era, when graph paper and the straightedge were the tools for fitting models to experimental data. Such primitive techniques have been replaced by the implementation of *linear* and *nonlinear regression* methods on the computer.

The theory of linear regression has been expounded by statisticians and econometricians, and a rigorous statistical analysis of the regression results has been developed. Nonlinear regression is an extension of the linear regression methods used iteratively to arrive at the values of the parameters of the nonlinear models. The statistical analysis of the nonlinear regression results is also an extension of that applied in the linear analysis but does not possess the rigorous theoretical basis of the latter.

In this chapter after giving a brief review of statistical terminology, we develop the basic algorithm of linear regression and then show how this is extended to nonlinear regression. We develop the methods in matrix notation so that the algorithms are equally applicable to fitting single or multiple variables and to using single or multiple sets of experimental data.

8.2 REVIEW OF STATISTICAL TERMINOLOGY USED IN REGRESSION ANALYSIS

It is assumed that the reader has a rudimentary knowledge of statistics. This section serves as a review of the statistical definitions and terminology needed for understanding the application of linear and nonlinear regression analysis and the statistical treatment of the results of this analysis. For a complete discussion of statistics, the reader should consult a standard text on statistics, such as Montgomery et al. [1] and Metcalfe et al. [2].

8.2.1 POPULATION AND SAMPLE STATISTICS

A *population* is defined as a group of similar items, or events, from which a sample is drawn for test purposes; the population is usually assumed to be very large—sometimes infinite. A *sample* is a random selection of items from a population, usually made for evaluating a variable of that population. The *variable* under investigation is a characteristic property of the population.

A *random variable* is defined as a variable that can assume any value from a set of possible values. A *statistic* or *statistical parameter* is any quantity computed from a sample; it is characteristic of the sample and is used to estimate the characteristics of the population variable.

Degree of freedom can be defined as the number of observations made in excess of the minimum theoretically necessary to estimate a statistical parameter or any unknown quantity.

Let us use N to denote the total number of items in the population under study, where $0 \leq N \leq \infty$, and n to specify the number of items contained in the sample taken from that population, where $0 \leq n \leq N$. The variable being investigated will be designated as X; it may have discrete values, or it may be a continuous function, in the range $-\infty < x < \infty$. For specific populations, these ranges may be more limited, as will be mentioned later.

For the sake of example and in order to clarify these terms, let us consider studying the entire human population and examine the age of this population. The value of N, in this case, would be approximately 8 billion. The age variable would range from 0 to possibly 150 years. Age can be considered either as a continuous variable because all ages are possible or, more commonly, as a discrete variable because ages are usually grouped by year. In the continuous case, the age variable takes an infinite number of values in the range $0 < x \leq 150$, and in the discrete case it takes a finite number of values x_j, where $j = 1, 2, 3, \ldots, M$ and $M \leq 150$. Assume that a random sample of n persons is chosen from the total population (say $n = 1$ million) and the age of each person in the sample is recorded.

The frequency with which each value of the variable (age, in the earlier example) may occur in the population is not the same; some values (ages) will occur more frequently than others.

Designating m_j as the number of times the value of x_j occurs, we can define the concept of *probability of occurrence* as:

$$Pr\{X = x_j\} = \left\{ \begin{array}{l} \text{Probability of} \\ \text{occurrence of} \\ x_j \end{array} \right\} = \frac{\text{number of occurrences of } x_j}{\text{total number of observations}}$$

$$= \lim_{n \to N} \frac{m_j}{n} = p(x_j)$$

(8.6)

For a discrete random variable, $p(x_j)$ is called the *probability function*, and it has the following properties:

$$0 \le p(x_j) \le 1$$
$$\sum_{j=1}^{M} p(x_j) = 1$$

(8.7)

The shape of a typical probability function is shown in Fig. 8.2a.

For a continuous random variable, the probability of occurrence is measured by the continuous function $p(x)$, which is called the *probability density function*, so that:

$$Pr\{x < X \le x + dx\} = p(x)dx$$

(8.8)

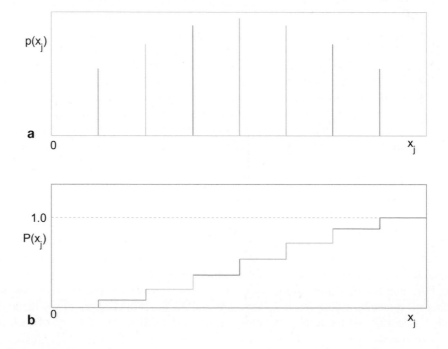

FIG. 8.2

(a) Probability function and (b) cumulative distribution function for a discrete random variable.

The probability density function has the following properties:

$$0 \leq p(x) \leq 1$$
$$\int_{-\infty}^{\infty} p(x)dx = 1 \tag{8.9}$$

The smooth curve obtained from plotting $p(x)$ versus x (Fig. 8.3a) is called the continuous probability density distribution.

The cumulative distribution function is defined as the probability that a random variable X will not exceed a given value x, that is:

$$Pr\{X \leq x\} = P(x) = \int_{-\infty}^{x} p(x)dx \tag{8.10}$$

The equivalent of Eq. (8.10) for a discrete random variable is:

$$Pr\{X \leq x_i\} = P(x_i) = \sum_{j=1}^{i} p(x_i) \tag{8.11}$$

The cumulative distribution functions for discrete and continuous random variables are illustrated in Figs. 8.2b and 8.3b, respectively.

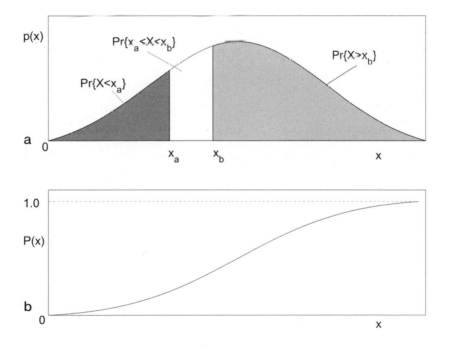

FIG. 8.3

(a) Probability function and (b) cumulative distribution function for a continuous random variable.

It is obvious from the integral of Eq. (8.10) that the cumulative distribution function is obtained from calculating the area under the density distribution function. The three area segments shown in Fig. 8.3a correspond to the following three probabilities:

$$Pr\{X \leq x_a\} = \int_{-\infty}^{x_a} p(x)dx \tag{8.12}$$

$$Pr\{x_a \leq X \leq x_b\} = \int_{x_a}^{x_b} p(x)dx \tag{8.13}$$

$$Pr\{X > x_b\} = \int_{x_b}^{\infty} p(x)dx \tag{8.14}$$

The *population mean*, or *expected value*, of a discrete random variable is defined as:

$$\mu = E[X] = \sum_{j=1}^{M} x_j p(x_j) \tag{8.15}$$

and that of a continuous random variable as:

$$\mu = E[X] = \int_{-\infty}^{\infty} xp(x)dx \tag{8.16}$$

The usefulness of the concept of *expectation*, as defined earlier, is that it corresponds to our intuitive idea of *average*, or equivalently to the center of gravity of the probability density distribution along the x-axis. It is easy to show that combining Eqs. (8.6) and (8.15) yields the arithmetic average of the random variable for the entire population:

$$\mu = E[X] = \frac{1}{N} \sum_{i=1}^{N} x_i \tag{8.17}$$

In addition, the integral of Eq. (8.16) can be recognized from the field of mechanics as the *first noncentral moment of X*.

The *sample mean*, or *arithmetic average*, of a sample of observations is the value obtained by dividing the sum of observations by their total number:

$$\bar{x} = \frac{1}{n} \sum_{i=1}^{n} x_i \tag{8.18}$$

The expected value of the sample mean is given by:

$$E[\bar{x}] = E\left[\frac{1}{n}\sum_{i=1}^{n} x_i\right] = \frac{1}{n}\sum_{i=1}^{n} E[x_i] = \frac{1}{n}\sum_{i=1}^{n} \mu = \mu \tag{8.19}$$

that is, the sample mean is an *unbiased estimate* of the population mean.

The *population variance* is defined as the expected value of the square of the deviation of the random variable X from its expectation:

$$\begin{aligned}
\sigma^2 &= V[X] \\
&= E\left[(X - E[X])^2\right] \\
&= E\left[(X - \mu)^2\right]
\end{aligned} \tag{8.20}$$

For a discrete random variable, Eq. (8.20) is equivalent to:

$$\sigma^2 = \sum_{j=1}^{M} (x_j - \mu)^2 p(x_j) \tag{8.21}$$

When combined with Eq. (8.6), Eq. (8.21) becomes:

$$\sigma^2 = \frac{1}{N} \sum_{i=1}^{N} (x_j - \mu)^2 \tag{8.22}$$

which is the arithmetic average of the square of the deviations of the random variable from its mean. For a continuous random variable, Eq. (8.20) is equivalent to:

$$\sigma^2 = \int_{-\infty}^{\infty} (x - \mu)^2 p(x) dx \tag{8.23}$$

which is the *second central moment of X* about the mean.

It is interesting and useful to note that Eq. (8.20) expands as follows:

$$\begin{aligned}
V[X] &= E\left[(X - E[X])^2\right] = E\left[X^2 + (E[X])^2 - 2XE[X]\right] \\
&= E\left[X^2\right] + E\left[(E[X])^2\right] - 2E[XE[X]] \\
&= E\left[X^2\right] + (E[X])^2 - 2(E[X])^2 \\
&= E\left[X^2\right] - (E[X])^2 \\
&= E\left[X^2\right] - \mu^2
\end{aligned} \tag{8.24}$$

Note that the expected value of a constant is that constant. In the previous equation, since the expected value of X is constant, $E[E[X]] = E[X]$.

The positive square root of the population variance is called the *population standard deviation*:

$$\sigma = +\sqrt{\sigma^2} \tag{8.25}$$

The *sample variance* is defined as the arithmetic average of the square of the deviations of x_i from the population mean μ:

$$s^2 = \frac{1}{n} \sum_{i=1}^{n} (x_i - \mu)^2 \tag{8.26}$$

However, because μ is not usually known, x is used as an estimate of μ, and the sample variance is calculated from:

$$s^2 = \frac{1}{n-1} \sum_{i=1}^{n} (x_i - \bar{x})^2 \tag{8.27}$$

where the degrees of freedom have been reduced to $(n - 1)$ because the calculation of the sample mean consumes one degree of freedom. The sample variance obtained from Eq. (8.27) is an unbiased estimate of population variance, that is,

$$E[s^2] = \sigma^2 \tag{8.28}$$

The positive square root of the sample variance is called the sample *standard deviation*:

$$s = +\sqrt{s^2} \tag{8.29}$$

In MATLAB® the built-in function $std(x)$ calculates the standard deviation of the vector x (Eq. 8.29). If x is a matrix, $std(x)$ returns a vector of standard deviations of each column.

The covariance of two random variables X and Y is defined as the expected value of the product of the deviations of X and Y from their expected values:

$$Cov[X, Y] = E[(X - E[X])(Y - E[Y])] \tag{8.30}$$

Eq. (8.30) expands to:

$$\begin{aligned} Cov[X, Y] &= E[XY - YE[X] - XE[Y] + E[X]E[Y]] \\ &= E[XY] - E[X]E[Y] \end{aligned} \tag{8.31}$$

The covariance is a measurement of the association between the two variables. If large positive deviations of X are associated with large positive deviations of Y, and likewise large negative deviations of the two variables occur together, then the covariance will be positive. Furthermore, if positive deviations of X are associated with negative deviations of Y and vice versa, then the covariance will be negative. On the other hand, if positive and negative deviations of X occur equally frequently with positive and negative deviations of Y, then the covariance will tend to zero.

The variance of X, defined earlier in Eq. (8.20), is a special case of the covariance of the random variable with itself:

$$\begin{aligned} Cov[X, X] &= E[(X - E[X])(X - E[X])] \\ &= E\left[(X - E[X])^2\right] = V[X] \end{aligned} \tag{8.32}$$

The magnitude of the covariance depends on the magnitude and units of X and Y, and could conceivably range from $-\infty$ to ∞. To make the measurement of covariance more manageable, the two dimensionless standardized variables are formed:

$$\frac{X - E[X]}{\sqrt{V[X]}} \quad \text{and} \quad \frac{Y - E[Y]}{\sqrt{V[Y]}}$$

The covariance of the standardized variables is known as the *correlation coefficient*:

$$\rho_{XY} = Cov\left[\frac{X - E[X]}{\sqrt{V[X]}}, \frac{Y - E[Y]}{\sqrt{V[Y]}}\right] \tag{8.33}$$

Using the definition of covariance reduces the correlation coefficient to:

$$\rho_{XY} = \frac{Cov[X, Y]}{\sqrt{V[X]V[Y]}} \tag{8.34}$$

If $\rho_{XY} = 0$, we say that X and Y are *uncorrelated*, and this implies that:

$$Cov = [X, Y] = 0 \tag{8.35}$$

We know from probability theory that if X and Y are *independent variables*, then,

$$p\{x, y\} = p_x(x)p_y(y) \tag{8.36}$$

from which it follows that:

$$E[XY] = E[X]E[Y] \qquad (8.37)$$

Combining Eqs. (8.37) and (8.31) shows that:

$$Cov[X, Y] = 0 \qquad (8.38)$$

and from Eq. (8.34):

$$\rho_{XY} = 0 \qquad (8.39)$$

Thus independent variables are uncorrelated.

The population and sample statistics discussed earlier are summarized in Table 8.1.

Table 8.1 Summary of population and sample statistics.

Statistics	Population		Sample
	Continuous variable	**Discrete variable**	
Mean	$\mu = E[X] = \int_{-\infty}^{\infty} xp(x)dx$	$\mu = E[X] = \sum_{j=1}^{M} x_j p(x_j)$	$\bar{x} = \frac{1}{n}\sum_{i=1}^{n} x_i$
Variance	$\sigma^2 = V[X] = E[(X - E[X])^2]$ $= \int_{-\infty}^{\infty} (x-\mu)^2 p(x)dx$	$\sigma^2 = V[X] - E[(X - E[X])^2]$ $= \sum_{j=1}^{M}(x_j - \mu)^2 p(x_j)$	$s^2 = \frac{1}{n}\sum_{i=1}^{n}(x_i - \mu)^2$ or $s^2 = \frac{1}{n-1}\sum_{i=1}^{n}(x_i - \bar{x})^2$
Standard deviation	$\sigma = +\sqrt{\sigma^2}$	$\sigma = +\sqrt{\sigma^2}$	$s = +\sqrt{s^2}$
Covariance Correlation coefficient	$Cov[X, Y] = E[(X - E[X])(Y - E[Y])]$ $\rho_{XY} = \frac{Cov[X,Y]}{\sqrt{V[X]V[Y]}}$		

EXAMPLE 8.1 PARTICLE SIZE DISTRIBUTION

The size distribution of a powder, shown in Table E8.1, was measured by sieve analysis. Determine:

(a) Probability density and cumulative distribution of size for this powder.
(b) Calculate the mean particle size and standard deviation of the distribution.

Method of solution

Part (a): The probability at each point j is calculated from:

$$p(D_j) = \frac{m_j}{\sum_{i=1}^{M} m_i \Delta D_i} \qquad (1)$$

(*Continued*)

Table E8.1 Sieve analysis.

Size, D (μm)	45–53	53–63	63–75	75–90	90–106	106–125	125–150	150–180	180–212
Weight, m (g)	1	6	22	44	54	45	25	3	0

EXAMPLE 8.1 (CONTINUED)

and the cumulative distribution is evaluated from:

$$P(D_j) = \frac{\sum_{i=1}^{j} m_i \Delta D_i}{\sum_{i=1}^{M} m_i \Delta D_i} \tag{2}$$

In these equations, D_j is the arithmetic mean of the size range.

Part (b): Arithmetic mean size and variance of the size distribution are evaluated based on Eqs. (8.16) and (8.23), respectively, from:

$$\overline{D} = \sum_{i=1}^{M} D_i p(D_j) \Delta D_i \tag{3}$$

$$s^2 = \sum_{i=1}^{M} (D_i - \overline{D})^2 p(D_j) \Delta D_i \tag{4}$$

Input and results

Calculation steps and results are shown in the Excel spreadsheet shown in Fig. E8.1a. The first three columns in this spreadsheet are the inputs as given in Table E8.1. The following steps were followed in creating this spreadsheet:

- The arithmetic mean of the size range is given in column D. The formula in cell D2 is " = (A2 + B2)/2" and then copied down to the final row 10.
- For computing the denominator of Eqs. (1) and (2), we write the formula " = C2*(B2 − A2)" in cell E2 and copy this formula down to cell E10.
- The summation in the denominator is then evaluated in cell E12 by typing in " = SUM(E2:E10)".
- The value of the probability distribution function at each particle size is calculated by typing " = C2/E12" in cell F2 and copying it in the following rows. Note that the symbol $ before the coordinates of the cell allows copying the formula without changing the reference cell, which is the cell E12 here (the summation for the whole distribution is the same for all points).
- The cumulative distribution is given in column G. The formulas for the cells of this column were typed row by row because of the fact that in each cell the value should be obtained, according to Eq. (2), by summation of all previous intervals. For example, the formula in cell F4 is: " = SUM(E2:E4)/E12".
- For calculation of the average particle size from Eq. (3), we type " = D2*F2*(B2 − A2)" in cell H2 and then copy it to the following rows. The average particle size or the arithmetic mean value, is then evaluated in cell H12 by the formula " = SUM(H2:H10)". It can be seen that the average particle size is 105.30 μm for this distribution.

(Continued)

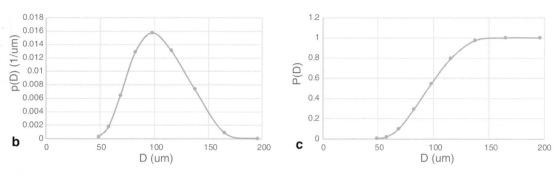

FIG. E8.1

Particle size distribution: (b) Probability density, (c) Cumulative distribution.

EXAMPLE 8.1 (CONTINUED)

- For calculation of the variance from Eq. (4), we type " = (D2-H12)^2*F2*(B2 − A2)" in cell I2 and then copy it to the following rows. The variance is then evaluated in cell I12 by the formula " = SUM(I2:I10)", and the standard deviation is given in cell I14 from the formula " = SQRT(I12)". The standard deviation of the distribution in this problem is 23.88 μm.

  Probability and cumulative distributions can be visualized in the same spreadsheet by following these steps:

$$\text{Insert} \rightarrow \text{Graphs} \rightarrow \text{Scatter} \rightarrow \text{Right-click on the chart} \rightarrow \text{Select Data}$$

  For the probability density, you should choose values in column D for Series X and those in column F for Series Y. For showing the cumulative distribution, these columns are D and G, respectively. These graphs are shown in Figs. E8.1b and c.

8.2.2 PROBABILITY DENSITY FUNCTIONS AND PROBABILITY DISTRIBUTIONS

Many different probability density functions are encountered in statistical analysis. Of particular interest to regression analysis are the *normal*, χ^2, *t*, and *F* distributions, which will be discussed in this section. The *normal* or *Gaussian* density function has the form:

$$p(x) = \frac{1}{\sigma\sqrt{2\pi}} \exp\left[-\frac{1}{2}\left(\frac{x-\mu}{\sigma}\right)^2\right] \tag{8.40}$$

where $-\infty < x < \infty$. The cumulative distribution function of the normal density function is:

$$P(x) = \frac{1}{\sigma\sqrt{2\pi}} \int_{-\infty}^{x} \exp\left[-\frac{1}{2}\left(\frac{x-\mu}{\sigma}\right)^2\right] dx \tag{8.41}$$

which involves an integral that does not have an explicit form and must be integrated numerically (see Chapter 4). The normal probability distributions are illustrated in Fig. 8.4.

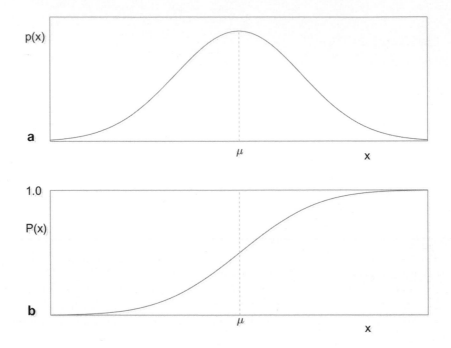

FIG. 8.4

(a) Normal probability density and (b) normal cumulative distribution for a continuous random variable.

The expected value of a variable that has a normal distribution is:

$$E[X] = \mu \tag{8.42}$$

and the variance is:

$$V[X] = \sigma^2 \tag{8.43}$$

For this reason, the normal density function is usually abbreviated as $N(\mu, \sigma^2)$, and the notation:

$$X \sim N(\mu, \sigma^2) \tag{8.44}$$

means that the variable X has a normal distribution with expected value μ and variance σ^2.

A normal density function can be transformed to the *standard normal density function* by the substitution:

$$u = \frac{x - \mu}{\sigma} \tag{8.45}$$

which transforms Eq. (8.40) to:

$$\phi(u) = \frac{1}{\sqrt{2\pi}} \exp\left(-\frac{u^2}{2}\right) \tag{8.46}$$

The expected value of the standardized variable u is:

$$E[u] = 0 \tag{8.47}$$

and the variance is:

$$V[u] = 1 \tag{8.48}$$

Therefore

$$u \sim N(0, 1) \tag{8.49}$$

The standard normal density function $\phi(u)$ and its cumulative distribution function:

$$\Phi(u) = \frac{1}{\sqrt{2\pi}} \int_{-\infty}^{u} \exp\left(-\frac{u^2}{2}\right) du \tag{8.50}$$

are shown in Fig. 8.5.

The function $\phi(u)$ is symmetrical about zero; therefore, the area in the left tail, below $-u$, is equal to the area in the right tail, above $+u$ (shaded area in Fig. 8.5a). The unshaded area between -1.960 and 1.960 is equivalent to 95% of the total area under the density function. This area is

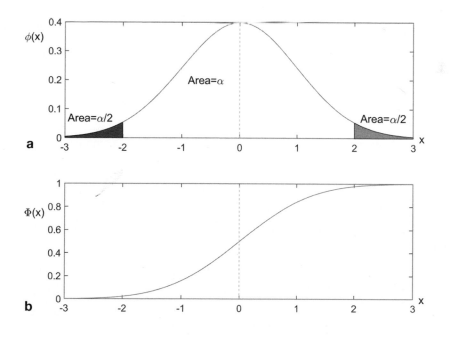

FIG. 8.5

(a) Standardized normal probability density and (b) standardized normal cumulative distribution.

designated as $(1 - \alpha)$ and the area under each tail as $\alpha/2$. Application of Eqs. (8.12) to (8.14) shows that:

$$Pr\{u \leq -1.960\} = \frac{\alpha}{2} = 0.025$$
$$Pr\{-1.960 < u \leq 1.960\} = 1 - \alpha = 0.95 \qquad (8.51)$$
$$Pr\{u > 1.960\} = \frac{\alpha}{2} = 0.025$$

If a set of normally distributed variables X_j, where

$$X_k \sim N\left(\mu_k, \sigma_k^2\right) \qquad (8.52)$$

is linearly combined to form another variable Y, where

$$Y = \sum_k a_k X_k \qquad (8.53)$$

then Y is also normally distributed, that is,

$$Y \sim N\left(\sum_k a_k \mu_k, \sum_k a_k^2 \sigma_k^2\right) \qquad (8.54)$$

The sample mean (Eq. 8.18) of a normally distributed population is a linear combination of normally distributed variables; therefore, the sample mean itself is normally distributed as follows:

$$\bar{x} \sim N\left(\mu, \frac{\sigma^2}{n}\right) \qquad (8.55)$$

It follows then, from Eqs. (8.45) and (8.49), that,

$$\frac{\bar{x} - \mu}{\sqrt{\sigma^2/n}} \sim N(0, 1) \qquad (8.56)$$

If we wish to test the hypothesis that a sample whose mean is x could come from a normal distribution of mean μ and known variance σ^2, the procedure is easy, because the variable $(x - \mu)/\sqrt{\sigma^2/n}$ is normally distributed as $N(0, 1)$ and can readily be compared with tabulated values. However, if σ^2 is unknown and must be estimated from the sample variance s^2, then *Student's t distribution*, which is described later in this section, is needed.

Now consider a sequence X_j of identically distributed, independent random variables (not necessarily normally distributed) whose second-order moment exists. Let

$$E[X_k] = \mu \qquad (8.57)$$

and

$$E\left[(X_k - \mu)^2\right] = V[X_k] = \sigma^2 \qquad (8.58)$$

for every k. Consider the random variable Z_n defined by:

$$Z_n = X_1 + X_2 + \cdots + X_n \tag{8.59}$$

where

$$E[Z_n] = n\mu \tag{8.60}$$

and, by the independence of X_k,

$$E\left[(Z_n - n\mu)^2\right] = n\sigma^2 \tag{8.61}$$

Let

$$\hat{Z}_n = \frac{Z_n - n\mu}{\sigma\sqrt{n}} \tag{8.62}$$

then the distribution of Z_n approaches the standard normal distribution, that is,

$$\lim_{n \to \infty} P_n(z) = \frac{1}{\sqrt{2\pi}} \int_{-\infty}^{z} \exp\left(-\frac{z^2}{2}\right) dz \tag{8.63}$$

This is the *central limit theorem*, a proof of which can be found in Seinfeld and Lapidus [3]. This is a very important theorem of statistics, particularly in a regression analysis where experimental data are being analyzed. The experimental error is a composite of many separate errors whose probability distributions are not necessarily normal distributions. However, as the number of components contributing to the error increases, the central limit theorem justifies the assumption of normality of the error.

Suppose we have a set of ν independent observations, $x_1, \ldots x_\nu$ from a normal distribution $N(\mu, \sigma^2)$. The standardized variables:

$$u_i = \frac{x_i - \mu}{\sigma} \tag{8.64}$$

will also be independent and have distribution $N(0, 1)$. The variable $\chi^2(\nu)$ is defined as the sum of the squares of u_i as follows

$$\chi^2(\nu) = \sum_{i=1}^{\nu} u_i^2 = \sum_{i=1}^{\nu} \frac{(x_i - \mu)^2}{\sigma^2} \tag{8.65}$$

The $\chi^2(\nu)$ variable has the so-called χ^2 (*chi-square*) *distribution function*, which is given by:

$$p(\chi^2) = \frac{1}{2^{\nu/2}\Gamma(\nu/2)} e^{-\chi^2/2} (\chi^2)^{(\nu/2)-1} \tag{8.66}$$

where $\chi^2 \geq 0$ and

$$\Gamma\left(\frac{\nu}{2}\right) = \int_0^\infty e^{-x} x^{(\nu/2)-1} dx \tag{8.67}$$

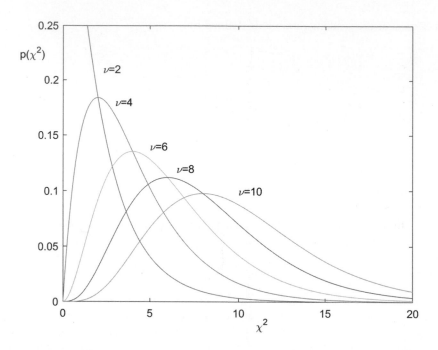

FIG. 8.6

The χ^2 distribution.

The χ^2 distribution is a function of the degrees of freedom ν, as shown in Fig. 8.6. The distribution is confined to the positive half of the χ^2-axis because the u_i^2 quantities are always positive.

The expected value of χ^2 variable is:

$$\mu = E[\chi^2] = \int_0^\infty \chi^2 p(\chi^2) d\chi^2 = \nu \tag{8.68}$$

and its variance is:

$$\sigma^2 = V[\chi^2] = \int_0^\infty (\chi^2)^2 p(\chi^2) d\chi^2 = 2\nu \tag{8.69}$$

The χ^2 distribution tends toward the normal distribution $N(\nu, 2\nu)$ as ν becomes large. The χ^2 distribution is widely used in statistical analysis for testing the independence of variables and the fit of probability distributions to experimental data.

We saw earlier that the sample variance was obtained from Eq. (8.27):

$$s^2 = \frac{1}{n-1} \sum_{i=1}^n (x_i - \bar{x})^2 \tag{8.27}$$

with $(n-1)$ degrees of freedom. When $\bar{x}$ is assumed to be equal to μ then,

$$s^2 = \frac{1}{n-1} \sum_{i=1}^n (x_i - \mu)^2 \tag{8.70}$$

Combining Eqs. (8.65) and (8.70) shows that:

$$\chi^2 = (n-1)\frac{s^2}{\sigma^2} \tag{8.71}$$

with $\nu = (n-1)$ degrees of freedom. This equation will be very useful in Section 8.2.3 in constructing confidence intervals for the population variance.

Let us define a new random variable t, so that:

$$t = \frac{u}{\sqrt{\chi^2/\nu}} \tag{8.72}$$

where $u \approx N(0, 1)$ and χ^2 is distributed as chi-square with ν degrees of freedom. It is assumed that u and χ^2 are independent of each other. The variable t is called *Student's t* and has the probability density function:

$$p(t) = \frac{1}{\sqrt{\nu\pi}}\frac{\Gamma[(\nu+1)/2]}{\Gamma(\nu/2)}\left(1+\frac{t^2}{\nu}\right)^{-(\nu+1)/2} \tag{8.73}$$

with ν degrees of freedom. The shape of the t density function is shown in Fig. 8.7.

The expected value of the t variable is:

$$\mu_t = E[t] = \int_{-\infty}^{\infty} tp(t)d - 0 \quad \text{for } \nu > 1 \tag{8.74}$$

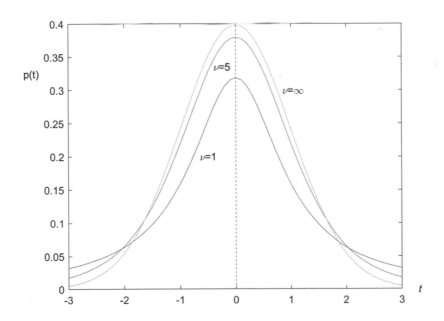

FIG. 8.7

The Student's t distribution.

and the variance is:

$$\sigma_t^2 = V[t] = \int_{-\infty}^{\infty} t^2 p(t) dt = \frac{\nu}{\nu - 2} \quad \text{for } \nu > 2 \tag{8.75}$$

The t distribution tends toward the normal distribution as v becomes large.

Combining Eq. (8.72) with Eqs. (8.56) and (8.71) gives:

$$t = \frac{(\bar{x} - \mu)/\sqrt{\sigma^2/n}}{\sqrt{s^2/\sigma^2}} = \frac{\bar{x} - \mu}{\sqrt{s^2/n}} \tag{8.76}$$

The quantity on the right-hand side of Eq. (8.76) is independent of σ and has a t distribution. Therefore, the t distribution provides a test of significance for the deviation of a sample mean from its expected value when the population variance is unknown and must be estimated from the sample variance.

Finally, we define the ratio:

$$F(\nu_1, \nu_2) = \frac{\chi_1^2/\nu_1}{\chi_2^2/\nu_2} \tag{8.77}$$

where χ_1^2 and χ_2^2 are two independent random variables of the chi-square distribution with ν_1 and ν_2 degrees of freedom, respectively. The variable $F(\nu_1, \nu_2)$ has the F distribution density function with ν_1 and ν_2 degrees of freedom as follows:

$$p(F) = \frac{\left(\frac{\nu_1}{\nu_2}\right)^{\nu_1/2} F^{(\nu_1/2)-1} \left(1 + \frac{\nu_1}{\nu_2} F\right)^{-(\nu_1+\nu_2)/2}}{\int_0^1 x^{(\nu_1/2)-1} (1-x)^{(\nu_2/2)-1} dx} \tag{8.78}$$

The F distribution is very useful in the analysis of variance of populations. Consider two normally distributed independent random samples:

$$x_{1,1}, \, x_{1,2}, \, \ldots, \, x_{1,n_1}$$

and

$$x_{2,1}, \, x_{2,2}, \, \ldots, \, x_{2,n_1}$$

The first sample, which has a sample variance s_1^2, is from a population with mean μ_1 and variance σ_1^2. The second sample, which has a sample variance s_2^2, is from a population with mean μ_2 and variance σ_2^2. Using Eq. (8.71), we see that:

$$\chi_1^2 = (n_1 - 1) \frac{s_1^2}{\sigma_1^2} \tag{8.79}$$

and

$$\chi_2^2 = (n_2 - 1) \frac{s_2^2}{\sigma_2^2} \tag{8.80}$$

Combining Eq. (8.77) with Eqs. (8.79) and (8.80) shows that:

$$F(n_1 - 1, n_2 - 1) = \frac{\chi_1^2/(n_1 - 1)}{\chi_2^2/(n_2 - 1)} = \frac{s_1^2/\sigma_1^2}{s_2^2/\sigma_2^2} \tag{8.81}$$

with $(n_1 - 1)$ and $(n_2 - 1)$ degrees of freedom. Furthermore, if the two populations have the same variance, that is, if $\sigma_1^2 = \sigma_2^2$, then:

$$F(n_1 - 1, n_2 - 1) = \frac{s_1^2}{s_2^2} \tag{8.82}$$

Therefore, the F distribution provides a means of comparing variances, as will be seen in Section 8.2.3.

8.2.3 CONFIDENCE INTERVALS AND HYPOTHESIS TESTING

The concept of the *confidence interval* is of considerable importance in regression analysis. A confidence interval is a range of values defined by an upper and a lower limit, the *confidence limits*. This range is constructed in such a way that we can say with a certain confidence that the true value of the statistic being examined lies within this range. The level of confidence is chosen at $100(1 - \alpha)\%$, where α is usually small, say, 0.05 or 0.01. For example, when $\alpha = 0.05$, the confidence level is 95%. We demonstrate the concept of the confidence interval by first constructing such an interval for the standard normal distribution, extending the concept to other distributions, and then calculating specific confidence intervals for the mean and variance.

We saw earlier that the standard normal variable u has a density function $\phi(u)$ (Eq. 8.46) and a cumulative distribution function $\Phi(u)$ (Eq. 8.50) and is distributed with $N(0, 1)$. Applying Eqs. (8.12) and (8.13) to the standard normal distribution,

$$Pr\{u \le u_{\alpha/2}\} = \int_{-\infty}^{u_{\alpha/2}} \phi(u)du = \Phi(u_{\alpha/2}) = \frac{\alpha}{2} \tag{8.83}$$

$$Pr\{u \le u_{1-\alpha/2}\} = \int_{-\infty}^{u_{1-\alpha/2}} \phi(u)du = \Phi(u_{1-\alpha/2}) = 1 - \frac{\alpha}{2} \tag{8.84}$$

and

$$Pr\{u_{\alpha/2} < u \le u_{1-\alpha/2}\} = \int_{u_{\alpha/2}}^{u_{1-\alpha/2}} \phi(u)du = \Phi(u_{1-\alpha/2}) - \Phi(u_{\alpha/2}) = 1 - \alpha \tag{8.85}$$

The inequality:

$$u_{\alpha/2} < u \le u_{1-\alpha/2} \tag{8.86}$$

defines the $100(1 - \alpha)\%$ interval for the variable u. If $\alpha = 0.05$, then the 95% confidence interval for the standard normal variable is:

$$-1.96 < u \le 1.96 \tag{8.87}$$

Let us now determine a confidence interval for the mean of a normally distributed population. We saw earlier that the sample mean $\bar{x}$ of a normally distributed population is also normally distributed,

$$\bar{x} \sim N\left(\mu, \frac{\sigma^2}{n}\right) \tag{8.55}$$

and that this can be converted to the standard normal distribution so that:

$$\frac{\bar{x} - \mu}{\sqrt{\sigma^2/n}} \sim N(0, 1) \tag{8.56}$$

Because the quantity $(x - \mu)/\sqrt{\sigma^2/n}$ is equivalent to u, Eq. (8.85) can be written as:

$$Pr\left\{u_{\alpha/2} < \frac{\bar{x} - \mu}{\sqrt{\sigma^2/n}} \leq u_{1-\alpha/2}\right\} = 1 - \alpha \tag{8.88}$$

or rearranged to:

$$Pr\left\{\bar{x} - u_{1-\alpha/2}\sqrt{\frac{\sigma^2}{n}} \leq \mu \leq \bar{x} - u_{\alpha/2}\sqrt{\frac{\sigma^2}{n}}\right\} = 1 - \alpha \tag{8.89}$$

The inequality:

$$\bar{x} - u_{1-\alpha/2}\sqrt{\frac{\sigma^2}{n}} \leq \mu \leq \bar{x} + u_{1-\alpha/2}\sqrt{\frac{\sigma^2}{n}} \tag{8.90}$$

is the $100(1 - \alpha)\%$ confidence interval for the population mean. Note that the density distribution of u is symmetrical around $u = 0$ so that $u_{\alpha/2} = -u_{1-\alpha/2}$. This substitution has been made in obtaining Eq. (8.90). For $\alpha = 0.05$, the 95% confidence interval of the mean of a normally distributed population is:

$$\bar{x} - 1.96\sqrt{\frac{\sigma^2}{n}} \leq \mu \leq \bar{x} + 1.96\sqrt{\frac{\sigma^2}{n}} \tag{8.91}$$

where $\bar{x}$ is the sample mean and σ^2 is the population variance. This simply says that we can state with 95% confidence that the true value of the population mean is in the range defined by the inequality in Eq. (8.91).

If the population variance σ^2 is not known, it will be estimated from the sample variance s^2. Replacing σ^2 with s^2 in the quantity $(x - \mu)/\sqrt{\sigma^2/n}$, we obtain the variable:

$$t = \frac{\bar{x} - \mu}{\sqrt{s^2/n}} \tag{8.76}$$

which has a Student's t distribution with $v = (n - 1)$ degrees of freedom, as shown in Section 8.2.2. The confidence interval in this case is obtained from:

$$Pr\left\{t_{\alpha/2} < \frac{\bar{x} - \mu}{\sqrt{s^2/n}} \leq t_{1-\alpha/2}\right\} = 1 - \alpha \tag{8.92}$$

which rearranges to:

$$Pr\left\{\bar{x} - t_{1-\alpha/2}\sqrt{\frac{s^2}{n}} \leq \mu \leq \bar{x} - t_{\alpha/2}\sqrt{\frac{s^2}{n}}\right\} = 1 - \alpha \tag{8.93}$$

to yield the $100(1-\alpha)\%$ confidence interval as follows:

$$\bar{x} - t_{1-\alpha/2}\sqrt{\frac{s^2}{n}} \leq \mu \leq \bar{x} + t_{1-\alpha/2}\sqrt{\frac{s^2}{n}} \tag{8.94}$$

The density distribution of the t variable is symmetrical around $t = 0$ so that $t_{\alpha/2} = -t_{1-\alpha/2}$. This substitution has been made in obtaining Eq. (8.94).

In Section 8.2.2 we showed that the sample variance s^2 and the population variance σ^2 were related through the χ^2 distribution:

$$\chi^2 = (n-1)\frac{s^2}{\sigma^2} \tag{8.71}$$

with $\nu = (n-1)$ degrees of freedom. This relation can now be used to construct the confidence interval for the variance from:

$$Pr\left\{\chi^2_{\alpha/2} < (n-1)\frac{s^2}{\sigma^2} \leq \chi^2_{1-\alpha/2}\right\} = 1 - \alpha \tag{8.95}$$

which gives the $100(1-\alpha)\%$ confidence interval for σ^2 as:

$$\frac{(n-1)s^2}{\chi^2_{1-\alpha/2}} \leq \sigma^2 \leq \frac{(n-1)s^2}{\chi^2_{\alpha/2}} \tag{8.96}$$

This discussion leads us to the concept of *hypothesis testing*. This consists of making an assumption about the distribution function of a random variable, very often about the numerical values of the statistical parameters of the distribution function (mean and variance), and deciding whether those values of the parameters are consistent with our sample of observations on that random variable.

For example, suppose that a sample of $n_1 = 10$ observations has a sample mean $\bar{x}_1 = 2.0$ and a sample variance $s_1^2 = 4$. Let us make the assumption that this sample came from a population that has a normal distribution with $\mu = 0$ and σ_1^2 unknown, that is, $X \approx N(0, \sigma_1^2)$. In order to test this assumption, we formalize it by stating the *null hypothesis*,

$$H_0 : \mu = \mu_0 = 0 \tag{8.97}$$

and the alternative hypothesis,

$$H_A : \mu = \mu_A \neq 0 \tag{8.98}$$

We recall from Eq. (8.76) that the quantity $(\bar{x}_1 - \mu)/\sqrt{s_1^2/n_1}$ has a t distribution, and from Eq. (8.92) that:

$$Pr\left\{t_{\alpha/2} < t < t_{1-\alpha/2}\right\} = 1 - \alpha \tag{8.99}$$

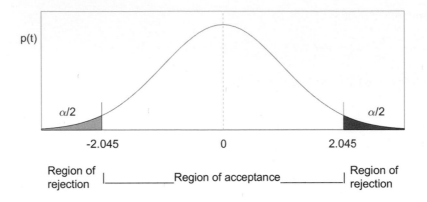

FIG. 8.8

Hypothesis test for the mean.

For 95% probability and $\nu = (n_1 - 1) = 9$ degrees of freedom, the previous equation is:

$$Pr\{-2.045 < t < 2.045\} = 0.95 \tag{8.100}$$

The region defined by Eq. (8.100) is shown in Fig. 8.8 as the *region of acceptance*, whereas the regions outside this range are labeled *regions of rejection*. Based on the assumption that the null hypothesis is true, if the statistic calculated from the experimental sample falls outside the region of acceptance, the null hypothesis is rejected and H_A is accepted. Otherwise, H_0 is accepted and H_A is rejected. In this example we calculate t_0:

$$t_0 = \frac{\bar{x}_1 - \mu_0}{\sqrt{s_1^2/n_1}} = \frac{2.0 - 0}{\sqrt{\frac{4}{10}}} = 3.16 \tag{8.101}$$

We see that t_0 is outside the region of acceptance defined by Eq. (8.100); therefore, we reject the null hypothesis.

We can generalize this test by saying that if:

$$\left| \frac{\bar{x} - \mu_0}{\sqrt{s^2/n}} \right| > t_{(1-\alpha/2)} \tag{8.102}$$

then the null hypothesis that $\mu = \mu_0$ must be rejected. This is the well-known *two-side t-test*, which is used extensively in regression analysis to test the values of regression parameters.

Let us now examine the variance of the sample. We draw a second sample of $n_2 = 21$ observations and find that $\bar{x}_2 = 2.0$ and $s_2^2 = 3$. We ask the question: "Is the second sample taken from the same population as the first sample, or from one which has a different variance than the first?" We state that the null hypothesis,

$$H_0 : \frac{\sigma_1^2}{\sigma_2^2} = 1 \tag{8.103}$$

and the alternative hypothesis,

$$H_A: \frac{\sigma_1^2}{\sigma_2^2} \neq 1 \tag{8.104}$$

We recall from Eq. (8.81) that the ratio $(s_1^2/\sigma_1^2)/(s_2^2/\sigma_2^2)$ has an F distribution with (ν_1, ν_2) degrees of freedom. From the probability distribution function:

$$Pr\{F_{\alpha/2}(\nu_1, \nu_2) < F \leq F_{1-\alpha/2}(\nu_1, \nu_2)\} = 1 - \alpha \tag{8.105}$$

To test the null hypothesis at the 95% confidence level, obtain the values of $F_{0.025}(9, 20)$ and $F_{0.975}(9, 20)$ for this example from the F distribution tables. The value of $F_{\alpha/2}(\nu_1, \nu_2)$ is obtained from the relationship: $F_{\alpha/2}(\nu_1, \nu_2) = 1/F_{1-\alpha/2}(\nu_2, \nu_1)$. Therefore, the interval of acceptance is given by:

$$Pr\left\{0.272 < \frac{s_1^2/\sigma_1^2}{s_2^2/\sigma_2^2} \leq 2.84\right\} = 0.95 \tag{8.106}$$

The null hypothesis assumes that $\sigma_1^2 = \sigma_2^2$; therefore, the inequality becomes:

$$0.272 < \frac{s_1^2}{s_2^2} \leq 2.84 \tag{8.107}$$

For this example:

$$\frac{s_1^2}{s_2^2} = \frac{4}{3} = 1.33 \tag{8.108}$$

Therefore the null hypothesis can be accepted.

This is the *two-side F test*, which is used in the analysis of variance of regression results to test the adequacy of a model in fitting the experimental data (see Section 8.5).

Hypothesis testing is an involved procedure, which we have briefly introduced here. It is outside the scope of this chapter to discuss hypothesis testing in more depth. The interested reader is referred to Montgomery et al. [1] and Metcalfe et al. [2] for further discussion.

8.3 LINEAR REGRESSION ANALYSIS

Most mathematical models in engineering and science are nonlinear in the parameters. However, for a complete understanding of nonlinear regression methods, it is necessary to first develop the linear regression case and show how this extends to nonlinear models.

The exact representation of a linear relationship may be shown as:

$$y = \alpha + \beta x \tag{8.109}$$

where y represents the true value of the dependent variable, x is the true value of the independent variable, β is the slope of the line, and α is the y-intercept of the line. This deterministic relationship is not useful in this form because it requires knowledge of the true values of y and x. Instead, the linear model is rewritten in terms of the observations of the values of the variables:

$$Y^* = \alpha + \beta X + u \tag{8.110}$$

where Y^* is the vector of observations of the dependent variable, X is the vector of observations of the independent variable, and u is the vector of *disturbance terms*. The purpose of the u term is to characterize the discrepancies that emerge between the true values and the observed values of the variables. These discrepancies can be attributed mainly to experimental error. Later in this section, u will be assumed to be a stochastic variable with some specified probability distribution.

Eq. (8.110) can be extended to include more than one independent variable:

$$Y^* = \beta_1 X_1 + \beta_2 X_2 + \cdots + \beta_k X_k + u \tag{8.111}$$

where $X_1, X_2, \ldots, X_k$ are the vectors of observations of k independent variables. To allow for a y-intercept, the vector X_1 can be taken as a vector whose components are all unity; thus β_1 becomes the parameter specifying the value of the y-intercept.

Eq. (8.111) can be condensed to matrix form:

$$Y^* = X\beta + u \tag{8.112}$$

where $Y^* = (n \times 1)$ vector of observations of the dependent variable

$X = (n \times k)$ matrix of observations of the independent variables
$\beta = (k \times 1)$ vector of parameters
$u = (n \times 1)$ vector of disturbance terms
$n =$ number of observations

Given a set of n observations in the Y variable and in each of the k independent variables, the problem now is to obtain an estimate of the β vector.

The basic assumptions made in the derivation of the method for estimating the parameters are the following:

1. The disturbance terms, represented by the vector u, are random variables with zero expectation, that is,

$$E[u] = \bar{u} = 0 \tag{8.113}$$

Because the variable u is the sum of errors from several sources, the central limit theorem implies that the distribution of u tends toward the normal distribution as the number of factors contributing to u increases.

2. The variance of the distribution of u is constant and independent of X, that is,

$$\begin{aligned} V[u_1] &= E\left[(u_1 - \bar{u}_1)^2\right] = \sigma^2 \\ V[u_2] &= E\left[(u_2 - \bar{u}_2)^2\right] = \sigma^2 \\ &\vdots \\ V[u_n][\,] &= E\left[(u_n - \bar{u}_n)^2\right] = \sigma^2 \end{aligned} \tag{8.114}$$

In addition, the values of u for each set of observations are independent of one another, that is,

$$E[u_i u_j] = E[u_i]E[u_j] \quad i \neq j \tag{8.115}$$

From assumption 1 and Eqs. (8.31) and (8.37), we also conclude that the covariance of u is zero:

$$Cov[u_i, u_j] = 0 \quad i \neq j \tag{8.116}$$

The variance-covariance matrix is defined as:

$$Var - Cov[u] = \begin{bmatrix} V[u_1] & Cov[u_1, u_2] & \cdots & Cov[u_1, u_n] \\ Cov[u_2, u_1] & V[u_2] & \cdots & Cov[u_2, u_n] \\ \cdots & \cdots & \cdots & \cdots \\ Cov[u_n, u_1] & Cov[u_n, u_2] & \cdots & V[u_n] \end{bmatrix} \tag{8.117}$$
$$= E[(u - E[u])(u - E[u])']$$

Combining Eqs. (8.113), (8.114), (8.116), and (8.117), we obtain:

$$Var - Cov[u] = E[uu'] = \begin{bmatrix} \sigma^2 & 0 & \cdots & 0 \\ 0 & \sigma^2 & \cdots & 0 \\ \cdots & \cdots & \cdots & \cdots \\ 0 & 0 & \cdots & \sigma^2 \end{bmatrix} = \sigma^2 I \tag{8.118}$$

In summary, Eq. (8.118) says that each u distribution has the same variance and that all distributions are pairwise uncorrelated.

3. The matrix X is a set of fixed numbers, that is, the values of X do not contain an error.
4. The rank of the matrix X is equal to k, and $k < n$. The first part of this assumption ensures that k variables are linearly independent. The second part requires that the number of observations exceeds the number of parameters to be estimated. This is essential in order to have the necessary degrees of freedom for parameter estimation.
5. The vector u has a multivariate normal distribution:

$$u \sim N(0, \sigma^2 I) \tag{8.119}$$

These assumptions are not overly restrictive. Because the value of u is the result of many factors acting in opposite directions, it should be expected that small values of u occur more frequently than large values and that u is a variable with a probability distribution centered at zero and having a finite variance σ^2. This is true when the form of Eq. (8.112) is close to the correct relationship. Because of the many factors involved, the central limit theorem would further suggest that u has a normal distribution, which gives the parameter estimates the desirable property of being maximum-likelihood estimates. Later on in the discussion, it will be shown that the regression method can handle cases where σ^2 is not constant and where u is not independent of X.

8.3.1 THE LEAST SQUARES METHOD

Let us consider the hypothesized linear model:

$$Y^* = X\beta + u \tag{8.112}$$

Let b denote a k-element vector, which is an estimate of the parameter vector β. We use this estimate to define a vector of residuals:

$$\varepsilon = Y^* - Xb = Y^* - Y \tag{8.120}$$

These residuals are the differences between the experimental observations Y^* and the calculated values of Y using the estimated vector b. A common way to evaluate the unknown vector b is the *least squares method*, which minimizes the sum of the squared residuals Φ:

$$\Phi = \varepsilon'\varepsilon = (Y^* - Xb)'(Y^* - Xb) \tag{8.121}$$

In order to calculate the vector b that minimizes Φ, we take the partial derivative of Φ with respect to b and set it equal to zero:

$$\frac{\partial \Phi}{\partial b} = (-X)'(Y^* - Xb) + (Y^* - Xb)'(-X) = 0 \tag{8.122}$$

We simplify this using the matrix-vector identity $A'y = y'A$:

$$-2X'(Y^* - Xb) = 0 \tag{8.123}$$

Eq. (8.123) can be further rearranged to yield:

$$(X'X)b = X'Y^* \tag{8.124}$$

This constitutes a set of simultaneous linear algebraic equations, called the *normal equations*. The matrix $(X'X)$ is a $(k \times k)$ symmetric matrix. Assumption 4 made earlier guarantees that $(X'X)$ is non-singular; therefore, its inverse exists. Thus, the normal equations can be solved for the vector b:

$$b = (X'X)^{-1}X'Y^* \tag{8.125}$$

The values of the elements of vector b can be obtained readily from Eq. (8.125) because the right-hand side of this equation contains the matrix of observations of the independent variables X and the vector of observations of the dependent variable Y^*, all of which are known.

Polynomial regression may be considered a special case of linear regression. In such a case the relationship between independent and dependent variables is expressed by the following $(k - 1)$ degree polynomial:

$$y = b_1 + b_2 x + b_3 x^2 + \cdots + b_k x^{k-1} \tag{8.126}$$

We may consider x^0, x^1, x^2,...,x^{k-1} independent variables X_1 to X_k and construct the matrix X for the polynomial regression as:

$$X = \begin{bmatrix} 1 & x_1 & x_1^2 & \cdots & x_1^{k-1} \\ 1 & x_2 & x_2^2 & \cdots & x_2^{k-1} \\ \vdots & \vdots & \vdots & \ddots & \vdots \\ 1 & x_n & x_n^2 & \cdots & x_n^{k-1} \end{bmatrix} \tag{8.127}$$

The vector of coefficients of the polynomial in Eq. (8.126) is then calculated from Eq. (8.125).

8.3.2 PROPERTIES OF THE ESTIMATED VECTOR OF PARAMETERS

The vector b is an estimate of β, which minimizes the sum of the squared residuals irrespective of any distribution properties of the residuals. In addition, b is an unbiased estimate of β. To show this, we combine Eqs. (8.125) and (8.112),

$$b = (X'X)^{-1}X'(X\beta + u)$$
$$= (X'X)^{-1}(X'X)\beta + (X'X)^{-1}X'u$$
$$= \beta + (X'X)^{-1}X'u \tag{8.128}$$

and take the expected value of b,

$$E[b] = E[\beta] + (X'X)^{-1}X'E[u] \tag{8.129}$$

but because $E[u] = 0$ (assumption 1) and β is constant, then,

$$E[b] = \beta \tag{8.130}$$

that is, the expected value of b is β.

Furthermore, the variance of b can be obtained as follows. Rearranging Eq. (8.128),

$$b - \beta = (X'X)^{-1}X'u \tag{8.131}$$

and using Eq. (8.130), we obtain:

$$b - E[b] = (X'X)^{-1}X'u \tag{8.132}$$

From the definition of the variance-covariance matrix (Eq. 8.117),

$$Var - Cov[b] = E[(b - E[b])(b - E[b])'] \tag{8.133}$$

Using Eq. (8.132) in Eq. (8.133),

$$Var - Cov[b] = E\left[(X'X)^{-1}X'uu'X(X'X)^{-1}\right]$$
$$= (X'X)^{-1}X'E[uu']X(X'X)^{-1} \tag{8.134}$$

But from Eq. (8.118), $E[uu'] = \sigma^2 I$. Therefore, the variance-covariance of b simplifies to:

$$Var - Cov[b] = \sigma^2(X'X)^{-1} \tag{8.135}$$

where σ^2 is the variance of u, as defined by Eq. (8.114).

The elements of the matrix $(X'X)^{-1}$ are designated as a_{ij}. Therefore, the variance of b_i is given by:

$$V[b_i] = \sigma^2 a_{ii} \tag{8.136}$$

and the covariance of b_i with b_j by:

$$Cov[b_i, b_j] = \sigma^2 a_{ij} \tag{8.137}$$

Therefore, if the variance of u is known or can be estimated, then the variance-covariance of the estimated parameter vector b can be calculated.

It can be seen from Eq. (8.134) that the variance-covariance matrix of b can still be calculated even if assumption 2 is not made. In that case, the matrix $E[uu']$ would not be a diagonal matrix.

We can now draw an important conclusion regarding the distribution of b. Eq. (8.128) shows that b is a linear combination of u. If u is a multivariate normal distribution (assumption 5, Section 8.3), then b is also a multivariate normal distribution, that is,

$$b \sim N\big(\beta, \sigma^2(X'X)^{-1}\big) \tag{8.138}$$

For each individual parameter:

$$b_i \sim N\big(\beta_i, \sigma^2 a_{ii}\big) \tag{8.139}$$

where a_{ii} is the ith element on the principal diagonal of $(X'X)^{-1}$. The normal distribution can be converted to the standard normal distribution:

$$\frac{b_i - \beta_i}{\sigma\sqrt{a_{ii}}} \sim N(0, 1) \tag{8.140}$$

The variance σ^2 of the distribution term is not usually known unless a large number of repetitive experiments have been performed. The value of σ^2 can be estimated from:

$$s^2 = \frac{\varepsilon'\varepsilon}{n - k} \tag{8.141}$$

where $\varepsilon'\varepsilon$ is the sum of squared residuals (see Eq. 8.121) and $(n - k)$ is the number of degrees of freedom. If there is no lack of fit of the model to the data (see analysis of variances in Section 8.5), then s^2 is an unbiased estimate of σ^2, that is,

$$E[s^2] = \sigma^2 \tag{8.142}$$

If lack of fit cannot be tested, the use of s^2 as an estimate of σ^2 implies an assumption that the model is correct.

We saw earlier that the ratio of s^2/σ^2 has a chi-square distribution,

$$\chi^2 = \nu \frac{s^2}{\sigma^2} \tag{8.71}$$

and that the t variable is given by:

$$t = \frac{u}{\sqrt{\chi^2/\nu}} = \frac{N(0, 1)}{\sqrt{\chi^2/\nu}} \tag{8.72a}$$

We can therefore combine Eqs. (8.140), (8.71), and (8.72a) to form the t variable:

$$t = \frac{(b_i - \beta_i)/\sigma\sqrt{a_{ii}}}{\sqrt{s^2/\sigma^2}} = \frac{b_i - \beta_i}{s\sqrt{a_{ii}}} \sim t(n - k) \tag{8.143}$$

Eq. (8.143) shows that the quantity $(b_i - \beta_i)/s\sqrt{a_{ii}}$ has a t distribution with $(n - k)$ degrees of freedom. This is a very important equation because it enables us to construct confidence intervals of the parameters from quantities that can be calculated from the regression analysis. For example, the $100(1 - \alpha)\%$ confidence interval for parameter β_i can be obtained from:

$$Pr\left\{t_{\alpha/2} < \frac{b_i - \beta_i}{s\sqrt{a_{ii}}} \leq t_{1-\alpha/2}\right\} = 1 - \alpha \tag{8.144}$$

which yields the interval:

$$b_i - t_{1-\alpha/2}s\sqrt{a_{ii}} \leq \beta_i < b_i + t_{1-\alpha/2}s\sqrt{a_{ii}} \tag{8.145}$$

These are *individual parameter confidence intervals*. Fig. 8.9 demonstrates these intervals for β_1 and β_2 in a two-parameter model.

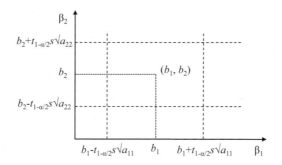

FIG. 8.9

Confidence intervals for parameters.

Furthermore, Eq. (8.143) enables us to perform the t-test for hypothetical values of β (see Section 8.2.3). For example, if it is suspected that the value of β_i is not significantly different from zero, the null hypothesis can be stated as:

$$H_0 : \beta_i = 0 \tag{8.146}$$

and the alternative hypothesis as:

$$H_A : \beta_i \neq 0 \tag{8.147}$$

When Eq. (8.146) is substituted in Eq. (8.143), the resulting expression:

$$t = \frac{b_i}{s\sqrt{a_{ii}}} \tag{8.148}$$

is calculated. If this value of t lies within the region of acceptance given by the t distribution for a two-sided test at the required confidence level, then the null hypothesis that $\beta_i = 0$ is accepted. This is a very useful test in deciding the significance of a parameter in a model and in helping the experimenter discriminate between competing models.

EXAMPLE 8.2 DETERMINING SOLUBILITY OF OXYGEN IN WATER WITH LINEAR REGRESSION

The solubility of oxygen in liquid water is a function of temperature, as given in Table E8.1.

(a) Determine the equation of solubility as a function of the square root of temperature. Estimate the solubility of oxygen in 0°C water.

(b) Calculate the 95% confidence interval for the parameters determined in (a).

(c) Compute the correlation coefficient between the equation and the experimental data.

Method of Solution

The proposed equation for solubility as a function of water is:

$$S = b_1 + b_2 \sqrt{T} \tag{1}$$

(Continued)

EXAMPLE 8.2 (CONTINUED)

where S is the solubility and T is temperature. The constants in this equation can be evaluated by the linear regression analysis.

Part (a): Considering $y = S$ and $x = \sqrt{T}$, Eq. (1) becomes the same as Eq. (8.109). Therefore, Eq. (8.125) will be used for computing the constants of the equation:

$$b = (X'X)^{-1}X'Y^* \tag{8.125}$$

where

$$X = \begin{bmatrix} 1 & \sqrt{5} \\ 1 & \sqrt{10} \\ 1 & \sqrt{15} \\ 1 & \sqrt{20} \\ 1 & \sqrt{25} \\ 1 & \sqrt{30} \end{bmatrix} \quad \text{and} \quad Y^* = \begin{bmatrix} 11.6 \\ 10.3 \\ 9.1 \\ 8.2 \\ 7.4 \\ 6.8 \end{bmatrix}$$

Putting these variables into Eq. (8.125), we get:

$$X'X = \begin{bmatrix} 6 & 24.2207 \\ 24.2207 & 105 \end{bmatrix}$$

$$(X'X)^{-1} = \begin{bmatrix} 2.4217 & -0.5586 \\ -0.5586 & 0.1384 \end{bmatrix} \tag{2}$$

$$(X'X)^{-1}X' = \begin{bmatrix} 1.1726 & 0.6552 & 0.2582 & -0.0765 & -0.3714 & -0.6380 \\ -0.2492 & -0.1210 & -0.0227 & 0.0602 & 0.1333 & 0.1993 \end{bmatrix}$$

$$b = (X'X)^{-1}X'Y^* = \begin{bmatrix} b_1 \\ b_2 \end{bmatrix} = \begin{bmatrix} 14.9853 \\ -1.5075 \end{bmatrix} \tag{3}$$

Therefore, the solubility in Eq. (1) becomes:

$$S = 14.9853 - 1.5075\sqrt{T}$$

Based on this equation, the solubility of oxygen in water at 0°C is 14.9853 ppm.

Let us also solve this example with Excel. The procedure is illustrated in Fig. E8.2. First, as shown in Fig. E8.2a, the data in Table E8.2 should be typed in the spreadsheet. The square root of the temperature also should be calculated in a separate column. Then, the plot of solubility against the square root of the temperature can be created from *Inset* → *Charts* → *Scatter*. Note that the x-series in the chart should be the square root of the temperature and the

(*Continued*)

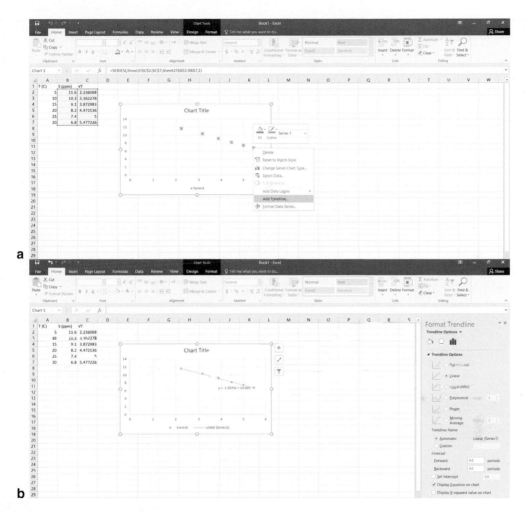

FIG. E8.2

Linear regression in Excel. (a) Creating the graph. (b) Adding the trend line and its equation.

Table E8.2 Solubility of oxygen at various temperatures.						
Temperature (°C)	5	10	15	20	25	30
Solubility (ppm)	11.6	10.3	9.1	8.2	7.4	6.8

EXAMPLE 8.2 (CONTINUED)

y-series is the solubility. On the chart, click on one of the symbols, then right-click and choose the *Add Trendline* option. The added trend line (the default is linear) can be seen in Fig. E8.2b. To show the fitted equation, click the box for *Display Equation on Chart* on the *Format Trendline* pane.

Part (*b*): The confidence interval should be determined from the inequality in Eq. (8.145):

$$b_i - t_{1-\alpha/2}s\sqrt{a_{ii}} \le \beta_i < b_i + t_{1-\alpha/2}s\sqrt{a_{ii}} \qquad i = 1, 2 \qquad (8.145)$$

In this example, $\alpha = 0.05$. The calculated parameters b_1 and b_2 are given in Eq. (3), and diagonal elements a_{11} and a_{22} are obtained from Eq. (2).

The standard deviation, *s*, should be evaluated from Eq. (8.141):

$$s^2 = \frac{\varepsilon'\varepsilon}{n - k} \qquad (8.141)$$

where

$$\varepsilon = Y^* - Xb = Y^* - Y \qquad (8.120)$$

In this problem, there are six data points ($n = 6$) and two calculated constants ($k = 2$). Therefore, we have:

$$Y = Xb = \begin{bmatrix} 1 & \sqrt{5} \\ 1 & \sqrt{10} \\ 1 & \sqrt{15} \\ 1 & \sqrt{20} \\ 1 & \sqrt{25} \\ 1 & \sqrt{30} \end{bmatrix} \begin{bmatrix} 14.9853 \\ -1.5075 \end{bmatrix} = \begin{bmatrix} 11.6145 \\ 10.2183 \\ 9.1469 \\ 8.2437 \\ 7.4480 \\ 6.7286 \end{bmatrix}$$

$$\varepsilon = Y^* - Y = \begin{bmatrix} 11.6 \\ 10.3 \\ 9.1 \\ 8.2 \\ 7.4 \\ 6.8 \end{bmatrix} - \begin{bmatrix} 11.6145 \\ 10.2183 \\ 9.1469 \\ 8.2437 \\ 7.4480 \\ 6.7286 \end{bmatrix} = \begin{bmatrix} -0.0145 \\ 0.0817 \\ -0.0469 \\ -0.0437 \\ -0.0480 \\ 0.0714 \end{bmatrix}$$

$$\varepsilon'\varepsilon = \begin{bmatrix} -0.0145 & 0.0817 & -0.0469 & 0.0437 & -0.0480 & 0.0714 \end{bmatrix} \begin{bmatrix} -0.0145 \\ 0.0817 \\ -0.0469 \\ -0.0437 \\ -0.0480 \\ 0.0714 \end{bmatrix} = 0.0184$$

(Continued)

EXAMPLE 8.2 (CONTINUED)

$$s^2 = \frac{\varepsilon'\varepsilon}{n-k} = \frac{0.0184}{6-2} = 0.0046 \text{ or } s = 0.0678 \tag{4}$$

The *Student's t distribution* is given by Eq. (8.73). In this example, the degree of freedom is $\nu = n - k = 4$ and $\alpha = 0.05$. Thus the value of $t_{0.975}$ can be obtained from:

$$0.975 = \int_{-\infty}^{t_{0.975}} \frac{1}{\sqrt{4\pi}} \frac{\Gamma(2.5)}{\Gamma(2)} \left(1 + \frac{t^2}{4}\right)^{-2.5} dt$$

From which we have:

$$t_{0.975} = 2.7765 \tag{5}$$

Now, we can present the 95% confidence interval for the parameters of Eq. (1) by inserting the values calculated in Eqs. (3), (4), and (5) into Eq. (8.145):

$$14.9853 - 2.7765 \times 0.0678\sqrt{2.4217} \le \beta_1 < 14.9853 + 2.7765 \times 0.0678\sqrt{2.4217}$$
$$-1.5075 - 2.7765 \times 0.0678\sqrt{0.1384} \le \beta_2 < -1.5075 + 2.7765 \times 0.0678\sqrt{0.1384}$$

or

$$14.6922 < \beta_1 < 15.2784$$
$$-1.5775 \le \beta_2 < -1.4374$$

Part (c): The correlation coefficient is defined by Eq. (8.34). In the regression analysis, we are interested in the correlation between calculated and experimental values of the dependent variable (Y and Y^*, respectively). Therefore, the correlation coefficient for this case can be shown as:

$$r = \frac{Cov[Y^*, Y]}{\sqrt{V[Y^*]V[Y]}} = \frac{\sum (Y^*_i - \overline{Y^*})(Y_i - \overline{Y})}{\sqrt{\sum (Y^*_i - \overline{Y^*})^2 \sum (Y_i - \overline{Y})^2}}$$

Using Y and Y^* vectors presented in the previous parts, the correlation coefficient becomes:

$$r = 0.9994$$

which indicates a very good fit of Eq. (1) to the experimental data points of Table E8.2.

8.4 NONLINEAR REGRESSION ANALYSIS

We have stated this earlier in the chapter, and we repeat it again: the mathematical models encountered in engineering and science are often nonlinear in their parameters. Consider, for example, the analysis of a chemical reaction such as:

$$A \xrightarrow{k_1} B \xrightarrow{k_2} C + D$$
$$C + A \xrightarrow{k_3} E + F$$

where the rate of formation of each component may be written as:

$$\frac{dC_A}{dt} = -k_1 C_A - k_3 C_A^n C_C^m$$

$$\frac{dC_B}{dt} = k_1 C_A - k_2 C_B$$

$$\frac{dC_C}{dt} = k_2 C_B - k_3 C_A^n C_C^m \qquad (8.149)$$

$$\frac{dC_E}{dt} = k_3 C_A^n C_C^m$$

This is only one possible formulation of the reaction mechanism. It contains five unknown parameters, k_1, k_2, k_3, n, and m, which must be calculated by fitting the model to experimental data. Suppose that experiments for this chemical system are carried out in a batch reactor and data of the form shown in Fig. 8.10 are collected. Because experimental data are available for all four dependent variables, C_A, C_B, C_C, and C_E, multiple nonlinear regression can be performed by simultaneously fitting all four equations of Eq. (8.149) to the data.

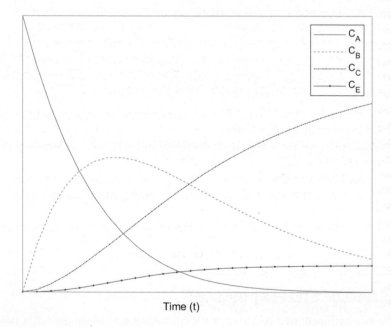

Time (t)

FIG. 8.10

Simulated data for batch reactor experiment.

A model consisting of differential equations, such as Eq. (8.149), may be shown in the form:

$$\frac{dY}{dx} = g(x, Y, b) \tag{8.150}$$

where

dY/dx = vector of derivatives of Y
   g = vector of functions
   x = independent variable
   Y = vector of dependent variables
   b = vector of parameters

We assume that if the boundary conditions are given and if the vector b can be estimated, then the differential equations in Eq. (8.150) can be integrated numerically or analytically to give the integrated results, which are:

$$Y = f(x, b) \tag{8.151}$$

For the simple case where the model consists of only one dependent variable, the sum of squared residuals is given by:

$$\Phi = \varepsilon' \varepsilon = (Y^* - Y)'(Y^* - Y) \tag{8.152}$$

where Y^* = vector of experimental observations of the dependent variable
      Y = vector of calculated values of the dependent variables obtained from Eq. (8.151).

There are several techniques for minimizing the sum of squared residuals described by Eq. (8.152). We review some of these methods in this section. The methods developed in this section will enable us to fit models consisting of multiple dependent variables, such as the one described earlier, to multiresponse experimental data in order to obtain the values of the parameters of the model, which minimize the *overall (weighted) sum of squared residuals*. In addition, a thorough statistical analysis of the regression results will enable us to:

1. Decide whether the model gives a satisfactory fit within the experimental error of the data.
2. Discriminate between competing models.
3. Measure the accuracy of the estimation of the parameters by constructing the confidence region in the parameter space.
4. Measure the correlation between parameters by examining the correlation coefficient matrix.
5. Perform tests to verify that repeated experimental data come from the same population of experiments.
6. Perform tests to verify whether the residuals between the data and the model are randomly distributed.

8.4.1 THE METHOD OF STEEPEST DESCENT

A simple method that has been used to arrive at the minimum sum of squares of a nonlinear model is that of *steepest descent*. We know that the gradient of a scalar function is a vector that gives the direction of the greatest increase of the function at any point. In the steepest descent method, we take advantage of this property by moving in the opposite direction to reach a lower function value.

Therefore, in this method, the initial vector of parameter estimates is corrected in the direction of the negative gradient of Φ:

$$\Delta b = -K\left(\frac{\partial \Phi}{\partial b}\right) \tag{8.153}$$

where K is a suitable constant factor and Δb is the correction vector to be applied to the estimated value of b to obtain a new estimate of the parameter vector:

$$b^{(m+1)} = b^{(m)} + \Delta b \tag{8.154}$$

where m is the iteration counter. Combining Eqs. (8.152) and (8.153) results in:

$$\Delta b = 2KJ'(Y^* - Y) \tag{8.155}$$

where J is the Jacobian matrix of partial derivatives of Y with respect to b evaluated at all n points where experimental observations are available:

$$J = \begin{bmatrix} \dfrac{\partial Y_1}{\partial b_1} & \cdots & \dfrac{\partial Y_1}{\partial b_k} \\ \vdots & \ddots & \vdots \\ \dfrac{\partial Y_n}{\partial b_1} & \cdots & \dfrac{\partial Y_n}{\partial b_k} \end{bmatrix} \tag{8.156}$$

The steepest descent method has the advantage in that it guarantees moving toward the minimum sum of squares without diverging, provided that the value of K, which determines the step size, is small enough. The value of K may be a constant throughout the calculations, changed arbitrarily at each calculation step, or obtained from optimization of the step size [4]. However, the rate of convergence to the minimum decreases as the search approaches this minimum, and the method loses its attractiveness because of this shortcoming.

8.4.2 THE GAUSS-NEWTON METHOD

Once again, we restate that in the least squares method, our objective is to find the vector of parameters b such that it minimizes the sum of squared residuals Φ. Thus, the vector b may be found by taking the partial derivative of Φ with respect to b and setting it to zero:

$$\frac{\partial \Phi}{\partial b} = 0 \tag{8.157}$$

Because Y is nonlinear with respect to the parameters, Eq. (8.157) will yield a nonlinear equation that would be difficult to solve for b. This problem was alleviated by Gauss, who determined that fitting nonlinear functions by least squares can be achieved by an iterative method involving a series of linear approximations. At each stage of the iteration, linear squares theory can be used to obtain the next approximation.

This method, known as the *Gauss-Newton method*, converts the nonlinear problem into a linear one by approximating the function Y by a Taylor series expansion around an estimated value of the parameter vector b:

$$Y(x, b) = Y(x, b^{(m)} + \Delta b) = Y(x, b^{(m)}) + \frac{\partial Y}{\partial b}\Big|_{b^{(m)}} \Delta b = Y + J\Delta b \tag{8.158}$$

where the Taylor series has been truncated after the second term, Eq. (8.158) is linear in Δb. Therefore, the problem has been transformed from finding b to that of finding the correction to b, that is, Δb, which must be added to an estimate of b to minimize the sum of squared residuals. To do this we replace Y in Eq. (8.152) with the right-hand side of Eq. (8.158) to get:

$$\Phi = (Y^* - Y - J\Delta b)'(Y^* - Y - J\Delta b) \tag{8.159}$$

Taking the partial derivative of Φ with respect to Δb, setting it equal to zero, and solving for Δb, we obtain:

$$\Delta b = (J'J)^{-1}J'(Y^* - Y) \tag{8.160}$$

The Gauss-Newton method applies to both the one-variable model and the multiple regression case (see Section 8.4.5). The algorithm of the Gauss-Newton method is shown as a flowchart in Fig. 8.11 and involves the following steps:

1. Assume initial guesses for the parameter vector b.
2. If the model is in the form of the differential equation(s), then use the vector b and the boundary condition(s) to integrate the equation(s) to obtain the profile(s) of Y. If the model is in the form of algebraic equation(s), then simply use the vector b to evaluate Y from the equation(s).
3. Evaluate the Jacobian matrix J from the equation(s) of the model.
4. Use Eq. (8.160) to obtain the correction vector Δb.
5. Evaluate the new estimate of the parameter vector from Eq. (8.154).

$$b^{(m+1)} = b^{(m)} + \Delta b \tag{8.154}$$

It is also possible to apply the relaxation factor in order to prevent the calculation from diverging (see Section 1.10).
6. Repeat steps 2 through 5 until either (or both) of the following conditions are satisfied:
   a. Φ does not change appreciably.
   b. Δb becomes very small.

The Gauss-Newton method is based on the linearization of a nonlinear model. Therefore, this method is expected to work well if the model is not highly nonlinear or if the initial estimate of the parameter vector is near the minimum sum squares. The contours of constant Φ in the parameter space of a linear model are ellipsoids (Fig. 8.12a). For a nonlinear model, these contours are distorted (Fig. 8.12b), but in the vicinity of the minimum Φ, the contours are very nearly elliptical. Therefore, the Gauss-Newton method is quite effective if the initial starting point for the search is in the nearly elliptical region. On the other hand, this method may diverge if the starting point is in the highly distorted region of the parameter hyperspace.

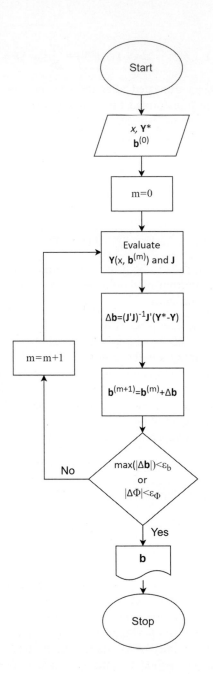

FIG. 8.11

Flowchart of the Gauss-Newton method.

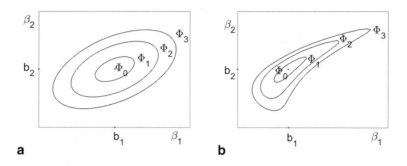

FIG. 8.12

Contours for a constant sum of squares in parameter space. (a) Linear model. (b) Nonlinear model.

8.4.3 NEWTON'S METHOD

Eq. (8.157) represents a set of nonlinear equations; therefore Newton's method may be applied to solve this set of nonlinear equations. First, let us expand Φ by Taylor series up to the third term:

$$\Phi(x,b) = \Phi(x,b^{(m)}) + \left(\frac{\partial \Phi}{\partial b}\right)^{(m)} \Delta b + \frac{1}{2}\Delta b' \left(\frac{\partial^2 \Phi}{\partial b^2}\right)^{(m)} \Delta b \tag{8.161}$$

Taking the partial derivative of both sides of Eq. (8.161) with respect to b gives:

$$\frac{\partial \Phi}{\partial b} = \left(\frac{\partial \Phi}{\partial b}\right)^{(m)} + \left(\frac{\partial^2 \Phi}{\partial b^2}\right)^{(m)} \Delta b \tag{8.162}$$

The first derivative of Φ with respect to b can be calculated by differentiating Eq. (8.152):

$$\left(\frac{\partial \Phi}{\partial b}\right)^{(m)} = -2J'(Y^* - Y) \tag{8.163}$$

and the second derivative of Φ with respect to b is called the *Hessian matrix* of the second-order partial derivatives of Φ with respect to b evaluated at all n points where experimental observations are available:

$$H = \begin{bmatrix} \dfrac{\partial^2 \Phi}{\partial b_1^2} & \dfrac{\partial^2 \Phi}{\partial b_1 \partial b_2} & \cdots & \dfrac{\partial^2 \Phi}{\partial b_1 \partial b_k} \\[2mm] \dfrac{\partial^2 \Phi}{\partial b_2 \partial b_1} & \dfrac{\partial^2 \Phi}{\partial b_2^2} & \cdots & \dfrac{\partial^2 \Phi}{\partial b_2 \partial b_k} \\[2mm] & \cdots & \cdots & \\[1mm] \dfrac{\partial^2 \Phi}{\partial b_k \partial b_1} & \dfrac{\partial^2 \Phi}{\partial b_k \partial b_2} & \cdots & \dfrac{\partial^2 \Phi}{\partial b_k^2} \end{bmatrix} \tag{8.164}$$

By applying the necessary condition of having a local minimum of Φ, Eq. (8.157), into Eq. (8.162), and combining with Eqs. (8.163) and (8.164), we can evaluate the correction vector Δb:

$$\Delta b = 2H^{-1}J'(Y^* - Y) \tag{8.165}$$

The calculation procedure for Newton's method is almost the same as that of the Gauss-Newton method, with the exception that the vector of corrections to the parameters is calculated from Eq. (8.165). If Φ is quadratic with respect to b (i.e., linear regression), Newton's method converges in only one step. Like all other methods applying Newton's technique for the solution of the set of nonlinear equations, a relaxation factor may be used along with Eq. (8.165) when correcting the parameters.

8.4.4 THE MARQUARDT METHOD[1]

Marquardt [5] developed an interpolation technique between the Gauss-Newton and the steepest descent methods. This interpolation is achieved by adding the diagonal matrix (λI) to the matrix ($J'J$) in Eq. (8.160):

$$\Delta b = (J'J + \lambda I)^{-1}J'(Y^* - Y) \tag{8.166}$$

The value of λ is chosen at each iteration so that the corrected parameter vector will result in a lower sum of squares in the following iteration. It can be seen that when the value of λ is small in comparison with the elements of the matrix ($J'J$), the Marquardt method approaches the Gauss-Newton method; when λ is very large, this method is identical to steepest descent, with the exception of a scale factor, which does not affect the direction of the parameter correction vector but which gives a small step size.

According to Marquardt, it is desired to minimize Φ in the largest neighborhood over which the linearized function will give an adequate representation of the nonlinear function. Therefore, the method for choosing λ must give small values of λ when the Gauss-Newton method would converge efficiently and large values of λ when the steepest descent method is necessary.

The Marquardt method may likewise be applied to Newton's method. In this case, the diagonal matrix λI is added to the Hessian matrix in Eq. (8.165):

$$\Delta b = 2(H + \lambda I)^{-1}J'(Y^* - Y) \tag{8.167}$$

The Marquardt method consists of the following steps:

1. Assume initial guesses for the parameter vector b.
2. Assign a large value, say 1000, to λ. This means that in the first iteration the steepest descent method is predominant and would assure that the method is moving toward the lower sum of squared residuals.
3. Evaluate the Jacobian matrix J from the equation(s) of the model. Also, evaluate the Hessian matrix H if using Newton's method.

---

[1]Also known as the Levenberg-Marquardt method.

4. Use either Eq. (8.166) or Eq. (8.167) to obtain the correction vector Δb.
5. Evaluate the new estimate of the parameter vector from Eq. (8.154):

$$b^{(m+1)} = b^{(m)} + \Delta b \tag{8.154}$$

6. Calculate the new value of Φ. If $\Phi^{(m+1)} < \Phi^{(m)}$, reduce the value of λ, by a factor of 4, for example. If $\Phi^{(m+1)} > \Phi^{(m)}$, keep the old parameters [$b^{(m+1)} = b^{(m)}$] and increase the value of λ, by a factor of 2, for example.
7. Repeat steps 3 through 6 until either (or both) of the following conditions are satisfied:
  a. Φ does not change appreciably.
  b. Δb becomes very small.

The flowchart of the Marquardt method is given in Fig. 8.13.

8.4.5 MULTIPLE NONLINEAR REGRESSION

In the previous four sections, the sum of squared residuals that was minimized was that given by Eq. (8.152). This was the sum of squared residuals determined from fitting one equation to measurements of one variable. However, most mathematical models may involve simultaneous equations in multiple dependent variables. For such a case, when more than one equation is fitted to multiresponse data, where there are v dependent variables in the model, the *weighted sum of squared residuals* is given by:

$$\Phi = \sum_{j=1}^{v} w_j \varepsilon_j' \varepsilon_j = \sum_{j=1}^{v} w_j \phi_j$$
$$= \sum_{j=1}^{v} w_j (Y_j^* - Y_j)'(Y_j^* - Y_j) \tag{8.168}$$

where w_j = weighting factor corresponding to the jth dependent variable:
  ϕ_j = sum of squared residuals corresponding to the jth dependent variable
To minimize Φ by the Gauss-Newton method, we first linearize the models using Eq. (8.158) and combine with Eq. (8.168) to obtain:

$$\Phi = \sum_{j=1}^{v} w_j (Y_j^* - Y_j - J_j \Delta b)'(Y_j^* - Y_j - J_j \Delta b) \tag{8.169}$$

Taking the partial derivative of Φ with respect to Δb, setting it equal to zero, and solving for Δb, we obtain:

$$\Delta b = \left[\sum_{j=1}^{v} w_j (J_j' J_j) \right]^{-1} \left[\sum_{j=1}^{v} w_j J_j' (Y_j^* - Y_j) \right] \tag{8.170}$$

Eq. (8.170) gives the correction of the parameter vector when fitting multiple dependent variables simultaneously. Eq. (8.170) becomes identical to Eq. (8.160) when $v = 1$, that is, when only

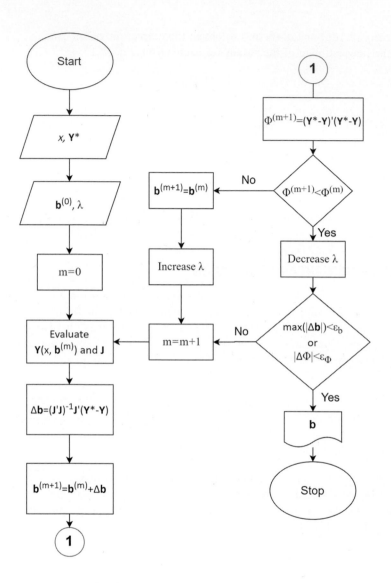

FIG. 8.13

Flowchart of the Marquardt method (based on the Gauss-Newton method).

one dependent variable is fitted. When using the Marquardt method, the correction of the parameter vector is calculated from:

$$\Delta b = \left[\lambda I + \sum_{j=1}^{v} w_j (J_j' J_j) \right]^{-1} \left[\sum_{j=1}^{v} w_j J_j' (Y_j^* - Y_j) \right] \qquad (8.171)$$

The weighting factors w_j are determined as follows: The basic assumption in the derivation of the regression algorithm was that the variance σ^2 of the distribution of the error in the measurements was constant throughout the profile of a single dependent variable. However, in the case of multiple regression, it is very unlikely that the variance σ_j^2 of all the curves will be the same. Therefore, in order to form an unbiased weighted sum of squared residuals, the individual sum of squares must be multiplied by a weighting factor that is proportional to $1/\sigma_j^2$. The equation for evaluating the weighting factors is given by:

$$w_j = \frac{1/\sigma_j^2}{\frac{1}{\sum\limits_{i=1}^{v} n_i} \left[\sum\limits_{i=1}^{v} \sum\limits_{l=1}^{n_i} \frac{1}{\sigma_j^2} \right]} \tag{8.172}$$

where σ_j^2 or σ_i^2 = variance for each curve

n_i = number of experimental points available for each curve
v = number of variables being fitted

The denominator of Eq. (8.172) accounts for the possibility that each curve may have a different number of experimental points n_i and weighs that accordingly. If the assumption that σ_j^2 is constant within one curve does not hold, then Eq. (8.172) can be extended so that the weighting factor can be calculated at each point with the appropriate value of σ_j^2.

In most cases, the values of σ_j^2 would not be known; however, the estimates of these variances s_j^2 can be obtained from repeated experiments, and the values of s_j^2 are then used in Eq. (8.172) to calculate the weighting factors. In the worst case, where no repeated experiments are made and no *a priori* knowledge of σ_j^2 is available, then the values of w_j must be guessed. Otherwise, the nonlinear regression algorithm would introduce a bias toward fitting more satisfactorily the curve with the highest ϕ_j and partially ignoring the curves with low ϕ_j.

The nonlinear regression can also be extended to fit multiple experimental values of the dependent variable at each value of the independent variable. This can be done by changing Eq. (8.168) so that the squared residuals are also summed up within each group of points. Finally, if the value of the variance of the error is proportional to the value of the dependent variable, the residual in the sum-of-squares calculation must be divided by the theoretical (calculated) value of the dependent variable at each point in the calculation.

8.5 ANALYSIS OF VARIANCE AND OTHER STATISTICAL TESTS OF THE REGRESSION RESULTS

The *t*-test on parameters, described in Section 8.3.2, is useful in establishing whether a model contains an insignificant parameter. This information can be used to make small adjustments to models and thus discriminate between models that vary from each other by one or two parameters. This test, however, does not give a criterion for testing the adequacy of this model. The residual sum of squares, calculated by Eq. (8.152), contains two components. One is the result of the scatter in the experimental data, and the other is the result of the lack of fit of the model. In order to test the adequacy of the fit of a model, the sum of squares must be partitioned into its components. This procedure is called the analysis of variance, which is summarized in Table 8.2. To maintain generality, we examine a set of nonlinear data and assume the availability of multiple values of the dependent variable y_{ij} at each point of the independent variable x_i (Fig. 8.14).

In Table 8.2, p is the number of points of the independent variable at which there are experimental (observed) values of the dependent variable; n_i is the numbers of repeated experiments available at each point of the independent variable; $\bar{y}$ is the mean value of each group of repeated experiments; y_i are the calculated values of the dependent variable; y^*_{ij} are the experimental values

Table 8.2 Analysis of variance.

Source of variance	Sum of squares	Degrees of freedom	Variance
Lack of fit	$\sum_{i=1}^{p} n_i (\bar{y}_i - y_i)^2$	$\nu_1 = p - k$	s_1^2
Experimental error	$\sum_{i=1}^{p} \sum_{j=1}^{n_i} \left(y^*_{ij} - \bar{y}_i \right)^2$	$\nu_2 = \left(\sum_{i=1}^{p} n_i \right) - p$	s_2^2
Total	$\sum_{i=1}^{p} \sum_{j=1}^{n_i} \left(y^*_{ij} - y_i \right)^2$	$\nu = \left(\sum_{i=1}^{p} n_i \right) - k$	s^2

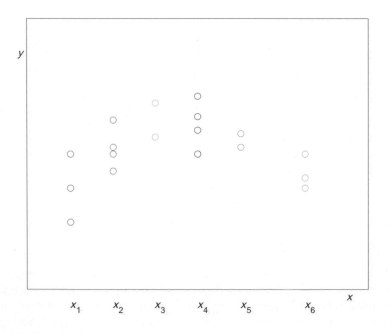

FIG. 8.14

Set of nonlinear data where multiple values of the dependent variable are available for repeated experiments ($p = 6$, $n_1 = 3$, $n_2 = 4$, $n_3 = 2$, $n_4 = 4$, $n_5 = 2$, $n_6 = 3$).

of the dependent variable; and k is the number of parameters being estimated. It should be realized that the total sum of squares shown in Table 8.2:

$$Total \ SS = \sum_{i}^{p} \sum_{j}^{n_i} \left(y_{ij}^* - y_i \right)^2 \tag{8.173}$$

is merely a generalization of Eq. (8.152) to apply to both linear and nonlinear models and an extension of that relationship to account for the presence of repeated experimental data.

The ratio of the variances s_1^2/s_2^2 has an F distribution with ν_1 and ν_2 degrees of freedom. This ratio must be tested against the F statistic in order to test the hypothesis that the experimental points are adequately represented by the predicted line. For a good fit, this ratio should be small, that is, to accept the hypothesis, the following must be true:

$$\frac{s_1^2}{s_2^2} < F_{1-\alpha}(\nu_1, \nu_2) \tag{8.174}$$

This would mean that the component of the variance resulting from the lack of fit is small when compared with the variance of the experimental error. In that case, the model adequately represents the data. It is obvious that if the experimental data have a large scatter, then s_2^2 is high and the requirements on the model are less stringent. Stating this more simply, almost anything can be fitted through very noisy or sloppy data, but the value of such a model would be marginal.

If more than one model is found to satisfy this test, the choice of the best one can be facilitated by performing an F test between the values s_1^2 of pairs of models. If the fit of any one of these models is significantly better than that of the others, it will be discovered by this test.

Furthermore, the F test may be used to determine if an experiment whose results deviate from those of other experiments performed under identical conditions should be grouped together with the other ones. To do this, the model is first fitted to each experiment separately. The individual sum of squares from each regression are pooled together as follows:

$$s_{pooled}^2 = \frac{\sum (\text{individual sum of squares})}{\sum (\text{degrees of freedom})} \tag{8.175}$$

Then the model is fitted to the grouped set of experiments to find the variance $s_{grouped}^2$. Finally, an F test is performed between the pooled and grouped variances. If inequality:

$$\frac{s_{grouped}^2}{s_{pooled}^2} < F_{1-\alpha} \tag{8.176}$$

is not satisfied, it means that the model fits the experiments better individually than when grouped together.

A final test can be performed to investigate the lack of fit of the model. In the least squares regression, the assumption is made that the model being fitted is the correct one and that the observations deviate from the model in a random fashion. The residuals between the observations and the model can be either positive or negative, but if these are truly random, the sign of the residuals should change in a random fashion. The randomness (or lack of fit) can be detected visually by plotting the residuals $(y^* - y)$ versus the independent variable x, and also versus the dependent variable y. Figs. 8.15–8.18 show several different cases of distribution of residuals. Fig. 8.15 demonstrates the case where the values of the residuals are randomly distributed around zero. This seems

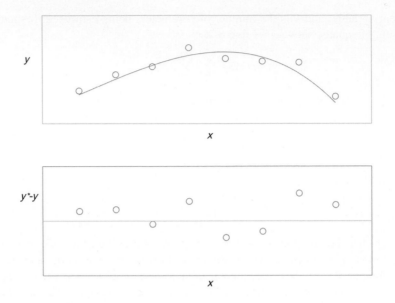

FIG. 8.15

Analysis of residuals showing a random trend.

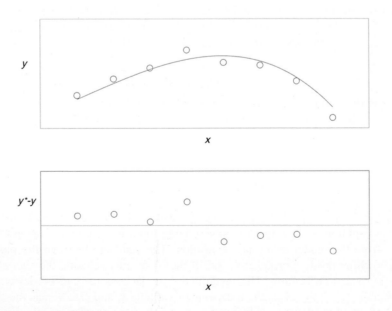

FIG. 8.16

Analysis of residuals showing a trend from positive to negative.

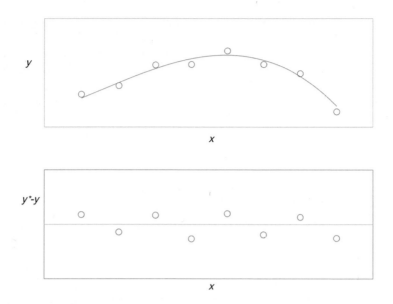

FIG. 8.17

Analysis of residuals showing an oscillatory trend.

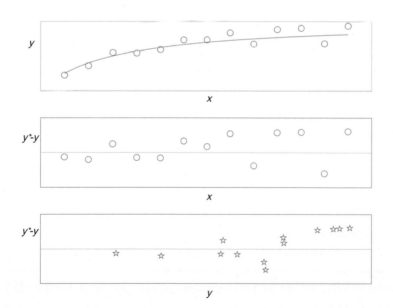

FIG. 8.18

Analysis of residuals showing a trend of increasing residuals in proportion to y.

to be a satisfactory fit that would probably pass the randomness test (*runs test*) described later in this section.

On the other hand, Fig. 8.16 shows a definite trend in the value of the residuals from positive to negative. The model gives a low prediction of y for low values of x and a high prediction of y for high values of x. A correction to the model to remedy this trend seems to be warranted. At first sight, Fig. 8.17 may seem to give a case of a well-fitting model, but careful examination of the residuals shows that there is an oscillation pattern in the distribution of these residuals around zero. The addition of a term that introduces oscillatory behavior in the model may improve considerably the fit of the model to the data.

Another case is demonstrated in Fig. 8.18, which shows that the value of the residuals grows proportionately to the value of y. In such a case, it would be more appropriate to normalize the residuals by dividing them by the appropriate value of y, that is,

$$\overline{\varepsilon} = \frac{\varepsilon}{y} \tag{8.177}$$

and then minimize the sum of normalized squared residuals:

$$\overline{\Phi} = \overline{\varepsilon}' \overline{\varepsilon} \tag{8.178}$$

The randomness of the distribution of the residuals can be quantified, measured, and tested by the so-called *runs test*. In this test, the total number of positive residuals is represented by n_1 and that of negative residuals by n_2. The number of times the sequence of residuals changes sign is r, which is called the number of runs. The distribution of r is approximated by the normal distribution. Brownlee [6] finds the mean and standard deviation of this variable to be:

$$\overline{r} = \frac{2n_1 n_2}{n_1 + n_2} + 1 \tag{8.179}$$

$$\sigma = \sqrt{\frac{2n_1 n_2 (2n_1 n_2 - n_1 - n_2)}{(n_1 + n_2)^2 (n_1 + n_2 - 1)}} \tag{8.180}$$

The standardized form of the variable is:

$$Z = \frac{r - \overline{r}}{\sigma} \tag{8.181}$$

which is distributed with zero mean and unit variance. To test the hypothesis that the deviations are random, Z is compared with the standard normal distribution. A two-sided test must be performed because if the value of Z is too low, the model is inadequate; and if Z is too high, then the data contain oscillations that must be accounted for by the model. On the other hand, if the value of Z falls in the region of acceptance for this test, then the hypothesis that the model is the correct one and that the residuals are randomly distributed can be accepted.

EXAMPLE 8.3 NONLINEAR REGRESSION USING THE MARQUARDT METHOD

Write a general MATLAB function to perform nonlinear regression analysis using the Marquardt method to fit the model equations to the given experimental data. You should

(Continued)

EXAMPLE 8.3 (CONTINUED)

evaluate the fitting parameters of the model and their statistical properties. The model may be presented in the form of either algebraic or differential equations.

(*a*) The fermentation process can be divided into vegetative growth and sporulation phases. For the vegetative phase, the logistic equation can be used to estimate the population of cells [7]:

$$\ln x_v = \frac{b_1}{1 + b_2 \exp(-b_3 t)}$$

where x_v is the vegetative cell density, t is time, and b_1 to b_3 are the constants of the equation.

The experimental data of vegetative cell density vs. time in the fermentation process of *Bacillus thuringiensis*, reported by Sarrafzadeh and Navarro [8], are given in Table E8.3a. Determine the parameters of the logistic equation by fitting these data to the model equation.

(*b*) In Problem 5.6 we described the kinetics of a fermentation process that manufactures penicillin antibiotics. When the microorganism *Penicillium chrysogenum* is grown in a batch fermenter under carefully controlled conditions, the cells grow at a rate that can be modeled by the logistic law:

$$\frac{dy_1}{dt} = b_1 y_1 \left(1 - \frac{y_1}{b_2}\right)$$

where y_1 is the concentration of the cell expressed as percent dry weight. In addition, the rate of production of penicillin has been mathematically quantified by the equation:

$$\frac{dy_2}{dt} = b_3 y_1 - b_4 y_2$$

where y_2 is the concentration of penicillin in units/mL.

The experimental data in Table E8.1b were obtained from two penicillin fermentation runs conducted at essentially identical operating conditions. Using the Marquardt method, fit these two equations to the experimental data and determine the values of the parameters b_1, b_2, b_3, and b_4, which minimize the weighted sum of squared residuals.

Method of Solution

The Marquardt method using the Gauss-Newton technique, described in Section 8.4.4, and the concept of multiple nonlinear regression, covered in Section 8.4.5, have been combined

(*Continued*)

Table E8.3a Vegetative cell density vs. time in the fermentation process of *Bacillus thuringiensis*.

t (hr)	3	12	14	16	18	20	23	24
x_v ($\times 10^9$ cell/mL)	0.04	2.11	3.13	3.19	3.55	3.34	3.62	3.62

Table E8.1b Experimental data for penicillin fermentation.

Time, hours	Run No. 1		Run No. 2	
	Cell concentration, percent dry weight	Penicillin concentration, units/mL	Cell concentration, percent dry weight	Penicillin concentration, units/mL
0	0.40	0	0.18	0
10		0	0.12	0
22	0.99	0.0089	0.48	0.0089
34		0.0732	1.46	0.0062
46	1.95	0.1446	1.56	0.2266
58		0.523	1.73	0.4373
70	2.52	0.6854	1.99	0.6943
82		1.2566	2.62	1.2459
94	3.09	1.6118	2.88	1.4315
106		1.8243	3.43	2.0402
118	4.06	2.217	3.37	1.9278
130		2.2758	3.92	2.1848
142	4.48	2.8096	3.96	2.4204
154		2.6846	3.58	2.4615
166	4.25	2.8738	3.58	2.283
178		2.8345	3.34	2.7078
190	4.36	2.8828	3.47	2.6542

EXAMPLE 8.3 (CONTINUED)

together to solve this example. Numerical differentiation by forward finite differences is used to evaluate the Jacobian matrix defined by Eq. (8.156).

In part (a), the independent variable is time (t) and the dependent variable is considered as:

$$y = \ln x_v$$

In part (b), the initial conditions of the model equations were chosen to be the average values of the corresponding experimental data at $t = 0$, that is, $y_1(0) = 0.29$ and $y_2(0) = 0.0$.

Program description

The programs and functions developed in this example can be found in: https://www.elsevier.com/books-and-journals/book-companion/9780128229613.

Two separate MATLAB functions are written for evaluating the fitting parameters and performing the statistical tests on the parameters. These functions, *NLR.m* and *statistics.m*, are described next.

(Continued)

EXAMPLE 8.3 (CONTINUED)

NLR.m: This function evaluates the fitting parameters by the Marquardt method. In the beginning, the function examines the length of the input arguments and sets the default value, if necessary. The experimental independent and dependent variables should be introduced to the function by matrices of the same size (column vectors in the case of a single independent variable), each column of which corresponds to a dependent variable. For example, the first column of x and y matrices contains the data points of the first variable, the second column of the second variable, and so on. It is not important to give the independent value in order (ascending or descending), but obviously it is important to give matrices of independent and dependent variables such that each element of the latter corresponds to the same element of the former. Because there may be a different number of experimental data points for each independent variable, these numbers should be given to the function as a vector. The structure of matrices x and y and relations between these matrices are illustrated in Fig. E8.3a.

The function *NLR.m* can handle both algebraic and ordinary differential equations as the model equations. If the model consists of only one type of equation, an empty matrix should be passed to the function in the place of the file name for the other one. It is important to note that both functions evaluating algebraic and differential equations return the function or derivative values as a column vector. The functions should also perform the calculations in an element-by-element manner, that is, using dotted operators (".*", "./", and ".^") where

(Continued)

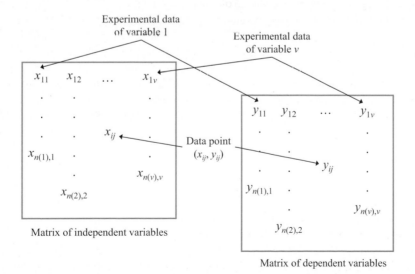

FIG. E8.3a

Structure of matrices x (independent variable) and y (dependent variable) and their relation. Each column of the matrix corresponds to a dependent variable. The number of dependent variables is shown by v, and the number of experimental data points for each dependent variable is given in the vector n.

EXAMPLE 8.3 (CONTINUED)

necessary. The boundary conditions passed to the function have to be the initial conditions of the dependent variables.

For multiple regression, the variances are used in Eq. (8.172) to determine the unbiased weighting factors, w_j, which are in turn used in Eq. (8.168) to determine the unbiased weighted sum of squared residuals. In the case where repeated experimental data are available, the program searches for repeated experimental points and evaluates the variance of each dependent variable by dividing the sum of squared differences by the number of degrees of freedom of each dependent variable (see Table 8.2). The degrees of freedom of each variable is calculated by the total number of experimental points available for the variable minus the number of means of groups of repeated data that have been calculated:

$$\text{Degrees of freedom} = \left(\sum_{i=1}^{p} n_i \right) - p$$

In the statistical analysis of the lack of fit, the degree of freedom is $(p - k)$, where k is the number of parameters associated with that variable. However, in the case where several dependent variables are fitted and the same parameter appears in the model equation of more than one variable, it is not so easy to decide exactly how many parameters correspond to each variable. For this reason, the *NLR.m* computer program apportions the parameters equally to each variable (i.e., k/v degrees of freedom are deducted from each curve).

The main iterative procedure consists of two nested loops. The inner loop evaluates the parameters while weighting factors, evaluated in the outer loop, are assumed constant. Starting with the initial guess of parameters, the function calculates the sum of squared residuals from Eq. (8.168). If the model contains ordinary differential equations, the function first solves the equations by the MATLAB function *ode23* and then evaluates by interpolation the calculated dependent variables corresponding to the vector of the independent variable. We use this procedure because the input may contain repeated experimental data, but *ode23* can handle only ascending or descending vectors of independent variables. The Jacobian matrix is then determined numerically, followed by evaluating the vector of corrections to the parameters from Eq. (8.171) for the Marquardt method or its equivalent for the Gauss-Newton method. If applicable, the value of λ is adjusted according to the new sum of squared residuals and the iterative procedure continues until:

$$\left| \frac{\Delta b_i}{b_i} \right| < \varepsilon_b \ \text{ if } \ b_i \neq 0$$

or

$$|\Delta b_i| < \varepsilon_b \ \text{ if } \ b_i = 0$$

(Continued)

EXAMPLE 8.3 (CONTINUED)

or

$$|\Delta\Phi_i| < \varepsilon_\Phi$$

When the inner iteration loop is complete, the function reevaluates the weighting factors from Eq. (8.141) and repeats the previous procedure until the increment of all the weighting factors satisfies the convergence condition:

$$|\Delta w_j| < \varepsilon_w$$

By default $\varepsilon_b = \varepsilon_w = 0.001$ and $\varepsilon_\Phi = 1 \times 10^{-6}$. These values may be changed through the 10th input argument (*tol*) to the function.

statistics.m: This function performs a statistical analysis of the data and the regression results calculated in *NLR.m*. First, the data are analyzed, and for each set of data, the following statistical information is calculated: total number of points, degrees of freedom, the sum of squares, variance, and standard deviation. Next, a statistical analysis of the regression results is performed. The standard deviations of the parameters are calculated from Eq. (8.136), where the value of σ^2 is approximated from Eq. (8.141) and a_{ii} are the diagonal elements of the matrix:

$$A = \left(\sum_{j=1}^{u} w_i J_j' J_j \right)^{-1}$$

In order to calculate the $100(1 - \alpha)\%$ confidence intervals of the parameters, first the value of $t_{1-\alpha/2}$ is calculated by the Newton-Raphson method using the following iteration formula:

$$t_{(1-\alpha/2)}^{(m+1)} = t_{(1-\alpha/2)}^{(m)} + \frac{\int_{t_{(1-\alpha/2)}^{(m)}}^{\infty} p(x, n-k)dx - \alpha/2}{p\left(t_{(1-\alpha/2)}^{(m)}, n-k \right)}$$

where $p(t, \nu)$ is the Student's t density function defined in Eq. (8.73). The confidence limits of the parameters then can be evaluated by Eq. (8.145). The Student's t density function is given in the function *stud.m*.

A significance test (t-test) is performed, as described in Section 8.2.3 (Eq. 8.102), on the parameters to test the null hypothesis that any one of the parameters might be equal to zero. The 95% confidence intervals of each measured variable are calculated. The variance-covariance matrix is calculated according to Eq. (8.135), and the matrix of correlation coefficients of the parameters is also evaluated. The analysis of variance of the regression results is performed as shown in Table 8.2. Finally, the randomness tests are applied to the residuals to test for the randomness of the distribution of these residuals.

(Continued)

EXAMPLE 8.3 (CONTINUED)

The results of the statistical analysis are stored in an output file so that they may be viewed and edited by the user. The file name can be supplied to the function by the user as the 9th input argument. Otherwise, the outputs will be saved in the file *NLRstatistics* (default name).

Example8_3a.m: The main program is written to get all the required inputs from the keyboard, perform regression analysis, and display the results both numerically and graphically. This program is written to solve the specific problem of evaluating the parameters of the logistic equation of Example 8.3a.

Ex8_3a_func.m: The logistic equation is given in this MATLAB function.

Example8_3b.m: The main program is written to get all the required inputs from the keyboard, perform regression analysis, and display the results both numerically and graphically. It is written in such a general form that it can be used in any other regression calculation, in addition to the problem in this example.

Ex8_3b_func.m: The model equations of this problem are given in this MATLAB function. Note that this function returns values of the derivatives as a column vector.

Input and results

Part (*a*)

>> **Example8_3a**

Evaluating parameters of the logistic equation

Vector of time (hr) = [3,12,14,16,18,20,23,24]

Vector of cell concentration (1/mL) = [0.04, 2.11, 3.13, 3.19, 3.55, 3.34, 3.62, 3.62]*1e9

Name of M-file containing the equation = 'Ex8_3a_func'

Vector of initial guess of fitting parameters = [20, 2, -0.5]

Method of solution

1 - Marquardt

2 - Gauss-Newton

Enter your choice: 1

Show results of each iteration (0/1)? 0

The results of the statistical analysis have been stored in the

specified output file. You may open this file using any editor.

Final Results

No. Parameter Standard 95% Confidence interval

deviation for the parameters

lower value upper value

1. 2.2044e + 01 4.4658e-02 2.1929e + 01 2.2159e + 01

2. 5.8389e-01 3.1617e-02 5.0261e-01 6.6516e-01

3. - 2.7034e-01 1.8385e-02 -3.1760e-01 -2.2308e-01

(*Continued*)

EXAMPLE 8.3 (**CONTINUED**)

   Part (*b*)

>> Example8_3b
Nonlinear Regression Analysis
Name of output file for storing results = 'Ex8_3b_results'
Experimental data input:
1 - Enter data from keyboard (point-by-point)
2 - Enter data from keyboard (in vector form)
3 - Read data from data file (prepared earlier)
Enter your choice: 1
Name of file for storing the data = 'Ex8_3b_data'
Number of dependent variables = 2
Variable 1
How many data sets for this variable? = 2
Data set 1
How many points in this set? = 9
Point 1
[x, y] = [0, 0.40]
Point 2
[x, y] = [22, 0.99]
Point 3
[x, y] − [46, 1.95]
Point 4
[x, y] = [70, 2.52]
Point 5
[x, y] = [94, 3.09]
Point 6
[x, y] = [118, 4.06]
Point 7
[x, y] = [142, 4.48]
Point 8
[x, y] = [166, 4.25]
Point 9
[x, y] = [190, 4.36]
Data set 2
How many points in this set? = 17
Point 1
[x, y] = [0, 0.18]
Point 2
[x, y] = [10, 0.12]

(*Continued*)

EXAMPLE 8.3 (CONTINUED)

Point 3
[x, y] = [22, 0.48]
Point 4
[x, y] = [34, 1.46]
Point 5
[x, y] = [46, 1.56]
Point 6
[x, y] = [58, 1.73]
Point 7
[x, y] = [70, 1.99]
Point 8
[x, y] = [82, 2.62]
Point 9
[x, y] = [94, 2.88]
Point 10
[x, y] = [106, 3.43]
Point 11
[x, y] = [118, 3.37]
Point 12
[x, y] = [130, 3.92]
Point 13
[x, y] = [142, 3.96]
Point 14
[x, y] = [154, 3.58]
Point 15
[x, y] = [166, 3.58]
Point 16
[x, y] = [178, 3.34]
Point 17
[x, y] = [190, 3.47]
Variable 2
How many data sets for this variable? = 2
Data set 1
How many points in this set? = 17
Point 1
[x, y] = [0, 0]
Point 2
[x, y] = [10, 0]
Point 3
[x, y] = [22, 0.0089]

(Continued)

EXAMPLE 8.3 (CONTINUED)

Point 4
[x, y] = [34, 0.0732]
Point 5
[x, y] = [46, 0.1446]
Point 6
[x, y] = [58, 0.5230]
Point 7
[x, y] = [70, 0.6854]
Point 8
[x, y] = [82, 1.2556]
Point 9
[x, y] = [94, 1.6118]
Point 10
[x, y] = [106, 1.8243]
Point 11
[x, y] = [118, 2.2170]
Point 12
[x, y] = [130, 2.2758]
Point 13
[x, y] = [142, 2.8096]
Point 14
[x, y] = [154, 2.6846]
Point 15
[x, y] = [166, 2.8738]
Point 16
[x, y] = [178, 2.8345]
Point 17
[x, y] = [190, 2.8828]
Data set 2
How many points in this set? = 17
Point 1
[x, y] = [0, 0]
Point 2
[x, y] = [10, 0]
Point 3
[x, y] = [22, 0.0089]
Point 4
[x, y] = [34, 0.0642]
Point 5
[x, y] = [46, 0.2266]

(Continued)

EXAMPLE 8.3 (CONTINUED)

Point 6
[x, y] = [58, 0.4373]
Point 7
[x, y] = [70, 0.6943]
Point 8
[x, y] = [82, 1.2459]
Point 9
[x, y] = [94, 1.4315]
Point 10
[x, y] = [106, 2.0402]
Point 11
[x, y] = [118, 1.9278]
Point 12
[x, y] = [130, 2.1848]
Point 13
[x, y] = [142, 2.4204]
Point 14
[x, y] = [154, 2.4615]
Point 15
[x, y] = [166, 2.2830]
Point 16
[x, y] = [178, 2.7078]
Point 17
[x, y] = [190, 2.6542]
Name of independent variable = 'Time, hours'
Name of dependent variable 1 = 'Cell concentration, % dry weight'
Name of dependent variable 2 = 'Penicillin concentration, units/mL'
Name of M-file containing differential equation(s) = 'Ex8_3b_func'
Value of independent variable at boundary condition = 0
Value(s) of dependent variable(s) at boundary condition = [0.29, 0]
Method of solution
1 - Marquardt
2 - Gauss-Newton
Enter your choice: 1
Vector of initial guess of fitting parameters = [0.1,4,0.02,0.02]
Show results of each iteration (0/1)? 0
The results of the statistical analysis have been stored in the
specified output file. You may open this file using any editor.
Final Results
No. Parameter Standard 95% Confidence interval

(Continued)

EXAMPLE 8.3 (CONTINUED)

deviation for the parameters
lower value upper value
1 - 4.2915e-02 2.3579e-03 3.8192e-02 4.7638e-02
2 - 3.9267e + 00 1.2089e-01 3.6846e + 00 4.1689e + 00
3 - 1.7680e-02 2.6762e-03 1.2319e-02 2.3041e-02
4 - 2.2162e-02 4.7105e-03 1.2726e-02 3.1598e-02

Discussion of results

Part (a): The fitted curve and the experimental data points are shown in Fig. E8.3b. The Marquardt method was used for calculations, and the constants of the logistic equation are computed as:

$$b_1 = 2.044 \quad b_2 = 0.5839 \quad b_3 = -0.2703$$

For the sake of brevity, calculation results at the end of each iteration are not shown here. However, it is strongly recommended that the reader repeat running the program *Example8_3a.m* and choose to show the results of each iteration. Also, detailed statistical results are saved in the file *NLRstatistics* (contents are not shown here). Note that because the name of the output file is not provided to the function *statistics.m*, the default name for the

(Continued)

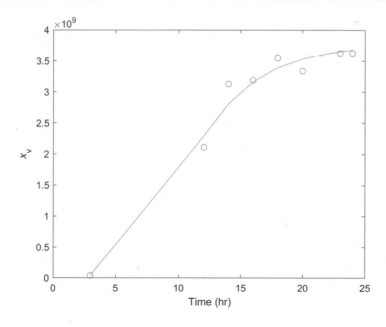

FIG. E8.3b

Results of fitting the logistic equation to the experimental data of vegetative cell density.

EXAMPLE 8.3 (CONTINUED)

output file is used by the function. The reader can open this file in any text editor to obtain more information about the properties of the calculated parameters, analysis of variance, and other statistical tests.

Part (*b*): The experimental and calculated values of y_1 and y_2 are shown in Fig. E8.3c. The final values of the parameters are calculated as:

$$b_1 = 0.0429 \quad b_2 = 3.9244 \quad b_3 = 0.0177 \quad b_4 = 0.0221$$

For the sake of brevity, calculation results at the end of each iteration are not shown here. However, it is strongly recommended that the reader repeat running the program *Example8_3b.m* and choose to show the results of each iteration. The program starts with the weighting factors calculated from the variances of the experimental data and the Marquardt method converges after eight iterations on the parameters. The weighting factors are then adjusted and the function repeats the iteration on the parameters and converges in three iterations. Finally, the weighting factors are changed for the third time and the method converges in one iteration. The reader is encouraged to repeat the calculations with different methods and different starting guesses of parameters. A bad guess may cause the Gauss-Newton method to diverge. If the Marquardt method is chosen, the method does not diverge.

(*Continued*)

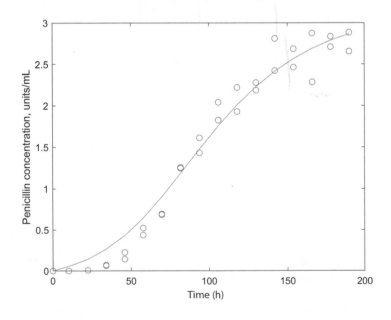

FIG. E8.3c

Results of the regression of cell and penicillin experimental data.

EXAMPLE 8.3 (CONTINUED)

However, a bad guess may require a large number of iterations to converge and may possibly exceed the maximum limit of iterations defined in the function.

The program gives a complete statistical analysis of the experimental data and the regression results. Statistical results are saved in the file *Ex8_3b_results* (contents are not shown here). The name of the file is provided in the inputs by the user when running *Example8_3b.m*. For the experimental data, the program calculates the following statistics for each dependent variable being fitted: total points, degrees of freedom, the sum of squares, variance, and standard deviation. These statistics are shown for both unweighted and weighted data. For the regression results, the program evaluates the standard deviation and 95% confidence intervals of the parameters, it performs a significance test (*t*-test) on each parameter to determine whether the value of that parameter is different than zero, it calculates the 95% confidence limit for each fitted variable, it lists the variance-covariance matrix and the matrix of correlation coefficients of the parameters, performs a complete analysis of variance, and does a randomness test on the distribution of the residuals.

8.6 USING BUILT-IN MATLAB® AND EXCEL FUNCTIONS

In MATLAB, the *Statistics and Machine Learning Toolbox* includes statistical functions and the *Curve Fitting Toolbox* provides various functions for linear and nonlinear regression analysis. Here, we only present some basic essentials.

The built-in functions *mean(x)* and *std(x)* calculate the mean value and standard deviation of the vector *x*, respectively. If *x* is a matrix, these functions return a vector of mean or standard deviation of each column. The function *cov(x, y)* calculates the covariance of the vectors of the same length *x* and *y* (Eq. 8.30). If *x* is a matrix where each row is an observation and each column a variable, *cov(x)* returns the covariance matrix.

The function *corrcoef(x, y)* in MATLAB calculates the matrix of the correlation coefficients of the vectors of the same length *x* and *y* (Eq. 8.34). If *x* is a matrix where each row is an observation and each column a variable, *corrcoef(x)* also returns the correlation coefficients matrix.

Various distributions are available in MATLAB. The probability density function can be created by the function *pdf* and the cumulative density function by *cdf*. The command $y = pdf(name, x, a)$ returns the array of values of the probability density function *y* for the given distribution specified by *name*, with parameters given in the vector *a* at the values in *x*. For example, $y = pdf('normal', x, mu, sigma)$ returns values of the probability density of a normal distribution with the expected value of *mu* and standard deviation of *sigma* at the values in *x*. The syntax for using the function *cdf* is the same and returns the cumulative distribution at values in *x*.

Polynomial regression in MATLAB can be carried out with the function *polyfit*. The statement *polyfit(x, y, n)* evaluates the coefficients of the *n*th-order polynomial fitted to the data points given in the vectors *x* (independent variable) and *y* (dependent variable). The same algorithm described in Section 8.3.1 is used in this function. Note that *polyfit* returns the coefficients in the descending order, which is the opposite of what is shown in Eq. (8.126).

Curve fitting in MATLAB can be done with the *fit* function. This function fits various types of curves or surfaces to the data. The command *cftool* opens the Curve Fitting Tool in a new window, in which you can perform linear and nonlinear regression in an interactive manner.

Nonlinear regression can also be carried out in the Optimization Toolbox of MATLAB. The statement $b = lsqcurvefit$ ('*file_name*', b_0, x, y) starts the regression calculations at the vector of initial guesses of the parameters b_0 and uses the least-squares technique to find the vector of parameters b that best fit the nonlinear expression, introduced in the MATLAB function *file_name.m*, to the data y. Inputs to the function *file_name* should be the vector of parameters b and the vector of independent variable x. The function *file_name* should return the vector of the dependent variable y. The default algorithm is Marquardt (see Section 8.4.4). Similarly, it is possible to use the MATLAB function *fitnlm* for fitting a nonlinear equation to the data. In this case, the statement $nlm = fitnlm(x, y, \text{'}file\_name\text{'}, b_0)$ fits the model equation specified in *file_name.m* to the data given in x and y, using b_0 as the initial value(s), and returns the nonlinear model *nlm*.

In Excel, line and curve fitting can be done using the Trendline command. The steps for adding a trend line are demonstrated in Example 8.2. The *Format Trendline* pane is shown in Fig. 8.19. As can be seen in this figure, there are different types of lines/curves that the user can choose from to fit the data. In the case of a polynomial trend line, you can specify the order of the polynomial. Also, at the bottom of the pane, the user can choose, by checking the appropriate box, to display the equation of the fitted line/curve in addition to the correlation coefficient on the chart.

8.7 SUMMARY

In many mathematical models in engineering, we encounter problems in which it is necessary to evaluate the parameters of the model and/or develop an empirical (or semi-empirical) equation based on the existing experimental data. Estimation of parameters of such equations can be carried out by regression analysis. In the regression analysis, the problem can be linear or nonlinear with respect to the unknown parameters.

Assessing the quality of the estimated parameters and comparing the performance of the obtained equations is possible through statistical methods. Therefore, this chapter starts with a review of basic statistical definitions and terminology employed in the regression analysis. Average, standard deviation, covariance, and correlation coefficient are the main statistics defined in this section.

Distribution functions applied in the regression analysis are introduced. Normal (or Gaussian) distribution is symmetric about the mean, showing that data near the mean occur more frequently than those far from the mean. If the average is estimated from a sample of small size, the Student's *t* distribution should be used instead of the normal distribution. The variance of a population of a normal distribution obeys the chi-square distribution when it is estimated from a sample. The *F* distribution provides a means of comparing variances.

In the next part, the linear regression analysis is developed and the least squares method is described for estimation of model parameters. In this method, the fitting parameters are determined such that the minimized sum of squared differences between the experimental and calculated data is obtained. Evaluating properties of the estimated parameters, including standard deviation and confidence interval, are also explained.

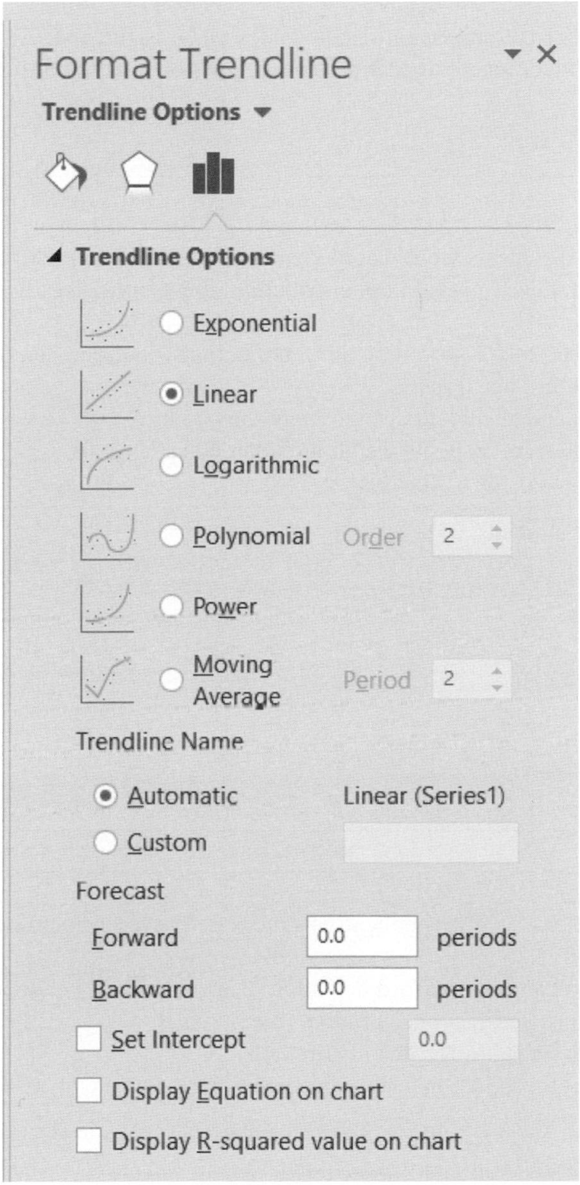

FIG. 8.19

The Trendline tab in Excel.

In the case of nonlinear regression, the steepest descent method, the Gauss-Newton method, Newton's method, and the Marquardt method for evaluating the model parameters are developed. When there is more than one dependent variable in the model, the weighted sum of squared errors should be considered in the calculations.

Analysis of variance of the regression results is described at the end. At the end of the chapter, built-in MATLAB and Excel functions that perform regression analysis are described.

PROBLEMS

8.1. The weights of 40 students were recorded and reported in Table P8.1.

    (a) Plot the normalized probability distribution and cumulative distribution of weights of students.

    (b) Assuming that this distribution is a normal distribution, determine the equations of these distribution functions.

    (c) What percentage of students weigh more than 75 kg?

    (d) What percentage of students weigh less than 60 kg?

Table P8.1 Weights of students.

Weight (kg)	Students
54–57	3
58–61	5
62–65	9
66–69	12
70–73	5
74–77	4
78–81	2
Total	40

8.2. Metal concentration, C, in a chemical was measured several times by atomic absorption, as reported in Table P8.2.

    (a) What is the mean and standard deviation of the measured concentration?

    (b) Knowing that these concentration measurements obey the normal distribution, determine the equation for this distribution.

    (c) What is the probability that the concentration of the metal in that sample is less than 2.05 ppm?

Table P8.2 Metal concentration.

Test #	1	2	3	4	5	6	7
C (ppm)	2.02	1.80	2.12	1.96	2.04	2.08	1.98

8.3. The heat capacity of gases and liquids is a function of temperature. Various forms of polynomial equations have been used to represent this functionality. Two such equations are shown as follows:

$$c_p = a + bT + cT^2 + dT^3$$
$$c_p = a + bT + cT^{-2}$$

Using the heat capacity data of Table P8.3, determine the coefficients of these equations. Discuss your results and recommend which equation gives the best representation of the data.

Table P8.3 Heat capacity data.

Temperature (°C)	Heat capacity (J/gmol °C)		
	Set No. 1	Set No. 2	Set No. 3
100	29.38	30.04	28.52
200	29.88	29.08	29.79
300	30.42	30.18	31.41
400	30.98	30.14	31.18
500	31.57	32.27	31.16
600	32.15	31.79	32.81
700	32.73	32.97	32.38
800	33.29	32.56	34.26
900	33.82	34.24	34.72
1000	34.31	35.27	33.69

8.4. The Clausius-Clapeyron equation for characterizing a discontinuous phase transition between two phases of a single-component material is given as:

$$\ln p^{sat} = A - \frac{\Delta H}{RT}$$

The data in Table P8.4 are given for saturated water. Determine the heat of evaporation and its 95% confidence interval based on these data points.

Table P8.4 Saturation temperature of water vapor as a function of pressure.

p^{sat} (kPa)	100	110	120	130	140	150	160	170	180
T (K)	373	375	378	380	382	385	386	388	390

8.5. Vapor pressure of n-butane is given in Table P8.5.

The relation between the temperature and vapor pressure of a pure substance can be correlated by the Antoine equation:

$$\log p^{sat} = A - \frac{B}{C + T}$$

(a) Estimate the constants of the Antoine equation for n-butane.
(b) Calculate the 95% confidence interval of the estimated parameters.

Table P8.5 Vapor pressure of *n*-butane as a function of temperature.

T (K)	270	280	290	300	310	320	330
p^{sat} (kPa)	91.55	132.97	187.65	258.11	347.06	457.31	591.79

8.6. The shrinking core model is the most used model for describing the kinetics in leaching processes. In this model, it is assumed that the reaction starts from the surface of the particle and continues until the reaction front reaches the center of the particle. Progress of leaching based on the shrinking core model in constant particle size involves the following steps in series: (1) diffusion of the leaching agent through the thin liquid film surrounding the particle, (2) diffusion of the leaching agent through the solid product layer, (3) the reaction on the surface of the unreacted core. If one of these steps is substantially slower than the other two, it becomes the controlling step, and the rate of leaching becomes the same as the rate of this step. Equations related to each controlling mechanism for spherical particles are [9]:

$$\text{Liquid film diffusion control: } X = f_F(X) = \frac{t}{\tau_F}$$

$$\text{Solid product layer diffusion control: } 1 - 3(1-X)^{\frac{2}{3}} + 2(1-X) = f_P(X) = \frac{t}{\tau_P}$$

$$\text{Chemical reaction control: } 1 - (1-X)^{\frac{1}{3}} = f_R(X) = \frac{t}{\tau_R}$$

In these equations, X is the fractional recovery of the metal ion, t is the time of leaching, and τ is the time constant for each controlling mechanism.

Detecting the controlling mechanism can be achieved by fitting the time-recovery experimental data to each of the previous equations. The formula of the correct mechanism exhibits a decent linear relationship with time. The best line can be recognized by its correlation coefficient to be greater than the other two. The mechanism whose equation fits best to a line is considered the rate-controlling step. The kinetics of leaching of cobalt from spent lithium-ion batteries can be described by the shrinking core model. Experimental recovery of vanadium against the time of leaching at various temperatures is given in Table P8.6 [10]. Determine the controlling mechanism in each temperature. For this purpose, do the following:

Table P8.6 Recovery of cobalt against leaching time at different temperatures.

Time, t (min)	Co Recovery, X (%)			
	@ 30°C	@ 45°C	@ 60°C	@ 80°C
30	18.74	23.36	32.94	45.50
60	28.56	53.43	57.53	68.20
90	47.42	67.38	78.59	87.06
120	56.99	76.39	83.50	89.80
210	85.14	88.70	92.79	93.89
300	93.33	94.71	96.35	97.70

(a) Use the data in Table P8.6 in each temperature and fit the data to each of the functions shown earlier (the line $f_i(X)$ vs. t, $i = F, P, R$).

(b) Calculate the correlation coefficient for each fitted line. The line with the highest correlation coefficient determines the controlling mechanism.

(c) Does the controlling mechanism change with temperature? Why?

Note that each line in this problem should pass from the origin (0, 0).

8.7. Calculate the coefficients of the equation:

$$y = b_1 + b_2 x^2 + b_3 x^3$$

by fitting it to the data in Table P8.7.

Table P8.7 Data points.

x	1	2	3	4	5	6	7
y	0.1	0.5	0.8	0.6	0.5	0.2	0.1

8.8. Concentration of a chemical during its decomposition is measured and reported in Table P8.8. Considering the first-order reaction, calculate the reaction rate constant, its standard deviation, and the correlation coefficient of the fitted equation.

Table P8.8 Decomposition data.

t (min)	0	2	5	10	20	40
C (mmol/L)	1.5	1.15	0.85	0.45	0.10	0.01

8.9. A mathematical model of the fermentation of the bacterium *Pseudomonas ovalis*, which produces gluconic acid, which describes the dynamics of the logarithmic growth phase, can be summarized as follows:

$$\text{Rate of cell growth: } \frac{dy_1}{dt} = b_1 y_1 \left(1.0 - \frac{y_1}{b_2} \right)$$

$$\text{Rate of gluconolactone formation: } \frac{dy_2}{dt} = \frac{b_3 y_1 y_4}{b_4 + y_4} - 0.9082 b_5 y_2$$

$$\text{Rate of gluconic acid formation: } \frac{dy_3}{dt} = b_5 y_2$$

$$\text{Rate of glucose consumption: } \frac{dy_4}{dt} = -1.011 \left(\frac{b_3 y_1 y_4}{b_4 + y_4} \right)$$

where y_1 = concentration of cell

y_2 = concentration of gluconolactone

y_3 = concentration of gluconic acid

y_4 = concentration of glucose

b_1 to b_5 = parameters of the system that are functions of temperature and pH

Using the batch fermentation data given in Table P8.9, determine the values of the parameters b_1 to b_5 at the three different temperatures of 25°C, 28°C, and 30°C.

Table P8.9 Data obtained by varying the temperature at pH 7.0 [13].

Time (hour)	Cell concentration (UOD/mL)	Gluconolactone concentration (mg/mL)	Gluconic acid concentration (mg/mL)	Glucose concentration (mg/mL)
Experiment No. 34 – Batch fermentation data at 25.0°C				
0.0	0.56	1.28	0.16	45.00
1.0	0.86	2.20	1.56	43.00
2.0	1.60	3.50	5.00	38.00
3.0	2.60	5.60	9.50	33.00
4.0	3.20	7.00	16.00	25.00
5.0	3.30	7.80	24.50	16.50
6.0	3.50	7.20	32.00	9.00
7.0	3.40	6.30	24.50	4.00
8.0	3.40	3.20	45.80	2.00
Experiment No. 33 – Batch fermentation data at 28.0°C				
0.0	0.66	–	–	48.00
1.0	1.00	1.96	0.15	45.00
2.0	1.60	6.67	7.00	37.50
3.0	2.60	10.50	15.00	28.00
4.0	3.20	10.50	25.00	18.00
5.0	3.30	7.58	35.00	8.00
6.0	3.30	2.05	42.50	3.00
7.0	3.30	1.90	45.50	–
Experiment No. 32 – Batch fermentation data at 30.0°C				
0.0	0.80	1.34	0.95	44.50
1.0	1.50	4.00	4.96	37.50
2.0	2.60	7.50	16.10	25.00
3.0	3.50	8.00	32.10	9.00
4.0	3.50	5.00	43.70	3.00
5.0	3.50	2.42	44.50	2.00

8.10. Accurate vapor-liquid equilibrium measurements can be used to compute liquid-phase activity coefficients and excess Gibbs free energies. Consider the data in Table P8.10 for benzene-2,2,4-trimethylpentane (B-TMP) mixtures at a constant temperature of 55°C.

(a) Assume that the gas phase is ideal and neglect any fugacity and Poynting corrections for the liquid phase. Calculate the activity coefficients for B and TMP and the molar excess Gibbs free energy at each temperature point. The vapor pressure of pure B at 55°C is 327.05 mm Hg and of pure TMP at 55°C is 178.08 mm Hg.

(b) If a *three-constant* Redlich-Kister expansion for the excess molar Gibbs free energy is assumed [11], evaluate the constants that appear using the data of part (a); that is, find "best fits" for A_0, A_1, and A_2.

(c) Calculate the activity coefficients from your expressions in part (b) for B and TMP. (*Hint*: First derive an expression for the activity coefficients assuming a three-constant Redlich-Kister expansion.)

(d) Plot the theoretical excess molar Gibbs free energy and the theoretical activity coefficients with the experimental data as well.

Table P8.10 Vapor-liquid equilibrium data [14].

Liquid-phase mole fraction x_B	Vapor-phase mole fraction y_B	Equilibrium total pressure P mm Hg
0.0819	0.1869	202.74
0.2192	0.4065	236.86
0.3584	0.5509	266.04
0.3831	0.5748	270.73
0.5256	0.6786	293.36
0.8478	0.8741	324.66
0.9872	0.9863	327.39

8.11. Use the data of Problem 5.8 to fit the Lotka-Volterra predator-prey equations:

$$\frac{dN_1}{dt} = \alpha N_1 - \beta N_1 N_2$$

$$\frac{dN_2}{dt} = -\gamma N_2 + \beta N_1 N_2$$

in order to obtain accurate estimates of the parameters of the model. Modify the Lotka-Volterra equations as recommended in Problem 5.8 and determine the parameters of your new models. Compare the results of the statistical analysis for each model, and choose the set of equations that gives the best representation of the data.

8.12. Svirbely and Blaner [12] modeled a set of chemical reactions represented by:

$$A + B \xrightarrow{k_1} C + F$$
$$A + C \xrightarrow{k_2} D + F$$
$$A + D \xrightarrow{k_3} E + F$$

using the following differential equations:

$$\frac{dA}{dt} = -k_1 AB - k_2 AC - k_3 AD$$

$$\frac{dB}{dt} = -k_1 AB$$

$$\frac{dC}{dt} = k_1 AB - k_2 AC$$

$$\frac{dD}{dt} = k_2 AC - k_3 AD$$

$$\frac{dE}{dt} = k_3 AD$$

Estimate the coefficients k_1, k_2, and k_3 (all positive) from the data of Table P8.12 and the following initial conditions:

$$C(0) = D(0) = 0.0$$
$$A(0) = 0.02090 \text{ mol/L}$$
$$B(0) = (1/3) \, A(0)$$

The estimates reported in the article were:

$$k_1 = 14.7$$
$$k_2 = 1.53$$
$$k_3 = 0.294$$

Could estimates be obtained if the initial conditions were not known?

Table P8.12 Experimental reaction data.

Time (min)	A (mmol/L)	Time (min)	A (mmol/L)
4.50	15.40	76.75	8.395
8.67	14.22	90.00	7.891
12.67	13.35	102.00	7.510
17.75	12.32	108.00	7.370
22.67	11.81	147.92	6.646
27.08	11.39	198.00	5.883
32.00	10.92	241.75	5.322
36.00	10.54	270.25	4.960
46.33	9.78	326.25	4.518
57.00	9.157	418.00	4.075
69.00	8.594	501.00	3.715

8.13. Choose one of the equations given in Problem 8.3 and perform the following:
(a) Fit the equation to the specific heat data given in Problem 8.3.
(b) Add random error (r.e.) to the data, where this error is in the range $-1.0 \leq$ r.e. ≤ 1.0, and fit the equation to the noisy data.
(c) Repeat part (b) with $-5.0 \leq$ r.e. ≤ 5.0.
(d) Repeat part (b) with $-10.0 \leq$ r.e. ≤ 10.0.
Compare the results of the statistical analysis in parts (a) to (d). What conclusions do you draw?

8.14. Solid mixing in fluidized beds is assumed to be described by the dispersion model, which is a diffusion-type model. Based on this model, the axial dispersion of solids is represented by the differential equation:

$$\frac{\partial C}{\partial t} = D \frac{\partial^2 C}{\partial z^2}$$

where C is the concentration of tagged particles at axial position z and time t, and D is the axial dispersion coefficient of the solids. In an experiment in a gas-solid fluidized bed, a certain amount of tagged particles (1000 units) is injected at the top of the bed ($z = 350$ mm) at the beginning of the experiment. The tagged particles are not added or taken out during the experiment. Therefore, the diffusion equation, in this case, should be solved subject to the following initial and boundary conditions:

$$\text{at } t = 0, \begin{cases} C = C_0 = 1000 \text{ for } z = L = 350 \text{ mm} \\ C = 0 \text{ for } z < L \end{cases}$$

$$\text{at } t \geq 0, \frac{\partial C}{\partial z}\bigg|_{z=L} = 0$$

$$\text{at } t \geq 0, \frac{\partial C}{\partial z}\bigg|_{z=0} = 0$$

The concentration of tagged particles at different heights and times are measured during the experiment and are given in Table P8.14.

Develop a MATLAB function to calculate the dispersion coefficient in the partial differential equation from $C(t, z)$ data by using a least-squares technique. Apply this function to the data of Table P8.14 to evaluate the dispersion coefficient at the conditions of this experiment.

Table P8.14 Experimental concentration of tagged particles as a function of time and height.

t (s)	$z = 300$ mm	$z = 250$ mm	$z = 200$ mm
0.0	0	0	0
0.1	20	5	0
0.2	63	22	6
0.3	72	40	24
0.4	57	53	36
0.5	30	56	38
0.6	29	32	48
0.8	26	25	40
1.0	25	23	25
1.5	17	21	16
2.0	29	20	23

REFERENCES

[1] Montgomery, D. C.; Runger, G. C.; Hubele, N. F. *Engineering Statistics*, 5th ed.; Wiley: New York, 2010.
[2] Metcalfe, A.; Green, D.; Greenfield, T.; Mansor, M.; Smith, A.; Tuke, J. *Statistics in Engineering*, 2nd ed.; Chapman and Hall/CRC: New York, 2019.
[3] Seinfeld, J. H.; Lapidus, L. *Mathematical Methods in Chemical Engineering*, vol. 3; Process Modeling, Estimation, and Identification, Prentice Hall: Englewood Cliffs, NJ, 1974.
[4] Edgar, T. F.; Himmelblau, D. M. *Optimization of Chemical Processes*, 2nd ed.; McGraw-Hill: New York, 2001.
[5] Marquardt, D. W. An Algorithm for Least Squares Estimation of Nonlinear Parameters. *J. Soc. Ind. Appl. Math.* **1963**, *11*, 431−441.

[6] Brownlee, K. A. *Statistical Theory and Methodology in Science and Engineering*, 2nd ed.; Wiley: New York, 1965.

[7] Soleymani, S.; Sarrafzadeh, M. H.; Mostoufi, N. Modeling of Fermentation Process of *Bacillus Thuringiensis* as a Sporulating Bacterium. *Chem. Prod. Process Model* **2019**, *14*, 20180007.

[8] Sarrafzadeh, M. H.; Navarro, J. M. The Effect of Oxygen on the Sporulation, δ-Endotoxin Synthesis and Toxicity of *Bacillus thuringiensis* H14. *J. World Microbiol. Biotechnol.* **2006**, *22*, 305−310.

[9] Faraji, F.; Alizadeh, A.; Rashchi, F.; Mostoufi, N. Kinetics of Leaching: A Review. *Rev. Chem. Eng.* **2022**, *38*, 113−148. Available from: https://doi.org/10.1515/revce-2019-0073.

[10] Golmohammadzadeh, R.; Rashchi, F.; Vahidi, E. Recovery of Lithium and Cobalt from Spent Lithium-Ion Batteries Using Organic Acids: Process Optimization and Kinetic Aspects. *Waste Manag.* **2017**, *64*, 244−254.

[11] Denbigh, K. *The Principles of Chemical Equilibrium*, 4th ed.; Cambridge University Press: Cambridge, 1981.

[12] Svirbely, W. J.; Blaner, J. A. The Kinetics of Three-Step Competitive Consecutive Second-Order Reactions. II. *J. Amer. Chem. Soc.* **1961**, *83*, 4115−4118.

[13] Rai, V. R. *Mathematical Modeling and Optimization of the Gluconic Acid Fermentation, Ph.D. Dissertation;* Rutgers—The State University of New Jersey: Piscataway, NJ, 1973.

[14] Weissman, S.; Wood, S. E. Vapor-Liquid Equilibrium of Benzene-2,2,4-Trimethylpentane Mixtures. *J. Chem. Phys.* **1960**, *32*, 1153.

ORTHOGONAL POLYNOMIALS

Orthogonal polynomials are a special category of functions that satisfy the following orthogonality condition with respect to a weighting function $w(x) \geq 0$, on the interval $[a, b]$:

$$\int_a^b w(x)g_n(x)g_m(x)dx = \begin{cases} 0 & if \quad n \neq m \\ c(n) > 0 & if \quad n = m \end{cases} \tag{A.1}$$

This orthogonality condition can be viewed as the continuous analog of the orthogonality property of two vectors:

$$x'y = 0 \tag{A.2}$$

in n-dimensional space, where n becomes very large and the elements of the vectors are represented as continuous functions of some independent variable.

There are many families of polynomials that obey the orthogonality condition. These are generally known by the name of the mathematician who discovered them. *Legendre, Chebyshev, Hermite*, and *Laguerre polynomials* are the most widely used orthogonal polynomials. In this section we list the Legendre, Chebyshev, and Laguerre polynomials.

The *Legendre polynomials* are solutions of the Legendre differential equation:

$$\frac{d}{dx}\left[(1 - x^2)\frac{dy}{dx}\right] + n(n + 1)y = 0 \tag{A.3}$$

These polynomials are orthogonal on the interval $[-1, 1]$ with respect to the weighting function $w(x) = 1$. The orthogonality condition is:

$$\int_{-1}^1 P_n(x)P_m(x)dx = \begin{cases} 0 & if \quad n \neq m \\ \dfrac{2}{2n + 1} & if \quad n = m \end{cases} \tag{A.4}$$

They also satisfy the recurrence relation:

$$(n + 1)P_{n+1}(x) - (2n + 1)xP_n(x) + nP_{n-1}(x) = 0 \tag{A.5}$$

Starting with $P_0(x) = 1$ and $P_1(x) = x$, the recurrence formula (A.5), or the orthogonality condition (A.4), can be used to generate the Legendre polynomials. These are listed in Table A.1 and drawn in Fig. A.1.

The *Chebyshev polynomials* are solutions of the Chebyshev differential equation:

$$(1 - x^2)\frac{d^2y}{dx^2} - x\frac{dy}{dx} + n^2y = 0 \tag{A.6}$$

477

n	$P_n(x)$
	Table A.1 Legendre polynomials.
0	$P_0(x) = 1$
1	$P_1(x) = x$
2	$P_2(x) = \frac{3x^2 - 1}{2}$
3	$P_3(x) = \frac{5x^3 - 3x}{2}$
4	$P_4(x) = \frac{35x^4 - 30x^2 + 3}{8}$
$\vdots$	$\vdots$
n	$P_n(x) = \sum_{m=0}^{[n/2]^a} (-1)^m \frac{(2n-2m)!}{2^n m!(n-m)!(n-2m)!} x^{n-2m}$

[a] The notation $[n/2]$ represents the integer part of $n/2$.

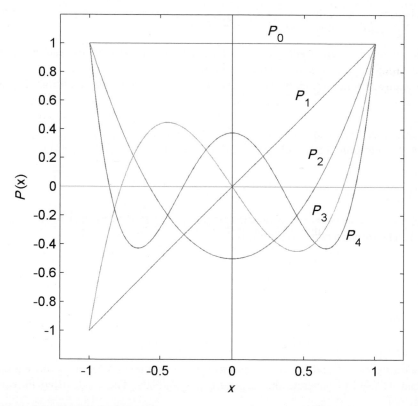

FIG. A.1 The Legendre orthogonal polynomials.

These polynomials are orthogonal on the interval $[-1, 1]$ with respect to the weighting function $w(x) = 1/\sqrt{1 - x^2}$. Their orthogonality condition is:

$$\int_{-1}^{1} \frac{1}{\sqrt{1 - x^2}} T_n(x)T_m(x)dx = \begin{cases} 0 & if & n \neq m \\ \pi & if & n = m = 0 \\ \pi/2 & if & n = m > 0 \end{cases} \tag{A.7}$$

and their recurrence relation is:

$$T_{n+1} - 2xT_n + T_{n-1} = 0 \tag{A.8}$$

Starting with $T_0(x) = 1$ and $T_1(x) = x$, the recurrence formula (A.7), or orthogonality condition (A.8), can be used to generate the Chebyshev polynomials which are listed in Table A.2 and drawn in Fig. A.2.

The *Laguerre polynomials* are solutions of the Laguerre differential equation:

$$x\frac{d^2y}{dx^2} - (1 - x)\frac{dy}{dx} + ny = 0 \tag{A.9}$$

These polynomials are orthogonal on the interval $[0, \infty)$ with respect to the weighting function $w(x) = e^{-x}$. Their orthogonality condition is:

$$\int_{0}^{\infty} e^{-x}L_n(x)L_m(x)dx = \begin{cases} 0 & if & n \neq m \\ (n!)^2 & if & n = m \end{cases} \tag{A.10}$$

and their recurrence relation is:

$$(n + 1)L_{n+1} - (2n + 1 - x)L_n + nL_{n-1} = 0 \tag{A.11}$$

Starting with $L_0(x) = 1$ and $T_1(x) = 1 - x$, the recurrence formula (A.10), or orthogonality condition (A.11), can be used to generate the Laguerre polynomials, which are listed in Table A.3 and drawn in Fig. A.3.

Table A.2 Chebyshev polynomials.	
n	$T_n(x)$
0	$T_0(x) = 1$
1	$T_1(x) = x$
2	$T_2(x) = 2x^2 - 1$
3	$T_3(x) = 4x^3 - 3x$
4	$T_4(x) = 8x^4 - 8x^2 + 1$
$\vdots$	$\vdots$
n	$T_n(x) = \sum_{m=0}^{[n/2]^a} (-1)^m \frac{(2n - 2m)!}{2^n m!(n - m)!(n - 2m)!} x^{n-2m}$

^{a} The notation $[n/2]$ represents the integer part of $n/2$.

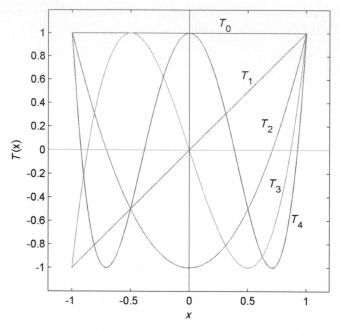

FIG. A.2 The Chebyshev orthogonal polynomials.

n	$L_n(x)$
	Table A.3 Laguerre polynomials.
0	$L_0(x) = 1$
1	$L_1(x) = -x + 1$
2	$L_2(x) = \frac{x^2 - 4x + 2}{2}$
3	$L_3(x) = \frac{-x^3 + 9x^2 - 18x + 6}{6}$
4	$L_4(x) = \frac{x^4 - 16x^3 + 72x^2 - 96x + 24}{24}$
$\vdots$	$\vdots$
n	$L_n(x) = \sum\limits_{m=0}^{n} \frac{(-1)^m}{m!} \binom{n}{m} x^m$

It should be noticed from Figs. A.1 and A.2 that these orthogonal polynomials have their zeros (roots) more closely packed near the ends of the interval of integration. This property can be used to improve the accuracy of interpolation of unequally spaced points. This can be done in the case where the choice of base points is completely free. The interpolation can be performed using the

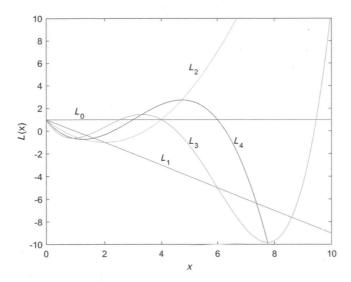

FIG. A.3 The Laguerre orthogonal polynomials.

Lagrange interpolation method described in Section 3.8.1 using the base points at the roots of the appropriate orthogonal polynomial. This concept is demonstrated in Chapter 4 in connection with the development of *Gauss quadrature* and Chapter 6 in the solution of boundary value problems by the *orthogonal collocation method*.

Index

Note: Page numbers followed by "*f*," "*t*," and "*b*" refer to figures, tables, and boxes, respectively.